DITA for Print: A DITA Open Toolkit Workbook

DITA Open Toolkit 2

Leigh W. White

http://xmlpress.net

DITA for Print: A DITA Open Toolkit Workbook

Credits

DITA and the DITA logo are used by permission from the OASIS open standards consortium.

Disclaimer

Trademarks

XML Press
Laguna Hills, CA
http://xmlpress.net

Second Edition
ISBN: 978-1-937434-54-0

Table of Contents

Chapter 18: List of Tables and List of Figures....................377

List of Figures

Preface

Preface

Why this book?

Several years ago I sat down with the DITA Open Toolkit for the first time. I wasn't new to XML or XSLT, but I wasn't an expert, either. Many cups of coffee and even more temper tantrums later, I had my first PDF customization up and running. Along the way, I had help from individuals and organizations. I attended webinars and conference sessions. Little by little, it all came together.

Back up to those last three sentences. It took a **lot** of different resources to enable me to put together that first PDF customization, and that's why I wrote this book. There was no single place where I, or anyone, could find all the information needed. Frustrating? Oh, a little. It's not that the information isn't out there. It's just that it's all over the place. Somewhere along the way it occurred to me that I ought to pull together everything I'd learned and write it down.

This book began as a general reference to all of the most common changes you'd need to make to the PDF plugin of the DITA Open Toolkit. I started down that road and realized it'd be like trying to describe every star in the universe. Trying to cover even a reasonable subset of general XSL/XSL-FO customizations would be almost impossible.

Then I considered that most readers—like you—would already have been publishing PDFs using something like FrameMaker or Microsoft Word. You already have a template that you need to reproduce in XSL-FO. So I decided to create this book as a tutorial, to walk you through migrating a template to FO.

After a brief explanation of the structure of the DITA Open Toolkit's PDF plugin, you'll start with a specification for the PDF you're going to work towards in this book: master pages, header and footer information, cover pages, heading formats, body text formats, image, list and table formats, table of contents and index creation and formatting, and more.

By working through specific tasks, you will create an actual PDF plugin that you can apply to your own template migration. So even though this book might not cover a specific kind of change, it probably covers something similar enough that you can apply the logic to whatever you need to do.

One important thing: I'm not a theorist. I'm a busy writer, information architect, and CMS implementor who's primarily concerned with the how, not the why. This book is mostly about the how, but I try to give enough background for you to understand what's happening and apply it to other situations. If you want to completely understand the why, there are good books and resources that explain XML, XSL, and XSL-FO thoroughly. I list a few of them in **Useful resources** *(p. 11).*

Who is this book for?

This book is for anybody who wants to learn how to create PDFs using the DITA Open Toolkit without learning everything there is to know about XSL-FO, XSLT, XPath, or even the DITA Open Toolkit itself.

You'll get the most out of this book if you:

- already have a firm grasp of authoring in DITA, understand the elements and attributes your organization uses, and can put together maps and bookmaps that represent the output you want;
- enjoy getting under the hood and are not afraid to try things, even if they don't work at first;
- want to know only as much detail as necessary to accomplish a task and don't need to know how the watch works to tell the time.

In other words, this book is for people who are already fluent in DITA and want to get custom PDFs up and running fast without a lot of technical background.

So, as you might expect, there's not a full explanation of the architecture of the DITA Open Toolkit in this book. There's not even a full explanation of how the PDF plugin works. Everything here is designed to tell you just as much as you need to know to do what you want to do. You don't need to understand the entire PDF transformation pipeline to understand how to change the header on even body pages.

There's also not much discussion about DITA itself and how it works. This book isn't a guide to DITA. In fact, it assumes you already have a firm understanding of authoring in DITA—including how to use the standard DITA formatting attributes and how to create maps and relationship tables that reflect appropriate relationships between topics. If you are new to DITA, you'll want to come up to speed on authoring before you try to follow these tutorials.

This book focuses on the "out of the box" elements in DITA 1.3. Specialization is one of the most important aspects of DITA: it lets you create a new element based on an existing element, where the new element inherits some or all of the characteristics of the original one. Processing specialized elements is a whole different ball of wax that isn't covered in this edition of the book.

The PDF plugin in the Open Toolkit is not pretty. It's had a lot of hands on it over the years, and it's been constantly updated to accommodate new features in DITA and the Open Toolkit while maintaining backward compatibility as much as possible. Which is to say, the XSL is less than pristine—but it works. However, one of the goals with Open Toolkit 2.x, besides incorporating processing for the new DITA 1.3

features, was to clean up the code and streamline some things. It's a big improvement over previous versions.

A lot of really smart people are working on alternatives, and many of them advise complete rewrites to sections of the PDF plugin. There's nothing wrong with that, but this book is focused on working with the plugin as-is, warts and all, and presenting the simplest, easiest-to-follow solutions for non-programmers.

Undoubtedly, there are technically better and more efficient ways to accomplish many of the tasks in this book. Especially with respect to the paths that select elements—many of them are inefficient and over-inclusive. However, they're based on the original code to keep your changes as simple and close to original examples as possible. If you come up with any better mousetraps, please post them to the Yahoo DITA Users list or the Google DITA OT Users list so we can all benefit from your ideas.

One last thing. I'm based in the United States, so this book is weighted towards U.S. standards such as inches, 8.5in x 11in paper, left-to-right word order, the English language, and so forth. In most cases, it should be pretty easy to substitute your own standards for the ones in the examples. Where that's not the case, let me know, and I'll consider it seed material for the next edition.

Keepin' it real

This book was written entirely in DITA using <oXygen/> XML Author and was formatted using the DITA Open Toolkit 2.4 using some of the same techniques you'll learn about here.

Even after you've written DITA content and developed a PDF plugin to format attractive PDF output, you still have to actually generate that output. The DITA Open Toolkit comes with a built-in tool to generate PDFs from DITA: the Apache FOP renderer. You can also use a commercial renderer like RenderX's XEP or Antenna House (this book was rendered using XEP).

There's absolutely nothing here that can't be done with DITA and the Open Toolkit out of the box—plus the custom PDF plugin this book will help you create. So let the book you hold in your hands (or look at on your screen) be the proof that all this stuff really does work!

About me

Here's the first thing I want you to know about me: I have degrees in English and Theoretical Linguistics. I've been a bookstore clerk, a bank teller, a graphic designer, a graduate teaching assistant, a technical writer, an information architect, and a CMS implementor. You'll notice that neither "programmer" nor "developer" is on that list. Aside from a few one-day classes in various software applications, I've never had any training in computer languages or programming.

I say that first because I want to stress: if I can do this, you can. You don't need to be a developer to get serious mileage out of the DITA Open Toolkit. I've been professionally creating and publishing content for over 20 years. During that time, I've worked with just about every desktop publishing application out there. No doubt, wrestling beautiful PDFs out of the DITA Open Toolkit is not as straightforward as designing a lovely FrameMaker template, but it's not rocket science either.

While I enjoy writing (though I may take a break now that this book is done...again!), I also enjoy teaching other writers. And, while I don't have a formal background in programming, I do like to tinker under the hood, which is a big advantage. I realize many writers just want to write, and when I teach, I keep that in mind. There has to be a woman (or a man) behind the curtain, but we don't have to see her (or him).

Aside from writing and teaching, I do love to talk. About nearly anything, really, but speaking at conferences and meeting others who do what we do is one of my favorite parts of this profession. Over the years, I've been fortunate enough to speak at DITA North America, DITA Europe, DITA OT Day, the STC Summit, NLDITA, WritersUA, LavaCon, Intelligent Content, SPECTRUM, DITA/Tech Comm, Real World DITA, and the FrameMaker Chautauquas. The energy and motivation, not to mention the knowledge, I've taken away from those conferences has been incredible. Do everything you can to get to at least one conference a year. It'll be amazing. I especially recommend DITA OT Day, held annually in Munich, Germany (at least, the previous three years). It's a full day of presentations by and networking with the best and brightest folks working with the OT today.

So enough about me. Well, not quite. I do have a life outside of the DITA Open Toolkit. When I can wrench myself away from the computer, I enjoy hiking and photography (often together), running, biking, team trivia, choral singing, reading, and building stuff. I also enjoy inventing languages, though my success at convincing anyone else to speak them has been spotty.

Contact me

Unfortunately, I'm not able to provide technical help with any of the exercises in this book, but I do want to hear from you! Hopefully there will be yet another edition of this book for a future DITA Open Toolkit release (3.0, anyone?). If there are other topics you'd like to see covered, if you have tips and tricks you'd like to share, or if you think I could have done a better job explaining something, post a comment on the **Yahoo! DITA Users Group**. I see the digest every day so I'll see your comment.

I may not be able to respond to you personally, but please know I appreciate every comment and suggestion.

If you do need help, please post your questions to the Yahoo! DITA Users Group so the hive mind can get you going and so other users can share the solution.

Acknowledgements

Since I started working with the DITA Open Toolkit, more people have generously helped me than I could possibly name here. But I'm going to name a few people, without whom this book wouldn't even exist. First, deepest gratitude to my publisher and editor, Richard Hamilton, who (again!) patiently waited for me to finish, offered invaluable advice on every aspect of this book and tested the Mac/Linux examples. Thanks again to Kristin Eberlein, who put me in touch with Richard in the first place, worked through an early draft, and gave me excellent feedback on what worked and what didn't. The wonderful folks at Suite Solutions have shared their considerable knowledge and expertise with me over the past few years and it's been a true pleasure to work with them.

I heard from quite a few folks using the first edition of this book, either through e-mail or via the DITA Users list. To all of you who let me know how helpful the book was to you, my humble thanks. That is exactly why I wrote it and your feedback made all of the hours of writing, testing, and re-writing more than worth it. To those of you who let me know about errors or passed along corrections or additional information, I thank you too. I tried to incorporate as much of that as I could. Any ongoing errors or misinterpretation are completely my own.

The DITA Users list has been my number one "go to" place for answers and advice. I can't count all the people who've responded to my posts, but I have to call out three of them: Kyle Schwamkrug for his contribution of the *GetChapterNumber* template to the DITA Users group and to this book; Severin Foreman for answering a gazillion questions back in the day and explaining even more carefully when I don't understand; and Eliot Kimber for spending hundreds of hours answering questions on the list, including mine—even though some of his posts are completely over my head, there's always a kernel of wisdom that gets through to me. They are my DITA gurus and I'd like to be any one of them when I grow up. Eliot also gave this book a good once-over to make sure some of the finer points are accurate, as did Dustin Clark, best known as one of the **Ditanauts** and developer of the QA plugin for DITA OT. Any errors are my own.

My eternal gratitude to my former colleagues on the Documentation Center of Excellence team at Allscripts Healthcare, LLC. When I first drove up in the DITA bus, they took a leap of faith and climbed on board. In the years after, they challenged me in every way, throwing me use cases and requirements that I would never have imagined. I know most of what I know because they asked me to figure it out. Thanks, y'all!

My current colleagues and customers at IXIASOFT Technologies are an amazing brain trust. Customer questions have challenged me to explore aspects of DITA and the Open Toolkit that I never would have looked at otherwise. The IXIASOFT Development, Services, and Support teams constantly supply me with ideas better than my own.

This list would not be complete without a bow to Jarno Elovirta, the developer responsible for most of the work on the DITA Open Toolkit over the last few years. Most of the work he does is on his own time, purely for love of the code and a desire to provide DITA users with the functionality we need to do our own work and publish our own documents. We are in your debt, Jarno! If I give the nod to Jarno, I have to give one to Robert Anderson as well. Robert is the project lead for the DITA-OT at IBM; he and Jarno are each other's right hands.

Finally, I'd like to recognize the amazing efforts of the OASIS DITA Technical Committee. No one has worked harder than they to bring the DITA 1.3 specification to reality, thus laying the foundation for even more wider adoption and productive use of DITA.

Lastly, I want to thank my assistant Zippy for his peerless company and feedback, constructive or otherwise.

This book is (again) for Nana, Daddy, and Sue. All my love.

Chapter 1

Introduction

What's different in this edition?

This edition of *DITA For Print* updates all examples to work with the DITA Open Toolkit 2. In addition, it includes new exercises that explain how to do the following:

- use localization strings along with localization variables
- move away from font-mapping
- customize the new built-in back cover page functionality
- [DITA 1.3] add templates and stylesheets to process and format the new troubleshooting topic type and elements
- format the index for FOP, Antenna House, and XEP
- use the new codeblock line numbering functionality
- specify the table heading—column or row
- determine the content of links to figures and tables
- generate bookmarks for index letter headings

In addition, existing exercises have been updated to explain the new or better way to:

- create a PDF plugin
- create a batch file to run the Open Toolkit
- create (not copy) attribute set file and stylesheets in your plugin
- retain the **topic.fo** file
- set up master pages
- set up headers and footers

Some of the changes are minor and some are pretty big, but they're all covered.

PDF changes in DITA OT 2.5

DITA Open Toolkit 2.5 will be released later in 2017, and among other new features, the PDF plugin will undergo some changes. While information about all of those changes is not yet available, there is one important change that you should know about now.

As you're aware (or as you'll see while using this book), it's not always as simple as it seems it should be to make basic changes to the `org.dita.pdf2` plugin. Even simple template or style overrides sometimes require working with many lines of code. In DITA OT 2.5, Jarno Elovirta has tried to simplify some of those basic overrides and make them more accessible. Doing so means re-writing certain sections of the legacy code (meaning some of the `org.dita.pdf2` code that this book works with).

However, there are lots of PDF plugins out there that are based on the existing templates and stylesheets, and some of these changes could break those plugins. It's not realistic to require people to drop everything and re-write their plugins simply to conform to the new templates and stylesheets if they want to use DITA OT 2.5.

As a compromise, the legacy templates and stylesheets will be bundled into a PDF plugin named `org.dita.pdf2.legacy`. To retain complete backwards compatibility between your existing PDF plugins and DITA OT 2.5, you simply install this legacy plugin and continue to run PDF transforms as before. The `org.dita.pdf2.legacy` plugin restores the older default code paths into the default PDF2 process so that customizations continue to operate based on that older code. (Thanks to Robert Anderson for this description.)

The legacy plugin can be found on GitHub at **https://github.com/dita-ot/org.dita.pdf2.legacy**.

Just to be clear, the changes in this book pertain to the `org.dita.pdf2` plugin as found in DITA-OT 2.4. The things you choose to customize for your PDF plugin using these instructions might or might not be affected by the changes coming in DITA OT 2.5. If you find that your PDF plugins break in OT 2.5, then you will need to install `org.dita.pdf2.legacy` to make DITA OT 2.5 PDF processing backwards-compatible with PDF plugins developed using earlier versions of the DITA OT.

What you'll need

To work with the PDF plugin of the DITA Open Toolkit, you will need a few tools.

DITA Open Toolkit

You need to have the DITA Open Toolkit installed, of course. This book is written for the DITA Open Toolkit 2.4, which is the current stable release as of the publication of this book.

Note There are lots of possible setups you could be working with. XML editors such as XMetaL or <oXygen/> come with their own installation of the DITA Open Toolkit. Because there's really no way to cover all the possibilities without creating a hopeless tangle, this book assumes a separate, standalone installation of the DITA Open Toolkit 2.4. If you're using the DITA Open Toolkit in some other configuration, you'll need to make corresponding adjustments to the instructions in these exercises.

PDF renderer

The DITA Open Toolkit comes with Apache FOP, a free program that generates PDF from an intermediate format (called "fo," for "formatting objects," and that's all you need to know about it). In many cases, FOP is adequate, but it may not be robust enough for a full production environment. Commercial alternatives are RenderX's XEP and Antenna House. Both have free trial versions. This book was created using XEP, but all of the exercises assume you are using FOP.

XML editor

You also need an XML editor that can validate your XSLT stylesheets while you're editing them. This will help you spot typos, missing elements, and other errors more quickly. If you've been authoring in DITA, you likely already have an XML editor such as <oXygen/>, XMetaL, Arbortext, XMLSpy, Syntext Serna, or XMLMind's XML Editor, to name a few. Alternatively, you can use a robust text editor such as the free Notepad++ or emacs using the NXML mode. With a free plug-in, Notepad++ can do basic syntax checking of XML files to help you spot some errors, and it color-codes elements, attributes, values, and plain text to make reading XML markup easier. Notepad++ also has some useful features such as file compare. If you're confident editing XML and XSL files without full validation, Notepad++ is a quick, lightweight alternative to a full-featured XML editor.

Search and replace utility

You'll need a good utility to search across multiple files and folders. Until you learn where things are in the PDF plugin, you'll do a lot of searching for specific attribute sets and templates. The free TextCrawler is a good option for Windows. For Macintosh or Linux systems, take a look at Find & Replace It! or the grep command.

XSL-FO reference

An absolute must is a good XSL-FO reference that lists and explains the different attributes available for formatting. When you get comfortable with the basics, you might want to explore other things you can do to add functionality to your PDFs and make your publishing process more powerful and flexible. Two books I can recommend are *XSL Formatting Objects Developer's Handbook*, by Doug Lovell and *XSL-FO*, by Dave Pawson. Both are excellent. Of the two, *XSL-FO* has a more instructional tone. Both of these books are complete references to XSL-FO, explaining how to create an FO file from the ground up (or rather, from the root up). Because the DITA OT creates the FO file for you, you probably won't use the entire book, but you will absolutely use the lists of FO elements and attributes that both books provide.

Alternatively, you can use an online FO reference. **W3Schools** has a pretty good one.

It's always a good idea to have a standard's specification around to refer to as well. The XSL-FO specification is at **www.w3.org**. Note that it was last fully updated in 2006, and at least for the time being, the standard is not being actively developed.

Some content

Finally, you need content you can use for testing. If you don't already have DITA topics and maps, you can use the garage samples in the samples subfolder of the DITA Open Toolkit or the source files in the doc subfolder of the Open Toolkit.

If you find that you need some more robust content for testing, you can download the Thunderbird source content from the **dita-demo-content-collection** on GitHub, contributed by Joe Gollner and others. The collection contains topics and maps using most of the elements and structures available in DITA 1.3.

The **DIM (Dynamic Information Model)** project on GitHub also provides sample content (which can also serve as great starter content for your own style guide) as well as a library of Schematron patterns that enforce the rules outlined in the style guide, XSLT scripts to generate a Schematron file, an oXygen configuration file, and an oXygen framework. The style guide content was contributed by Comtech Services and the remainder by Syncro Soft, makers of the oXygen XML editor.

Tools websites

DITA Open Toolkit: **http://www.dita-ot.org/** or **https://github.com/dita-ot/dita-ot/releases/** (for previous versions and source code)

FOP: **http://xmlgraphics.apache.org/fop/**

XEP: **http://www.renderx.com/tools/xep.html**

Antenna House: **http://www.antennahouse.com**

: **http://www.oxygenxml.com/download.html**

XMetaL: **http://na.justsystems.com/index.php**

XMLMind: **http://www.xmlmind.com**

Notepad++: **http://notepad-plus-plus.org**

TextCrawler: **http://www.digitalvolcano.co.uk/content/textcrawler**

Find & Replace It!: the Mac App Store

Useful resources

It takes a village...

Because DITA is an open standard, thousands of people use it, and most of those people are happy to help you. If you haven't already done so, consider joining the Yahoo! DITA Users list and the Google DITA-OT Users list. Both lists are active and well-monitored. The Search feature isn't the best, but you should still try to search the archives as thoroughly as you can to make sure you don't ask a question that's already been asked and answered many times.

The OASIS DITA Users list is an alternative to the Yahoo! DITA Users list. It's currently not as active as the Yahoo list, but there is a large, searchable archive.

If you need a brush-up on DITA itself, there are several very good books available that can get you going. There's a list at the end of this section.

These lists are by no means exhaustive. There are many other excellent websites and blogs...too many to list. Just enter "DITA" into a search engine, and you'll see that you could easily spend a lifetime exploring all the information available. Don't be distracted by Ms. Von Teese.

Websites

DITA 1.3 specification: **http://docs.oasis-open.org/dita/dita/v1.3/dita-v1.3-part0-overview.html**

DITA Community on GitHub: **https://github.com/dita-community**

oXygen XML's GitHub site: **https://github.com/oxygenxml/**

GitHub site for DITA-OT download: **https://github.com/dita-ot/dita-ot/releases/**

DITA-OT site for download and documentation: **http://www.dita-ot.org/**

Yahoo! DITA Users: **https://groups.yahoo.com/neo/groups/dita-users/info**

Google DITA-OT Users: **https://groups.google.com/forum/#!forum/dita-ot-users**

Eliot Kimber's specialization tutorials: **http://dita4practitioners.github.io/dita-specialization-tutorials/**

Guide to ANT build properties:**http://www.dita-ot.org/2.2/parameters/index.html**

List of entity codes: **http://www.entitycode.com**

Apache ANT: **http://ant.apache.org/**

Apache ANT manual: **http://ant.apache.org/manual/index.html**

Blogs

DITA Writer blog: **http://www.ditawriter.com/**. This blog in turn contains one of the more comprehensive lists of DITA resources, including books, websites, and blogs. A great starting point for research.

Scriptorium: **http://www.scriptorium.com/blog/**

oXygen XML: **http://blog.oxygenxml.com/**

Metadita.org: **http://metadita.org/toolkit/**. A few of Robert Anderson's thoughts on DITA and the OT.

Jarno Elovirta: **http://www.elovirta.com/**. Jarno's thoughts on DITA, the DITA OT, akido, and...whatever else occurs to him.

Books

Introduction to DITA: A User Guide to the Darwin Information Typing Architecture (Jennifer Linton, Kylene Bruski; Comtech Services, Inc.)

DITA 101: Fundamentals of DITA for Authors and Managers (Ann Rockley, Steve Manning and Charles Cooper; The Rockley Group, Inc.)

Practical DITA (Julio Vazquez; Lulu)

DITA Best Practices: A Roadmap for Writing, Editing, and Architecting in DITA (Laura Bellamy, Michelle Carey, Jenifer Schlotfeldt; IBM Press)

DITA for Practitioners, Volume I: Architecture and Technology (Eliot Kimber; XML Press)

DITA Style Guide (Tony Self; Scriptorium Publishing Services, Inc.)

XSL Formatting Objects Developer's Handbook (Doug Lovell; Sams Publishing)

XSL-FO (Dave Pawson; O'Reilly Media, Inc.)

Tools/Utilities

KeyAnalyzer: **http://www.maxprograms.com/products/keyanalyzer.html**. A sleek utility by Rodolfo Raya that analyzes keys in a map and delivers a report of all keys, their usage, their resolutions in context (keyscope-aware), and lists any undefined keys.

Jarno Elovirta's web-based DTD and plugin generator: **https://dita-generator-hrd.appspot.com/**

Online training

Scriptorium's LearningDITA series: **http://www.learningdita.com/**

DITA Open Toolkit plugins

DITA community plugins: **https://github.com/dita-community**

QA plugin for DITA OT: **https://sourceforge.net/projects/qa-plugin-dot/**. An extremely useful plugin that scans an input map and produces a report of errors such as incorrect terminology, questionable structure and more. You can easily customize and extend what the plugin searches for and reports on. Also available at **https://github.com/dita-community/**.

Extended DITA-OT Plugin: **http://www.hscherzer.de/dita.html**. A set of improvements to several DITA-OT output types, including PDF. The plugin offers improved processing for: paragraph, section, table, fig/image, note, links, lists, page control, title, and mini-TOC. Offered by Helmut Scherzer.

DITA4Publishers: **http://www.dita4publishers.org/**. Eliot Kimber's set of additional Open Toolkit plugins and transformation frameworks, including DITA to InDesign and the EPUB transform, as well as extensions to the PDF2 and HTML transforms.

The DITA Open Toolkit itself

After you work your way through this book, if you're eager to learn more about customizing the PDF plugin (or any other kind of plugin), you can turn to the Developer Reference documentation within the DITA Open Toolkit folder itself.

This documentation is in DITA-OT/doc, and it provides a wealth of information about the different processing stages of a DITA Open Toolkit build, as well as details on extending and further customizing plugins. Many thanks to Roger Fienhold Sheen for his tireless work updating this documentation and making it exponentially more usable.

This information is delivered as a map and source topics, so you'll probably want to build a PDF to browse it easily, which you will be able to do after you finish the first few chapters of this book!

Some conventions

Book organization

Each chapter addresses one particular aspect of a print publication. Most chapters contain exercises you need to complete to set up a PDF plugin that meets the specifications outlined in **Specifications used in these exercises** *(p. 393)*. Some chapters have other exercises to do other cool things with your PDF plugin. These exercises can be found in the **What you need to know** section in those chapters.

Folder names

For the sake of simplicity, in all the examples and paths throughout this book, the DITA Open Toolkit folder is called DITA-OT. If you download the DITA Open Toolkit 2.4, you'll see that by default, the folder is named dita-ot-2.4. You can leave that name as-is or change it. Just keep in mind that DITA-OT refers to whatever you have actually named your folder, so edit the paths you use accordingly.

Paths

Most of the code samples and paths in the book use the URL syntax convention of forward slashes because DITA and XSL-FO are Web applications. Windows systems should accept the forward slashes as well in almost all cases, though you will see backslashes in some Windows-only command lines.

Typographical conventions

The following type styles are used to make certain items more obvious.

- **Bold**: files, emphasized sections in code samples.
- *Italic*: attribute sets, attributes, variables, parameters, and properties.
- ***Bold italic***: template names.
- `Monospace`: file paths, folders, code samples, plugins, and elements (DITA, XSL, and FO).
- Double straight quotes: attribute values and marker names.
- ▶: indicates a line return has been inserted in a code example for clarity (usually because without the line break, the code would not fit on the printed page); the actual code does not include a line return at that location.

Order of exercises

You can do most of the exercises in any order. You might not always get exactly the results you expect, especially with respect to formatting, but your build will work. There are a few sets of exercises that must be done in a specific order because each exercise depends on variables, attribute sets, or page layouts created in a previous exercise. These exercises are grouped together with a note recommending that you complete them in order.

Comment, comment, comment...test, test, test

Anytime you change your code, there are two things you should do that will make life a lot easier.

Comment your code

First, comment all your changes, even minor ones like changing an attribute value or a localization variable value. Even for these simple changes, it's good to have a record of what you touched and what the original value was.

For more complicated changes, it almost goes without saying that you need to explain what you did and why so anyone who comes across the code can understand why the change was made. A year from now, even you might not remember.

Ideally, start all your comments with something unique, to make it easy to find them later. Here is a good example of an XML comment:

```
<!--LWW (20160518): Switched the output order of the figure and its title-->
```

This comment includes the initials of the person making the change, the date the change was made, and an explanation of the change.

Comments can go almost anywhere in your XSLT stylesheets, attribute set files, or localization variables files. The only place you can't put a comment is within another comment.

All comments must start with <!-- and end with -->.

Test your changes frequently

Second, test your changes as you go. Test each change as you make it. If you delay testing until you've made several changes and then something goes wrong, it will be more difficult for you to find exactly where the mistake occurred. Even if the changes seem simple and straightforward, test them. Then, when you encounter a problem, you will know it occurred within the last change you made.

From painful experience, I can tell you that even a small error, such as forgetting to close an element tag or leaving out a quote character, can set off a cascade of error messages that make it look like there are dozens of mistakes, when in fact only one error messed up everything after it.

Most of the exercises in this book end with, "Save your changes, run a build, and test your work." Seriously, do that.

Use source control

Anytime you work with code (or any kind of file, for that matter), it's a good idea to keep your files under source control. With source control, if you hopelessly mess up, you can always go back to a previous, good revision of anything you've placed under source control. You can keep a revision history of all your changes and know what files you have updated It's easy to compare two revisions of a file to see exactly what has changed, and you can restore your files if your system should crash.

There are many free source control utilities available. **Apache subversion (SVN)** is a widely used, free utility that will let you set up a source control repository. SVN provides a command-line interface, and most XML editors provide a built-in graphical interface to SVN. You can also use the free utility **Tortoise SVN**, which provides a graphical interface. As an alternative to SVN, you can also use **Git**. Each of these source control tools has advantages and disadvantages; it really comes down to personal preference and tools availability.

Chapter 2

Custom PDF plugin creation

The whole point of this book is to show you how to create and use your own PDF plugin with the DITA Open Toolkit. This chapter gets you started.

First, I explain what a PDF plugin is and how the one that comes with the DITA Open Toolkit is organized.

Next, I explain how to download and install the DITA Open Toolkit (if you haven't already done that).

Finally, I explain how to create your own PDF plugin and tell the DITA Open Toolkit to use it instead of the default plugin.

> **Note** You should perform the exercises in this section in the order they're given. Each exercise builds on the previous one, so skipping around is not a good idea.

What is a PDF plugin and why do you need one?

Out of the box, the DITA Open toolkit provides ANT build files, XSLT stylesheets, and Java executables that let you convert a collection of DITA topics into a variety of output formats, including PDF. The processing for each of these output formats comes bundled in a set of plugins that you can extend and customize. In this book, you'll explore how to create a plugin that customizes PDF processing.

"Do I need my own PDF plugin," you ask. Unless you think the out-of-the-box PDFs produced by the DITA Open Toolkit are just dandy, then yes—you need a PDF plugin. No question about it.

Now that we've gotten that out of the way, here's the story with PDF plugins.

To customize PDF output, you could just edit the files in the default plugin (`org.dita.pdf2`). That would be fine, until your boss comes to you and says, "We also need to produce a Quick Reference Guide. It should be 5.5 inches by 8.5 inches. And all the fonts need to be smaller. And the margins need to be different. And the headers and footers need to be different...." What do you do? If you're using the default files, you have to keep editing them over and over. Pretty soon, you won't be able to keep up. And maintaining more than one type of PDF output at the same time will be nearly impossible.

Instead, the right strategy is to create your own separate PDF plugin. That way, you bundle up all of your changes into one nice, neat package that you can easily maintain.

Going back to the example above, it makes a lot more sense to create one PDF plugin for User Guides and another for Quick Reference Guides. That way, you simply call the plugin you need when you create a PDF. It's like having different templates in Adobe FrameMaker or Microsoft Word for different kinds of documents.

When you have a nicely bundled-up PDF plugin, you can easily move it from one installation of the DITA Open Toolkit to another. (Or you can keep it in another location entirely; it doesn't have to be within the DITA OT folder.) You can also send the plugin to other writers on your team, to contractors, to marketing, or to whomever. It's self-contained and portable.

So now that you're convinced you need a PDF plugin, exactly what is one?

To create PDFs from a map or bookmap, the DITA Open Toolkit uses many different files. These files are in various places in the Open Toolkit, but most of them, and almost all of the ones you'll need to work with, are found in `DITA-OT/plugins/org.dita.pdf2`. A PDF plugin contains only the files you changed for your customization. (In this path, and throughout the book, DITA-OT refers to whatever you've named your DITA Open Toolkit folder.)

These files fall mainly into three categories: attribute set files, XSLT stylesheets, and variables files.

- *Attribute set files* control the way each element looks in the PDF, including indentation, font characteristics, line spacing, hyphenation, alignment, and so forth. Think (very generally): "attribute set file equals appearance."

- *Stylesheets* control the way each element is processed. For example, if you want to number a paragraph or assign a custom attribute set to an element based on the value of the *@outputclass* attribute, you would use the appropriate stylesheet. Think (again, very generally): "stylesheet equals behavior."

- *Variables files* do several things, but one of the most important is to dynamically insert language-specific boilerplate text during processing. For example, if you want to add a label in front of a particular element and that label differs depending on the language, you would set that up in a variables file. Variables files also play an important role in setting up headers and footers. Think: "variables file mostly equals translation and headers and footers."

In addition to these files, a plugin can also include its own ANT build file. If you want a plugin that simply overrides attribute sets, XSLT stylesheets, and localization variables, you can create the plugin without an ANT build file. However, if you need additional functionality, you need to include an ANT build file. It is easy to create a plugin with an ANT build file, and doing so will make it easier to add functionality in the future, so we will take that approach in this book.

Upgrading existing plugins

Objective: A list of things to look out for when upgrading a plugin created for an earlier version of the DITA Open Toolkit.

This book assumes you're creating a brand-new PDF plugin from scratch. However, if you're upgrading an existing plugin that you created for an earlier version of the DITA Open Toolkit, here are some things to look out for:

- *insertVariable* template

  This template (which was used all over the place) has been deprecated. Instead, use the *getVariable* template.

- *$ditaVersion*

  This variable, previously used in templates in **bookmarks.xsl**, **front-matter.xsl**, **index.xsl**, and **toc.xsl**, is no longer defined in 2.4. From here forward, the assumption is that your content is compliant with at least DITA 1.1.

- Stylesheets specific to Apache FOP, Antenna House Formatter, and RenderX XEP

  Rather than all being mixed into the org.dita.pdf2 plugin, they are now in separate plugins for each processor. If your plugin uses any of these processor-specific stylesheets, you might need to extend the new org.dita.pdf2.fop, org.dita.pdf2.axf, or org.dita.pdf2.xep plugins.

- Localization variables

  Many of these that are no longer used in PDF processing have been deprecated and will be removed in an upcoming release. If your plugin uses them, you should refactor those usages.

  In addition to the localization variables files, the org.dita.pdf2 plugin also uses the strings files found in DITA-OT/xsl/common/. Many variables that were previously in the localization variables files are now found in the strings files.

This is not a comprehensive list of gotchas, but it does cover some of the major differences. To keep this list manageable, I've included only the differences between version 1.8.5 (which is what the previous edition of this book was written for) and the current version, 2.4.

If you're upgrading a plugin created for an earlier version of the DITA OT, the differences are going to be more drastic. If your plugin was created for a DITA OT version prior to 1.6, I advise starting over from scratch. Seriously.

For a more complete list of changes between each version since 1.5.4, see the **DITA OT documentation**.

Organization of the org.dita.pdf2 plugin

Before you get started creating your own plugin, you need to understand a little about the `org.dita.pdf2` plugin, because your plugin will have almost the same organization.

IBM originally developed DITA and the Open Toolkit to create online output. IBM has its own print solution, but they did not include it with the Open Toolkit components they contributed to open source. After DITA became a public standard, everybody recognized the need to publish PDFs as well.

The `org.dita.pdf2` plugin was originally developed by Idiom and is still sometimes referred to as the "Idiom PDF plugin." However, the plugin is now maintained by a different group of people. Like the rest of the DITA Open Toolkit, the PDF plugin is open source, meaning that no one really owns it. In this case, "maintains" just means that certain groups make the code changes that are included in the official plugin as it ships with the DITA Open Toolkit.

Out-of-the-box, the DITA Open Toolkit is organized like this:

Figure 1: Default OT organization

In addition to these folders, there are also build files, catalog files, the integrator file, and the `startcmd` batch and shell files. More on those later.

The `org.dita.pdf2` plugin, not so shockingly, is found in the `plugins` subfolder.

> **Note** It wasn't always that way. In DITA Open Toolkit versions before 1.6, the plugin lived in the `demo` subfolder because the capability was presented more or less as a demonstration of future possibilities. As of DITA Open Toolkit 1.6 (released in 2012), it finally found its way to its logical location.

Plugins folder organization

Here's what the `plugins` folder looks like out of the box:

```
📂 plugins
   📁 com.sophos.tocjs
   📁 org.dita.base
   📁 org.dita.docbook
   📁 org.dita.eclipsecontent
   📁 org.dita.eclipsehelp
   📁 org.dita.html5
   📁 org.dita.htmlhelp
   📁 org.dita.javahelp
   📁 org.dita.odt
   📁 org.dita.pdf2
   📁 org.dita.pdf2.axf
   📁 org.dita.pdf2.fop
   📁 org.dita.pdf2.xep
   📁 org.dita.specialization.dita11
   📁 org.dita.specialization.eclipsemap
   📁 org.dita.troff
   📁 org.dita.wordrtf
   📁 org.dita.xhtml
   📁 org.oasis-open.dita.v1.2
   📁 org.oasis-open.dita.v1.3
```

Figure 2: Plugins folder organization

It's kind of scary at first. Block out everything but the `org.dita.pdf2` subfolder; that's where the PDF plugin lives.

Keep in the back of your mind the other PDF plugin folders: `org.dita.pdf2.axf`, `org.dita.pdf2.fop`, and `org.dita.pdf2.xep`. These plugins correspond to the three PDF renderers mentioned earlier. Because of the subtle, and not-so-subtle, ways each renderer handles PDF formatting, each one has some renderer-specific stylesheets.

If you're designing a plugin to be used with Antenna House, for example, you'd start with the generic stylesheets available in `org.dita.pdf2` and supplement them as necessary with the Antenna House-specific stylesheets in `org.dita.pdf2.axf`. These three plugins are structured like `org.dita.pdf2`, so once you learn your way around that plugin, you'll understand the other three.

Note Why is there also a folder named `org.dita.pdf`? Well, because the PDF plugin has been through a lot of changes in its life. At one point, the developers replaced the original one with a snazzier new one. But because everything in the DITA Open Toolkit is designed to be backwards-compatible, the old one is still hanging around. The two folders are legacies of that history. There's nothing much in the `org.dita.pdf` folder, though, and these days, when you create a PDF, you're using the `org.dita.pdf2` plugin.

`org.dita.pdf2` **folder organization**

Here's what the `org.dita.pdf2` folder contains out of the box.

Figure 3: `org.dita.pdf2` **organization**

You won't need to work with any of those **.xml** files for a typical PDF plugin, so forget about them. You'll invoke some of them behind the scenes, but you won't directly edit them.

There are the following sub folders in the `org.dita.pdf2` folder:

cfg   Contains the sub folders `common` and `fo`. More on these below.

Customization   This folder used to contain a starter set for your own PDF customization. It became obsolete in version 1.6 of the DITA Open Toolkit.

lib   Contains **fo.jar**, a Java executable needed for PDF processing.

xsl   Contains the XSLT stylesheets that process DITA content into PDFs. You'll learn how to copy these XSLT stylesheets to your own PDF plugin and modify them.

The `cfg` subfolder is where the magic happens, as far as customization is concerned. Well, most of the magic, anyway. The `cfg` subfolder has this structure.

Here's a description of each of the sub folders of cfg.

artwork   Contains images and icons used by the default PDF plugin. If you have custom images for your plugin, such as icons or logos, store them in the artwork subfolder.

index   Contains the index sorting files for various languages. These files determine the order in which the DITA Open Toolkit sorts the index for different languages.

vars   Contains files with boilerplate text that can be dynamically inserted in your PDF, avoiding the need for redundant translation. Boilerplate text includes content such as the labels for figure and table titles, header and footer information, and more.

attrs   Contains the attribute set files that determine the appearance of DITA elements in a PDF.

i18n   Contains the character sets for various languages. These character sets specify which subset of the Unicode character set the language uses.

xsl   Empty by default (except for the **custom.xsl** file). You don't really do anything with this folder. But, in your plugin folder, you'll have an xsl subfolder just like this one. You copy any XSLT stylesheets you want to modify for your plugin into this folder.

The fo subfolder also contains **font-mappings.xsl** and **layout-masters.xsl**. **Font-mappings.xsl** determines the fonts that appear in your PDF. **Layout-masters.xsl** determines the master pages available in your PDF—body master pages, index master pages, cover pages, and so forth. If this doesn't make a lot of sense right now, don't worry. As you work through the exercises in this book, you'll understand these folders better.

DITA Open Toolkit 1.5 and earlier

In case you don't know, there was a significant re-architecting of the PDF plugin in the DITA Open Toolkit beginning with version 1.6. For one thing, before DITA Open Toolkit 1.6, PDF wasn't really a plugin at all. It was a customization that lived in DITA-OT/demo/fo. Although the customization folder was organized almost identically to the plugin, the method for creating and calling a PDF customization was different from the method explained in this book.

Also, many files had a "_1.0" suffix. These files supplemented the original files with code specific to new functionality introduced with DITA 1.1. All of those files went away with version 1.6 of the DITA Open Toolkit. However, if you're working with older versions, you need to be aware of them.

Many people still use older versions of the DITA Open Toolkit, and it would have been nice to include instructions for those versions in this book. However, it was impractical to do so. The good news is that almost everything in this book, except this chapter, applies to older Open Toolkit releases. You might have to tweak things a little, but you should be able to follow the steps and get the results you want.

DITA Open Toolkit post-2.0

If you've worked with previous versions of the DITA Open Toolkit, you'll also notice significant changes between, say version 1.8.5 (which was the focus of the first edition of this book) and version 2.4.

Beginning with DITA Open Toolkit 2.0 (released in late 2014), there has been an ongoing clean-up effort. This includes removing obsolete or redundant Java and XSLT code, converting stylesheets to XSLT 2.0, implementing a new command line, and adding processing for the new DITA 1.3 features such as scoped keys, branch filtering, and so forth.

To see a comprehensive list of changes for each version, go to: **http://www.dita-ot.org/download**, where you can see a quick overview of changes and new features for any version and access a link to the full release notes.

One reason the DITA Open Toolkit had become an accumulation of old and inefficient code was the commitment to complete backwards compatibility. While this commitment made upgrades a lot easier, it eventually resulted in a mess that was hard to maintain and impossible to optimize. Beginning with version 2.0, there is no more promise of backwards compatibility.

This means that if you have existing plugins that were developed for versions of the DITA Open Toolkit before 2.0, it's likely you are going to have to update them to some degree. The changes may be minor or they may be so involved that you might as well start over. In any case, be prepared for your pre-2.0 plugins to sputter and backfire, and possibly crash and burn. (Naturally, this metaphor would in no way be an allusion to jet packs, which Robert Anderson, one of the principal architects of the DITA Open Toolkit, promised as an upcoming DITA OT feature. No, really, he did: **DITA OT Jet Packs**.)

Exercise: Download and install the DITA Open Toolkit

Objective: Download the DITA Open Toolkit and install it.

The first step in creating a PDF plugin is to make sure you have a working DITA Open Toolkit installation. In case you don't already have one, here are brief instructions for installing and testing it. These instructions assume you have Java installed on your computer. If you don't, you need to install it separately because the DITA Open Toolkit requires Java to run. Installing Java is beyond the scope of this book.

1. Download the latest stable version of the DITA Open Toolkit from **http://www.dita-ot.org/download**
2. After downloading the zip file, expand it to a folder on your computer.

A good practice, and the practice that this book assumes, is to expand it to your C: drive on Windows or the Mac/Linux equivalent.

You'll end up with a folder named something like dita-ot-2.4.

> **Note** Look carefully. After you expand the downloaded zip, there is a dita-ot-2.4 folder with *another* dita-ot-2.4 folder inside of it. You want to move the innermost one to your C: drive and discard the now-empty one.

In versions prior to 2.0, there was a demo build you could run to verify you had a working Open Toolkit installation. Unfortunately, this demo build no longer exists.

Exercise: Create your own PDF plugin

Objective: Create a custom PDF plugin that you can use to build PDFs with your own look and feel.

Before you start

Complete the exercise **Download and install the DITA Open Toolkit** *(p. 24).*

Creating your own plugin makes it easier to move your changes to a new version of the DITA Open Toolkit and also ensures the originals (in org.dita.pdf2) remain intact in case you ever need clean copies or just want to use the default PDF processing.

I. In DITA-OT/plugins, create a new folder named com.company.pdf.

You could name this folder anything you want, depending on what you want to name the plugin. For the purposes of this book, I'll use this name.

> **Note** You can download a ready-to-use copy of this folder from **http://xmlpress.net/publications/dita/dita-for-print/plugin** to use as a comparison as you work through these exercises.

The plugins in the plugins folder use Java-style package naming. Most of the plugins in the plugins folder start with org. The typical naming convention for plugins is to start the folder name with org for non-commercial plugins (such as those developed for the DITA Open Toolkit by various non-commercial organizations) or com for commercial plugins (such as those developed by companies for their own use). The plugin name normally uses the real Internet domain of the person or organization that owns the plugin. The third part of the plugin typically names the kind of plugin and should be clear enough to enable you to distinguish between different plugins that you develop.

For example, if you work for BigCorp, Inc., (www.bigcorp.com) all of your plugin folder names would probably start with `com.bigcorp`. The third part of the folder name is what keeps them separate: `com.bigcorp.pdf`, `com.bigcorp.xhtml`, `comp.bigcorp.eclipse`. But keep in mind that you might develop several PDF plugins, for different styles or PDF. In that case, you'd want the third part of the plugin folder names to be more specific: `com.bigcorp.pdf-ug` (for User Guides), `com.bigcorp.pdf-qr` (for Quick Reference Guides), and so forth.

2. In `com.company.pdf`, create a subfolder named `cfg`.

3. In the `cfg` subfolder, create subfolders `common` and `fo`.

4. In the `common` subfolder, create subfolders `artwork`, `index`, `strings` and `vars`.

5. In the `fo` subfolder, create subfolders `attrs` and `xsl`.

When you're finished, the folder structure should look like this:

Figure 4: Plugin folder structure

6. In the `cfg` subfolder, create a file named **catalog.xml**.

7. Add this content to **catalog.xml**.

```
<?xml version="1.0" encoding="UTF-8"?>
<catalog prefer="system" xmlns="urn:oasis:names:tc:entity:xmlns:xml:catalog">
  <uri name="cfg:fo/attrs/custom.xsl" uri="fo/attrs/custom.xsl"/>
  <uri name="cfg:fo/xsl/custom.xsl" uri="fo/xsl/custom.xsl"/>
</catalog>
```

The **catalog.xml** file in your PDF plugin is like a tourist information booth; it tells the Open Toolkit where to look for attribute set files, XSLT stylesheets, index files, and more in your plugin. The two lines in this file tell the Open Toolkit to look first for attribute files and XSLT stylesheets in `com.company.pdf` and use those rather than the corresponding default ones in `org.dita.pdf2`.

8. In the `fo/attrs` subfolder, create a file named **custom.xsl**.

9. Add this content to **custom.xsl**.

```
<?xml version="1.0"?>
<xsl:stylesheet xmlns:xsl="http://www.w3.org/1999/XSL/Transform"
    xmlns:fo="http://www.w3.org/1999/XSL/Format"
    version="2.0">

</xsl:stylesheet>
```

10. Copy **custom.xsl** to the `fo/xsl` subfolder.

Even though the **custom.xsl** files in the `fo/xsl` and `fo/attrs` subfolders have the same name and structure, they serve different (though similar) purposes. In this book, I'll refer to them as **custom.xsl** (attrs) and **custom.xsl** (xsl) to distinguish them.

11. In the `com.company.pdf` folder, create a file named **plugin.xml**.

12. Add the following content to the file.

```
<?xml version='1.0' encoding='utf-8'?>
<plugin id="com.company.pdf">
    <require plugin="org.dita.pdf2" />
    <feature extension="dita.conductor.transtype.check" value="custpdf" />
    <feature extension="dita.transtype.print" value="custpdf" />
    <feature extension="dita.conductor.target.relative" file="integrator.xml" />
</plugin>
```

The *dita.conductor.transtype.check* and *dita.transtype.print* feature extensions both have the value **custpdf**, which is the name of the transform type that calls this plugin.

13. In the `com.company.pdf` folder, create a file named **integrator.xml**.

14. Add the following content to the file:

```
<?xml version='1.0' encoding='utf-8'?>
<project name="com.company.pdf">
    <target name="dita2custpdf.init">
        <property name="customization.dir" location="${dita.plugin.com.company.pdf.dir}/cfg"/>

        <property name="pdf2.i18n.skip" value="true"/>
    </target>
    <target name="dita2custpdf" depends="dita2custpdf.init, dita2pdf2"/>
</project>
```

Notice that the project name (`com.company.pdf`) corresponds to the name of the plugin folder. When you create new plugin folders, be sure to change the project name to match.

Notice also that the names of both targets in this file (**dita2custpdf.init** and **dita2custpdf**) incorporate the name of the custom transtype (**custpdf**). This is essential. The name of the first target should always be **dita2**_[transtype]_**.init**. The name of the second target should always be **dita2**_[transtype]_.

The second target should always depend on the first target and on any other targets that it extends. In this case, you're extending the main target of the `org.dita.pdf2` plugin, **dita2pdf2**, so you must also include **dita2pdf2** in the *depends* list.

This means that the DITA Open Toolkit runs **dita2pdf2** first and then overrides aspects of it as appropriate with corresponding targets and properties, as specified in your `com.company.pdf` plugin.

The first target (**dita2custpdf.init**) has two properties. The first, *customization.dir*, specifies the location of all your stylesheets, attribute set files, and variables files. The second property, *pdf2.i18n.skip*, is set to "true" which overrides a default setting in the Open Toolkit and disables i18n font processing. To learn more about that, see **About fonts in the PDF plugin** *(p. 81)*.

15. Copy **layout-masters.xsl** from `DITA-OT/plugins/org.dita.pdf2/cfg/fo/` to `DITA-OT/plugins/com.company.pdf/cfg/fo/`.

I recommend copying this entire file rather than creating an empty file and copying individual page masters from `DITA-OT/plugins/org.dita.pdf2/cfg/fo/`**layout-masters.xsl** into your copy.

Reality check

Reality check: Your plugin folder should look exactly like this fully expanded example:

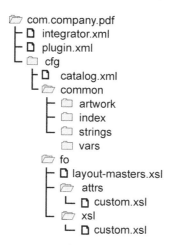

Figure 5: Plugin folder structure - expanded

If it doesn't, go back and review these steps and make changes until it does. If you get sideways here, nothing else will work correctly!

You now have the essential parts of an PDF plugin folder set up. The next step is to integrate your plugin into the DITA Open Toolkit.

Exercise: Integrate your plugin into the DITA-OT

Objective: Make the DITA Open Toolkit recognize your custom PDF plugin.

Before you start

Complete the exercise **Create your own PDF plugin** *(p. 25)*.

You've created a plugin, but the Open Toolkit can't detect it automatically; You need to tell the Open Toolkit about your plugin. You do this with the plugin integrator (**integrator.xml**), which is an ANT file that integrates your plugins with the Open Toolkit. In the past, you could use `startcmd.bat` (or `startcmd.sh`) to run the integrator, but both of those files have been deprecated. For now, they are still included in the DITA Open Toolkit, and they still work (though with protest), but at some point they will go away. If you've used them in the past and adapted them to make your own batch files, now is the time to wean yourself.

> **Note** If you're wondering why these ever-so-useful files have been deprecated, there is a good reason. Robert Anderson has written an article explaining why: **Where, oh where, is my startcmd going?** In a nutshell, Jarno Elovirta (one of the principal architects of the DITA Open Toolkit) has eliminated the need for a batch file that tells the DITA OT where all of its parts are. It's a big improvement but can be disconcerting at first if you've been used to using startcmd in the past.

There is a new dita command line tool in the DITA Open Toolkit that can be used to install and uninstall plugins. However, because plugin integration is something you might do fairly often, why not create a batch file to do it rather than having to type on a command line each time?

1. In the DITA-OT folder, create a file named **integrate_plugins.bat**.

2. Add this content to **integrate_plugins.bat**:

```
set DITA_DIR=C:\DITA-OT\
call %DITA_DIR%bin\ant -f integrator.xml
pause
```

> **Note** This is just one of the formats that Robert Anderson's article explains. You might prefer to follow another of his examples.

The `pause` command enables you to see whether the build was successful before the command prompt window closes.

3. Save the file.

4. Double-click it.

The integrator scampers through the plugins subfolder, finds all the installed plugins, detects their transtypes and other information, and marks them present and accounted for to the DITA Open Toolkit. You can (and probably will) run the integrator multiple times; it doesn't hurt anything to re-integrate a plugin.

You won't see much in the command prompt window. Most likely, just the following:

```
strict:

integrate:
```

And then (hopefully) BUILD SUCCESSFUL, just before the command prompt disappears (unless you added `pause`). When you see that, your plugin is ready to go.

5. In your DITA-OT folder, open the **build.xml** file and search for **custpdf** (or whatever you named your custom transtype).

You should find it in a section something like this:

```
<target name="init">
  <dita-ot-fail id="DOTA001F">
    <condition>
      <and>
        <not><equals arg1="${transtype}" arg2="pdf" casesensitive="false"/></not>
       ▶<not><equals arg1="${transtype}" arg2="pdf2" casesensitive="false"/></not>
       ▶<not><equals arg1="${transtype}" arg2="custpdf" casesensitive="false"/></not>
        ...
```

This section outlines fail conditions for the DITA OT. In this case, it's listing all of the transtypes that the DITA OT should recognize and saying, "If the transtype specified for the current build is not in this list, fail the build." So obviously, you want **custpdf** to be in this list. If it's missing, your plugin was not correctly integrated.

You should also find a target in **build.xml** with the same name as your target in the **integrator.xml** file in your plugin:

```
<target name="dita2custpdf.init">
    <property name="customization.dir" location="${dita.plugin.com.company.pdf.dir}/cfg"/>
    <property name="pdf2.i18n.skip" value="true"/>
</target>
```

If this target is missing from **build.xml**, your plugin was not correctly integrated. If that happens, go back through the previous exercise, **Create your own PDF plugin** *(p. 25)*, and carefully check every detail to be sure there are no typos and that you've followed all of the naming rules outlined in that exercise. Don't continue until your plugin is successfully integrated into the DITA Open Toolkit.

Exercise: Create an attribute set file in your plugin

Objective: Create an empty attribute set file in your custom plugin so you can copy and modify attribute sets in it without affecting the default attribute sets.

Each time you need to change the appearance of an element in your PDF, you need to copy the attribute set that formats that element to your plugin. I recommend that you create an empty attribute set file in your plugin and give it the same name as the file in `org.dita.pdf2` that contains the attribute set you want to modify. Then, copy the attribute set to this file. This section contains instructions for doing this.

> **Note** In the first edition of this book, I suggested that you copy the entire attribute set file that contains the attribute set you want to modify. I now recommend that you create an empty attribute set file in your customization and copy only the attribute sets that you need to modify. This should make it easier to handle upgrades to the Open Toolkit.

1. After you determine the attribute set you need to customize, note the name of the file it's in.

   For example, suppose you want to change the appearance of the `wintitle` element. That element is formatted by the *wintitle* attribute set, which you can find in the file named **ui-domain-attr.xsl** in `DITA-OT/plugins/org.dita.pdf2/cfg/fo/attrs/`.

2. Create an empty text file of the same name (in this example **ui-domain-attr.xsl**) in `DITA-OT/plugins/com.company.pdf/cfg/fo/attrs/`.

3. Add the following content to that file:

   ```
   <?xml version="1.0" encoding="UTF-8"?>

   <xsl:stylesheet xmlns:xsl="http://www.w3.org/1999/XSL/Transform"
                   xmlns:fo="http://www.w3.org/1999/XSL/Format"
                   version="2.0">

   </xsl:stylesheet>
   ```

4. Save the new file.

5. Add an `xsl:import` statement with the following content to `DITA-OT/plugins/com.company.pdf/cfg/fo/attrs/`**custom.xsl**:

   ```
   <xsl:import href="ui-domain-attr.xsl"/>
   ```

6. Save **custom.xsl** (attrs).

Complete the **Add an attribute set to your plugin** *(p. 32)* exercise to add an attribute set to your custom attribute set file.

Exercise: Add an attribute set to your plugin

Objective: Copy an attribute set from the default PDF plugin to your custom plugin so you can make changes to it without affecting the default attribute set.

Each time you need to change the appearance or formatting of an element in your PDF, you need to find the attribute set that controls that element and copy it to your plugin. Including only attribute sets that you have modified makes your plugin a little more future-proof and easier to maintain. This section contains instructions you can follow each time you modify an attribute set.

These steps assume you have followed the instructions in **Create an attribute set file in your plugin** *(p. 31)* and have an attribute set file in your plugin for this attribute set.

> **Important:** If you include **commons.xsl** in your plugin and then add an attribute set to your plugin, you must also copy the corresponding template from the corresponding XSLT stylesheet, even if you don't need to change anything in it. If you don't add the template, changes to the attribute sets in your attribute set file are not reflected in your PDF. This seems a little confusing, but there is a reason. To understand why, you need to understand DITA element classes. You can wait and find out about those a little later, or you can jump ahead a little and read about them in **DITA element classes** *(p. 126)*.

1. After you determine the attribute set you need to customize, copy it from `DITA-OT/plugins/org.dita.pdf2/cfg/fo/attrs/` to your custom attribute set file of the same name in `DITA-OT/plugins/com.company.pdf/cfg/fo/attrs/`.

   If there is no attribute set file of the same name in your plugin, create one following the instructions in **Create an attribute set file in your plugin** *(p. 31)*.

2. Save the changes to your attribute set file.

Exercise: Create an XSLT stylesheet in your plugin

Objective: Create an empty XSLT stylesheet in your custom plugin so you can make, copy, and modify templates without affecting the default templates.

Each time you need to change the behavior of an element in your PDF, you need to copy the template that processes that element to your plugin. I recommend that you create an empty stylesheet file in your plugin and give it the same name as the file in `org.dita.pdf2` that contains the template you want to modify. Then, add the modified template to this file. This section contains instructions for doing this.

> **Note** In the first edition of this book, I suggested that you copy the entire file that contains the template you want to modify. I now recommend that you create an empty stylesheet file in your customization and copy only the templates that you need to modify. This should make it easier to handle upgrades to the Open Toolkit.

1. After you determine the template you need to customize, note the name of the stylesheet it's in.

   You can search on the module/tagname value to find the templates. For example, to find the template that processes the `<p>` element, look for a template that begins with:

   ```
   <xsl:template match="*[contains(@class, ' topic/p ')]">
   ```

   This is standard syntax found throughout the DITA OT. Using your search utility, you can limit the search to the *match* attribute on the `xsl:template` and then run that search across all the `.xsl` files in the PDF plugin.

   For example, say you want to change the way the `<p>` element is processed. That element is processed by a template found in `DITA-OT/plugins/org.dita.pdf2/xsl/fo/`**commons.xsl**.

2. Create an empty text file of the same name (in this example **commons.xsl**) in `DITA-OT/plugins/com.company.pdf/cfg/fo/xsl/`.

3. Paste information similar to following into that file:

   ```
   <?xml version="1.0" encoding="UTF-8"?>

   <xsl:stylesheet xmlns:xsl="http://www.w3.org/1999/XSL/Transform"
                   xmlns:fo="http://www.w3.org/1999/XSL/Format"
                   version="2.0">

   </xsl:stylesheet>
   ```

   > **Important:** Be sure to copy the `xsl:stylesheet` element exactly as it appears in the original file. Each stylesheet might include different namespaces (*xmlns*) or other attributes such as *exclude-result-prefixes*, and that information must appear in your stylesheet, too.

4. Save the new file.

5. Add an `xsl:import` statement to `DITA-OT/plugins/com.company.pdf/cfg/fo/xsl/`**custom.xsl**.

   For example:

   ```
   <xsl:import href="commons.xsl"/>
   ```

6. Save **custom.xsl** (xsl).

Complete the **Add a template to your plugin** *(p. 34)* exercise to add the template to your plugin.

Exercise: Add a template to your plugin

Objective: Copy a template from the default PDF plugin to your custom plugin so you can make changes to the processing without affecting the default template.

Each time you need to change the behavior of an element in your PDF, you need to find the template that processes the element and copy it to your plugin. Including only templates that you have modified makes your plugin a little more future-proof and easier to maintain. As with attribute sets, you don't need to do this right now, but you'll do it a lot as you develop your plugin. This section contains instructions you can follow each time.

These steps assume you have followed the instructions in **Create an XSLT stylesheet in your plugin** *(p. 32)* and have a stylesheet file in your plugin for this template.

1. After you determine the template you need to customize, copy it from `DITA-OT/plugins/org.dita.pdf2/xsl/fo/` to your custom attribute set file of the same name in `DITA-OT/plugins/com.company.pdf/cfg/fo/xsl/`.

   If there is no stylesheet of the same name in your plugin, create one following the instructions in **Create an XSLT stylesheet in your plugin** *(p. 32)*.

2. Save the changes to your stylesheet.

Why not use a single custom file for all your changes?

The process just described for handling modified attribute sets and templates is straightforward and makes it easy to track your customizations. However, if you plan to customize only a few attribute sets or templates, and you want to keep the number of files in your plugin to a minimum, you could just copy those attribute sets or templates to **custom.xsl** (attrs) or **custom.xsl** (xsl).

For example, here's **custom.xsl** (attrs) with an attribute set copied from **commons-attr.xsl**:

```
<xsl:stylesheet xmlns:xsl="http://www.w3.org/1999/XSL/Transform"
    xmlns:fo="http://www.w3.org/1999/XSL/Format"
    version="2.0">

    <xsl:attribute-set name="fig.title" use-attribute-sets="base-font common.title">
        <xsl:attribute name="font-weight">bold</xsl:attribute>
        <xsl:attribute name="space-before">5pt</xsl:attribute>
        <xsl:attribute name="space-after">10pt</xsl:attribute>
        <xsl:attribute name="keep-with-previous.within-page">always</xsl:attribute>
    </xsl:attribute-set>

</xsl:stylesheet>
```

There's no functional reason not to do this. If you know you're going to edit only a handful of templates or attribute sets, this method makes sense. However, a full PDF plugin rarely has so few changes. Odds are, you're going to change dozens of attribute sets and templates, and eventually, you will find it hard to keep track of your templates and attribute sets if you throw them all into a single file. At that point, you'll have to bite the bullet and re-organize them into multiple files.

If you start out naming your files the same as the files in the `org.dita.pdf2` plugin, you can easily find the original templates and attribute sets when you need them.

On the other hand, if you create a template from scratch or use one that some generous person has given you, it makes sense to add it to **custom.xsl** (xsl), unless it fits neatly into one of the existing XSLT stylesheet files. The same is theoretically true of attribute sets, but most are related to an existing attribute set, so it may make more sense to add them to that attribute set file.

Exercise: Add a localization variables file to your plugin

Objective: Create a localization variables file in your custom plugin so you can make changes to variables without affecting the default values.

If you worked with previous versions of the Open Toolkit, you might notice that some variables that used to be present in the localization variables files are no longer there. The Open Toolkit developers are migrating many of the localization variables to the strings files in `DITA-OT/xsl/common/`. The goal is to maintain the strings that are common to all plugins in one location, and keep only PDF-specific variables in the existing localization variables files. For the moment, there are overlaps and gaps between the localization variables files and the strings files, so you need to work with both sets of files.

For simplicity, this example assumes that you author in English and translate to French. You need to account for both languages in your plugin.

1. In the folder `DITA-OT/plugins/com.company.pdf/cfg/common/vars/`, create two empty files and name them **en.xml** and **fr.xml**.

2. Paste the following into both files:

   ```
   <?xml version="1.0" encoding="UTF-8"?>
   <vars xmlns="http://www.idiominc.com/opentopic/vars">
   </vars>
   ```

3. Save the new files.

You don't need to do anything special to get your plugin to recognize these files; it does so automatically as long as they are in the `vars` folder.

Exercise: Add a localization strings file to your plugin

Objective: Create a localization strings file in your custom plugin so you can make changes to strings without affecting the default values.

For simplicity, this example assumes that you author in English and translate to French. You need to account for both languages in your plugin.

1. Create two empty text files named **strings-en-us.xml** and **strings-fr-fr.xml** in `DITA-OT/plugins/com.company.pdf/cfg/common/strings/`.

2. Paste the following into **strings-en-us.xml**:

```
<?xml version="1.0" encoding="utf-8"?>
<strings xml:lang="en-us">
</strings>
```

3. Paste the following into **strings-fr-fr.xml**:

```
<?xml version="1.0" encoding="utf-8"?>
<strings xml:lang="fr-fr">
</strings>
```

4. Save the new files.

Exercise: Add a strings.xml file to your custom plugin

Objective: Create a main **strings.xml** file in your custom plugin to serve as a central pointer to all of your localization strings files.

For simplicity, this example assumes that you author in English and translate to French. You need to account for both languages in your plugin. At this point you should have the following language-specific strings files in `DITA-OT/plugins/com.company.pdf/cfg/common/strings/`:

- **strings-en-us.xml**
- **strings-fr-fr.xml**

You now need to tell your plugin to use your copies of these files rather than the ones in `DITA-OT/xsl/common/`.

1. Create an empty text file named **strings.xml** file in the directory `DITA-OT/plugins/com.company.pdf/cfg/common/strings/`.

2. Paste the following into that file:

```
<?xml version="1.0" encoding="UTF-8"?>

<langlist>

</langlist>
```

3. Save the new file.

4. Add the following line to your **plugin.xml** file (at the root of your `com.company.pdf` folder):

```
<feature extension="dita.xsl.strings" file="cfg/common/strings/strings.xml"/>
```

5. Add the highlighted lines to your copy of **strings.xml**:

```
<langlist>

    <lang filename="strings-en-us.xml" xml:lang="en"/>
    <lang filename="strings-fr-fr.xml" xml:lang="fr"/>

</langlist>
```

These lines tell the Open Toolkit to use the strings in your copy of **strings-en-us.xml** whenever a map's or topic's *xml:lang* value is "en" and those in **strings-fr-fr.xml** whenever a map's or topic's *xml:lang* value is "fr".

6. Run the Open Toolkit integrator again to register this change in the DITA OT.

7. Save your changes.

Mapping strings files to xml:lang values in the strings.xml file

Objective: Edit the **strings.xml** file so that it points correctly to all of the localization strings files that you've included in your plugin.

The **strings.xml** file points to each of the localization strings files that you've included in your plugin, to ensure the DITA Open Toolkit finds them and uses your values rather than the default ones.

What do you do if your content is likely to use multiple locales of a language? For example, say your English content comes from different sources and is likely to have *xml:lang* values of "en", "en-ca" (Canadian English), "en-gb" (British English) or "en-us" (U.S. English).

You might choose to copy three strings files to your plugin (**strings-en-ca.xml**, **strings-en-gb.xml**, and **strings-en-us.xml**) and add lines like this to **strings.xml**:

```
<lang filename="strings-en-ca.xml" xml:lang="en-ca"/>
<lang filename="strings-en-gb.xml" xml:lang="en-gb"/>
<lang filename="strings-en-us.xml" xml:lang="en-us"/>
<lang filename="strings-en-us.xml" xml:lang="en"/>
```

In this example, each strings file maps to a unique *xml:lang* code (except for **strings-en-us.xml**, which is mapped to both "en-us" and the more generic "en"). You could maintain separate strings files for each locale, but the string values are likely to be identical across locales. That is a lot of duplication.

Instead, use the *xml:lang* attribute on the `<lang>` elements in your strings files to map a single strings file to multiple locales. For example, to tell the Open Toolkit to use your copy of **strings-en-us.xml** when `xml:lang` is "en", "en-ca", "en-gb", or "en-us", map **strings-en-us.xml** to all three locales and the generic "en" as follows:

```
<lang filename="strings-en-us.xml" xml:lang="en-ca"/>
<lang filename="strings-en-us.xml" xml:lang="en-gb"/>
<lang filename="strings-en-us.xml" xml:lang="en-us"/>
<lang filename="strings-en-us.xml" xml:lang="en"/>
```

‖ **Important:** Remember to run the integrator each time you change **strings.xml**.   ‖

Wrap-up

If you worked your way through the preceding pages, you now have your very own plugin folder `com.company.pdf`. You also understand how to add more files to your plugin for any formatting and processing changes you want to make. And you have just enough understanding of how the plugin is organized to be dangerous … er, to sound knowledgeable.

As you work through this book, you'll need to test your changes. One easy way to do that is to set up simple ANT build files to create a PDF. You'll look at that in the next chapter.

Each exercise in this book reminds you to comment the templates and attribute sets that you change or add in the course of the exercise. When your plugin is complete, you should go through each of the stylesheets and attribute set files that you copied into your plugin and delete all of the templates and attribute sets that you did not change or add—i.e., those with no comment. By doing so, you'll make it much easier to maintain and update your plugin through future DITA Open Toolkit versions.

Chapter 3

DITA Open Toolkit builds

I will use the term *build* to refer to the process of creating a PDF using the DITA Open Toolkit.

This chapter explains exactly what builds are and how to run them from the DITA Open Toolkit command line as well as from a custom ANT build file and batch, or shell, file.

What is an ANT build file?

If you just want to tell the DITA Open Toolkit to build a PDF from your ditamap, you can use a command line such as this:

```
dita -i samples/hierarchy.ditamap -f pdf -o output/pdf
```

The `dita` command invokes a **dita.bat** file in the `DITA-OT/bin` folder, *-i samples/hierarchy.ditamap* specifies the location and name of the ditamap to be built, *-f pdf* specifies the output type, and *-o output/pdf* specifies the output location. You can add additional parameters to this command line to specify other properties of the PDF you want to create, but who wants to type all this over and over?

Wouldn't it be great if you could put all this information into a single file that you could run with a click of the mouse? You can! You can create your own ANT build file to specify all the properties for your PDF and then run it with a batch file.

So far, I've referred to running builds, building output, PDF builds, ANT build files, and so on. At this point, you might be wondering just what a build or an ANT build file is. Simply put, a build is a set of processing steps that create an output. As you might guess, these steps are usually found in files whose names include the word *build*, such as the **build.xml** and **build_custpdf_template.xml** files that you created in your plugin folder, or the **build.xml** file found in the `org.dita.pdf2` folder.

The build files included with the DITA Open Toolkit and the ones that you'll create are ANT build files, meaning that they are written using the ANT scripting language.

Apache Ant is a Java library and command-line tool. (This is why you need to have Java installed to use the DITA Open Toolkit.) It's mainly used to build Java applications, but it can be used to build non-Java applications as well. In the context of the DITA Open Toolkit, ANT is used to run the processes that convert DITA content to various output formats. ANT is obviously a key part of the Open Toolkit, but you can use it outside of the Open Toolkit as well. For example, you could write scripts to move group of files from one location to another using ANT.

Each ANT build file contains one or more targets. A *target* is a set of tasks. Targets can depend on other targets and can call other targets. In other words, instead of having one gigantic set of instructions for processing output, the DITA Open Toolkit uses multiple smaller sets of instructions that interact with each other and are performed in a specific order. Some instructions are reused for many different output types, while other instructions are used for just one output type.

The DITA Open Toolkit includes an ANT installation, so you don't have to install it separately. However, because ANT's uses aren't limited to just the DITA Open Toolkit, you may find it helpful to have a second, standalone installation for running non-Open-Toolkit builds. For more information on downloading and using ANT, check out the links in **Useful resources** *(p. 11)*.

So let's create an ANT build file that includes all the parameters you need to create output from a map.

Exercise: Create an ANT build file

Objective: Create an ANT build file that contains properties for your custom PDF plugin.

The easiest way to create output directly from the DITA Open Toolkit—and avoid cumbersome command lines—is to create an ANT build file that includes all the output parameters. Using ANT build files and batch files, you can set up one- (well, two-) click builds that make testing easy.

> **Note** If you're using an XML editor to create DITA content, that editor probably includes the capability to interact with the DITA Open Toolkit to create output. How you set up that interaction varies from editor to editor, so it's not possible to cover all the possibilities in this book. If you prefer to create output from your XML editor, you're on your own as far as the setup.
>
> However, you can still add custom plugins to a DITA Open Toolkit within an XML editor using the techniques described in this book once your editor is set up. Just look for where the editor puts the Open Toolkit folder. For example, on Windows, oXygen uses the folder `C:\Program Files\`*[oXygen]*`\frameworks\dita\DITA-OT`, where *[oXygen]* is the name of the oXygen program file, which varies by version number.
>
> However, beginning with oXygen 17, you can point oXygen at any DITA-OT installation, not just the one installed with the application.

1. Create a folder for the ANT build file.

   Name the folder something intuitive like `Buildfiles`. This folder can be anywhere. To keep things simple, you can create the folder in your home folder (on Windows 10 that is `C:\Users\<username>`, and on Linux and Mac that is your home directory). ANT uses the variable *${user.home}* for your home folder, which makes it easy to create reusable ANT build files.

2. Copy the text from the **Sample ANT build file** *(p. 479)* appendix of this book into a text file.

3. Name the file something that matches your project, such as **WidgetUserGuide.ant**.

   You can use the **.xml** extension as well as **.ant**, but as you create more and more content, you'll end up with a lot of **.xml** files. Giving your ANT build files an **.ant** extension helps distinguish them from DITA topic files or other kinds of XML files.

4. Copy the file to the `Buildfiles` folder.

5. Add information for your specific build to the ANT build file as follows.

   a. Substitute your project name.

   ```
   project name="sample_pdf"
   ```

   changes to

   ```
   project name="WidgetUserGuide"
   ```

   The project name is used in ANT-aware development environments, such as the Eclipse Ant view, which lists build files and targets in a nice tree view. It uses the project *@name* value in the tree. It's also a good a memory device for you within the ANT build file.

   b. Replace the original path with the path to your DITA Open Toolkit installation.

   ```
   <property name="dita.dir" location="${basedir}/../.."/>
   ```

   changes to

   ```
   <property name="dita.dir" location="C:\DITA-OT"/>
   ```

   c. Replace the original input map or bookmap path and file name with the path and file name of the map or bookmap you are using in the build.

   ```
   <property name="args.input" location="${dita.dir}/samples/sequence.ditamap"/>
   ```

   changes to

   ```
   <property name="args.input"
      location="C:${file.separator}Widget${file.separator}WidgetUserGuide.ditamap"/>
   ```

If you don't have your own map, you can leave *args.input* pointing to **sequence.ditamap** or, for a sample bookmark, use **taskbook.ditamap**, which you can find in the same folder.

> **Note** *${file.separator}* is a variable that can resolve to either \ or /, depending on whether you are using a Windows system or a UNIX system. If you have users on both operating systems, using the variable keeps you from having to create duplicate ANT build files for different environments. If you prefer, you can also use /, as it will also resolve on both Windows and Mac/UNIX systems.

d. Replace the original output path with the full path to the location where you want the PDF created.

```
<property name="output.dir" location="${dita.dir}/out/samples/pdf"/>
```

changes to

```
<property name="output.dir"
    location="C:${file.separator}Widget${file.separator}output${file.separator}pdf"/>
```

e. Move the line in the previous step so that it's directly below the line that defines *dita.dir*, like this:

```
<property name="dita.dir" location="C:\DITA-OT"/>
<property name="output.dir"
    location="C:${file.separator}Widget${file.separator}output${file.separator}pdf"/>
```

When you define a property within a target, you can only use it within that target. By moving the definition of the output directory outside of a target, you can use it in any target.

f. Change the transtype to **custpdf**. This should match the value for transform type that you put in the file **plugin.xml** in **Create your own PDF plugin** *(p. 25)*.

g. Change the line that specifies the output directory to clean.

```
<delete dir="${dita.dir}/out/samples/pdf"/>
```

changes to

```
<delete dir="${output.dir}"/>
```

You can use the *output.dir* variable here because it is defined outside of a target—the move you made a few steps earlier.

> **Important:** Be careful here. Global properties in plugins (those that are defined outside of specific targets) that are integrated into other ANT projects become global to all the ANT projects. With something common like *${output.dir}*, make sure that you don't inadvertently override something you shouldn't.

6. Save the ANT build file.

Reality check

You now have an ANT build file that specifies 1) the location of the DITA Open Toolkit, 2) the map to process, 3) the folder in which to create the PDF, and 4) the plugin to use (based on the transtype). These are the only four required items for ANT build files. There are many more parameters and properties you can add to the ANT build file to fine-tune the output, but for now, this is enough to create a PDF.

Your ANT build file should look like this:

```
<?xml version="1.0" encoding="UTF-8" ?>
<!-- This file is part of the DITA Open Toolkit project hosted on
     Sourceforge.net. See the accompanying license.txt file for
     applicable licenses.-->
<!-- (c) Copyright IBM Corp. 2004, 2006 All Rights Reserved. -->

<project name="WidgetUserGuide" default="samples.pdf" basedir=".">

    <property name="dita.dir" location="C:${file.separator}DITA-OT"/>
    <property name="output.dir"
        location="C:${file.separator}Widget${file.separator}output${file.separator}pdf"/>

    <target name="samples.pdf" description="build the samples as PDF"
        depends="clean.samples.pdf">
        <ant antfile="${dita.dir}/build.xml">
            <property name="args.input"
                location="C:${file.separator}Widget${file.separator}WidgetUserGuide.ditamap"/>
            <property name="transtype" value="custpdf"/>
        </ant>
    </target>

    <target name="clean.samples.pdf" description="remove the sample PDF output">
        <delete dir="${output.dir}"/>
    </target>

</project>
```

Exercise: Create a batch file to launch an ANT build (Windows)

Objective: Create a batch file that launches your ANT build file with a single click.

Before you start

Complete the exercise **Create an ANT build file** *(p. 40)*.

Now that you have an ANT build file, you need to choose how to run it. As mentioned earlier, **startcmd.bat** is not going to be an option for much longer. There is also the new DITA command line option, but that requires typing and text-based folder navigation. Ick.

> **Note** If you'd like to understand why **startcmd.bat** is going away, read Robert Anderson's very good explanation: **http://metadita.org/toolkit/startcmd.html**

My favorite method remains a simple batch file. With a batch file, you can run a build with a double-click. You can also set up other processes to run multiple builds using batch files.

1. In your `Buildfiles` folder, create a text file.
2. Rename the file to reflect the purpose of the build or the title of the output.

   For example, **WidgetUserGuide.bat**.
3. Open the file using Notepad or Notepad++.
4. Paste the following into the file:

```
REM
set DITA_DIR=C:\DITA-OT\
call %DITA_DIR%bin\ant -f [your ANT build file]
```

   Where *[your ANT build file]* is the actual name of the ANT build file you created in a previous exercise—for example, **WidgetUserGuide.ant**.

   > **Note** This is just one of the formats that Robert Anderson's **article** explains. You might prefer to follow another of his examples.

5. After the word REM, type a short explanation of what this batch file is for.

   For example:

```
REM This builds the PDF output for the Widget User Guide.
```

6. If you want a log file, add `-l [logfilename].log` to the command line.

   For example

```
call %DITA_DIR%bin\ant -f WidgetUserGuide.ant -l WidgetUserGuide.log
```

7. Save the batch file.

Reality check

The batch file now looks like this:

```
REM This builds the PDF output for the Widget User Guide.
set DITA_DIR=C:\DITA-OT\
call %DITA_DIR%bin\ant -f WidgetUserGuide.ant -l WidgetUserGuide.log
```

From now on, you only need to double-click the batch file to launch the ANT build. If you did not specify a log file, the build progress appears in the command window, which closes as soon as the build completes. If you did specify a log file, the progress does not appear in the command window, but you can view the log file to confirm a successful build or troubleshoot a failed build.

> **Tip:** The free **BareTail** utility lets you watch a log being generated in real time while you simultaneously send the log to a file.

> **Note** If you see the following message, you can ignore it:
>
> ```
> Unable to locate tools.jar. Expected to find it in...
> ```
>
> Here's Don Day's explanation of the error from the DITA Users list:
>
> "Ant issues this message whenever it is executed in a Java Runtime Environment rather than in a full Java developer's environment. The message comes from Ant and to my knowledge cannot be configured off. It can be ignored because DITA-OT never calls the Java compiler—Ant is being used only for its ability to orchestrate processes and dependencies in the DITA processing pipeline."

Exercise: Create a shell file to launch an ANT build (Mac, Linux)

Objective: Create a shell file that launches your ANT build file.

Before you start

Complete the exercise **Create an ANT build file** *(p. 40)*.

Now that you have an ANT build file, you need to choose how to run it. The **startcmd.sh** script won't be an option much longer, and the new DITA command requires typing and textual folder navigation. Ick. My favorite method remains a simple shell file. With a shell file, you can run one or more builds by typing a single command.

1. In the DITA-OT folder, make a copy of **startcmd.sh** (Mac, Linux).
2. Rename the copy to reflect the purpose of the build or the title of the output.

   For example, **WidgetUserGuide.sh**.
3. Move the new shell file to the appropriate location.

   I suggest you put the shell file in the same location as your ANT build file. For example, `/home/abc/Buildfiles`.

4. Open the shell file using a text editor:

```
#!/bin/sh
# Generated file, do not edit manually"
echo "NOTE: The startcmd.sh has been deprecated, use the 'dita' command instead."

realpath() {
  case $1 in
    /*) echo "$1" ;;
    *) echo "$PWD/${1#./}" ;;
  esac
}

if [ "${DITA_HOME:+1}" = "1" ] && [ -e "$DITA_HOME" ]; then
  export DITA_DIR="$(realpath "$DITA_HOME")"
else #elif [ "${DITA_HOME:+1}" != "1" ]; then
  export DITA_DIR="$(dirname "$(realpath "$0")")"
fi

if [ -f "$DITA_DIR"/bin/ant ] && [ ! -x "$DITA_DIR"/bin/ant ]; then
  chmod +x "$DITA_DIR"/bin/ant
fi

export ANT_OPTS="-Xmx512m $ANT_OPTS"
export ANT_OPTS="$ANT_OPTS -Djavax.xml.transform.TransformerFactory
▶=net.sf.saxon.TransformerFactoryImpl"
export ANT_HOME="$DITA_DIR"
export PATH="$DITA_DIR"/bin:"$PATH"
NEW_CLASSPATH="$DITA_DIR/lib:$NEW_CLASSPATH"
NEW_CLASSPATH="$DITA_DIR/lib/ant-apache-resolver-1.9.4.jar:$NEW_CLASSPATH"
NEW_CLASSPATH="$DITA_DIR/lib/ant-launcher.jar:$NEW_CLASSPATH"
NEW_CLASSPATH="$DITA_DIR/lib/ant.jar:$NEW_CLASSPATH"
NEW_CLASSPATH="$DITA_DIR/lib/commons-codec-1.9.jar:$NEW_CLASSPATH"
NEW_CLASSPATH="$DITA_DIR/lib/commons-io-2.4.jar:$NEW_CLASSPATH"
NEW_CLASSPATH="$DITA_DIR/lib/dost-configuration.jar:$NEW_CLASSPATH"
NEW_CLASSPATH="$DITA_DIR/lib/dost.jar:$NEW_CLASSPATH"
NEW_CLASSPATH="$DITA_DIR/lib/guava-19.0.jar:$NEW_CLASSPATH"
NEW_CLASSPATH="$DITA_DIR/lib/icu4j-54.1.jar:$NEW_CLASSPATH"
NEW_CLASSPATH="$DITA_DIR/lib/saxon-9.1.0.8-dom.jar:$NEW_CLASSPATH"
NEW_CLASSPATH="$DITA_DIR/lib/saxon-9.1.0.8.jar:$NEW_CLASSPATH"
NEW_CLASSPATH="$DITA_DIR/lib/xercesImpl-2.11.0.jar:$NEW_CLASSPATH"
NEW_CLASSPATH="$DITA_DIR/lib/xml-apis-1.4.01.jar:$NEW_CLASSPATH"
NEW_CLASSPATH="$DITA_DIR/lib/xml-resolver-1.2.jar:$NEW_CLASSPATH"
NEW_CLASSPATH="$DITA_DIR/plugins/org.dita.pdf2/lib/fo.jar:$NEW_CLASSPATH"
NEW_CLASSPATH="$DITA_DIR/plugins/org.dita.odt/lib/odt.jar:$NEW_CLASSPATH"
NEW_CLASSPATH="$DITA_DIR/plugins/org.dita.pdf2.axf/lib/axf.jar:$NEW_CLASSPATH"
NEW_CLASSPATH="$DITA_DIR/plugins/org.dita.pdf2.xep/lib/xep.jar:$NEW_CLASSPATH"
NEW_CLASSPATH="$DITA_DIR/plugins/org.dita.wordrtf/lib/wordrtf.jar:$NEW_CLASSPATH"
if test -n "$CLASSPATH"; then
  export CLASSPATH="$NEW_CLASSPATH":"$CLASSPATH"
else
  export CLASSPATH="$NEW_CLASSPATH"
fi
cd "$DITA_DIR"
"$SHELL"
```

5. From the beginning of the file, remove the following lines (highlighted above):

```
if [ "${DITA_HOME:+1}" == "1" ] && [ -e "$DITA_HOME" ]; then
  export DITA_DIR="$(realpath "$DITA_HOME")"
else #elif [ "${DITA_HOME:+1}" != "1" ]; then
  export DITA_DIR="$(dirname "$(realpath "$0")")"
fi
```

6. In the same location, add the line export DITA_DIR=*DITA_HOME* where *DITA_HOME* is the full pathname for the DITA Open Toolkit directory. For example,

```
export DITA_HOME=/usr/share/dita-ot-2.4
```

7. Delete this line, which is near the bottom of the file: "$SHELL"

8. In the same location, add the following, using the absolute path for the ANT file:

```
ant -f /home/abc/Buildfiles/WidgetUserGuide.ant
```

9. If you want a log file, add -l [logfilename].log to the command line, for example:

```
ant -f /home/abc/Buildfiles/WidgetUserGuide.ant -l /home/abc/logs/WidgetUserGuide.log
```

10. Save the shell file.

Reality check

From now on, you can just type ./WidgetUserGuide.sh to launch the ANT build. The build progress appears in the shell window, which closes as soon as the build completes. You can view the log file to confirm a successful build or troubleshoot a failed build.

In addition, you have to set the file to be executable. It probably already is if you copied **startcmd.sh** from the Open Toolkit, but to set it executable, use: chmod +x WidgetUserGuide.sh. This makes the file executable for all users.

Exercise: Test your plugin

Objective: Make a small change in your PDF plugin to be sure the DITA Open Toolkit recognizes it correctly.

Before you start

Be sure you've created a batch file or a shell file to launch the ANT build file you created.

- **Create a batch file to launch an ANT build (Windows)** *(p. 43)*
- **Create a shell file to launch an ANT build (Mac, Linux)** *(p. 45)*

Before you continue, do a small test to be sure the DITA Open Toolkit is reading the values in your plugin rather than those in the cfg folder.

1. Copy **commons-attr.xsl** from `DITA-OT/plugins/org.dita.pdf2/cfg/fo/attrs/` to `DITA-OT/plugins/com.company.pdf/cfg/fo/attrs/`.

2. Open **commons-attr.xsl** (your copy) and locate the following line.

```
<xsl:attribute-set name="topic.title" use-attribute-sets="common.title common.border__bottom">
```

Note that this attribute set calls, among others, the *common.border__bottom* attribute set, which determines the appearance of the line underneath top-level headings.

3. In **commons-attr.xsl**, find the *common.border__bottom* attribute set:

```
<xsl:attribute-set name="common.border__bottom">
    <xsl:attribute name="border-after-style">solid</xsl:attribute>
    <xsl:attribute name="border-after-width">1pt</xsl:attribute>
    <xsl:attribute name="border-after-color">black</xsl:attribute>
</xsl:attribute-set>
```

(It's a few attribute sets before the *topic.title* attribute set.)

4. Change the value of the *border-after-color* attribute from **black** to **red**.

The Open Toolkit's out-of-the-box PDF formatting puts a thick black line underneath top-level headings. You're changing the color of that line (and all other lines under all other headings) to red, which should be easy to spot in the PDF.

5. Save and close **commons-attr.xsl**.

6. In `DITA-OT/plugins/org.dita.pdf2/cfg/fo/attrs/`, open **custom.xsl**.

7. Add the `xsl:import` statement that points to your copy of **commons-attrs.xsl**.

```
<xsl:import href="commons-attr.xsl"/>
```

8. Save and close **custom.xsl**.

9. Save and close the ANT build file.

10. Run the build and open the resulting PDF.

Reality check

If the top-level heading now has a red line below it, congratulations! Your plugin is set up correctly and the Open Toolkit is using it.

If the line is still black, or the build failed, carefully go back through all the set-up instructions to this point to find where you went astray.

Exercise: Use a specific PDF renderer

Objective: Edit your ANT build file to indicate the PDF renderer you're using.

If you're using FOP as your PDF renderer, you don't have to do a thing. The PDF plugin points to it by default. If you're using XEP or Antenna House, you need to do a little extra.

1. Open the ANT build file you're using for testing.
2. Add the following lines after the line:

```
<property name="transtype" value="custpdf"/>
```

For XEP, add:

```
<property name="pdf.formatter" value="xep"/>
<property name="xep.dir" value="C:\Program Files\RenderX\XEP"/>
```

For Antenna House, add:

```
<property name="pdf.formatter" value="ah"/>
<property name="axf.path" value="C:\Program Files\Antenna House\AHFormatterV62"/>
```

In both cases, the first property, *pdf.formatter*, tells the build what PDF renderer to use, and the second property, *xep.dir* or *axf.path*, tells the build where that PDF renderer is installed.

> **Note** Be sure to use the actual XEP or Antenna House installation location on your machine for the *xep.dir* or *axf.path* properties. The examples here show the default path on Windows, but your path might be different.

Other things you can do

Here are a few additional customizations you might want to make to your plugin. These customizations aren't part of the PDF specifications you're following, but they can be useful (and cool).

The topic.fo file

When you build a PDF from a bookmap or a map using the DITA Open Toolkit's PDF plugin, one of the first things that happens is that the plugin combines all of the topics in your map into one big XML file named **[map name]_MERGED_xml**. It then converts that file from the DITA tag set to a different tag set, the XSL-FO tag set. The result of that conversion is a file named **topic.fo**. This file, **topic.fo**, is sent to the PDF renderer. **topic.fo** is useful as a troubleshooting tool. Normally, the Open Toolkit deletes **topic.fo** after the build is complete, but it's a good idea to retain it. To set up your ANT build files to retain **topic.fo**, see **Retain the topic.fo file for troubleshooting** *(p. 52)*.

It's beyond the scope of this book to explain XSL-FO, but because **topic.fo** is an XML document, just like any of your DITA topics, it can be helpful for troubleshooting even if you don't know much about XSL-FO.

What would you use topic.fo to look for?

Lots of things! Mainly, you'll use it to track down the source of unexpected or weird formatting. Let's say that you're expecting to see something like this:

Keepin' it real

This book is written entirely in DITA, using <oXygen/> XML Author and formatted using the DITA Open Tookit 2.2, using some of the same techniques you'll learn about. There's absolutely nothing here that can't be done (or wasn't done) with DITA or the Open Tookit. So let the book you hold in your hands (or look at on your screen) be the proof that all this stuff really does work!

but instead you see this:

Keepin' it real

This book is written entirely in DITA, using <oXygen> XML Author and formatted using the DITA Open Tookit 2.2., using some of the same techniques you'll learn about. There's absolutely nothing here that can't be done (or wasn't done) with DITA or the Open Tookit. So let the book you hold in your hands (or look at on your screen) be the proof that all this stuff really does work!

Notice all the extra space above the paragraph? You check your *p* attribute set and the title-related attribute sets and don't see anything that is creating this extra space. This is when you want to take a peek at **topic.fo** to see exactly what formatting is being applied to the paragraph.

You do a search and find this paragraph within **topic.fo**. Looking at the tree or outline view, you see that the paragraph is contained in an `fo:block` element, and that `fo:block` element is contained in another `fo:block` element:

```
▲  ● fo:flow  "xsl-region-body"
   ▲  ● fo:block  concept/concept
         !-- concept/concept
      ▷  ● fo:block  "0pt solid black"
      ▲  ● fo:block  "#000000" This book is written entirely in
         ▷  ● fo:block  "0.6em" This book is written entirely in
```

As you work with **topic.fo**, you'll see a lot of nesting of blocks within blocks. It's common to find extra unexpected formatting coming from a parent block, especially if that block uses an attribute set that you typically wouldn't add a certain attribute to.

In this case, you look at the code in **topic.fo** and see the following:

```
<fo:block color="#000000" font-size="10pt" line-height="120%" margin-top=".5in" start-indent="0in">

    <fo:block space-after="0.6em" space-before="0.6em" text-indent="0em">
    This book is written entirely in DITA, using &lt;oXygen/&gt; XML Author and
    formatted using the DITA Open Toolkit <fo:inline border-left-width="0pt"
    border-right-width="0pt">2.2.5</fo:inline>, using some of the same techniques
    you'll learn about. There's absolutely nothing here that can't be done
    (or wasn't done) with DITA or the Open Toolkit. So let the book you hold in
    your hands (or look at on your screen) be the proof that all this stuff
    really does work!
    </fo:block>
</fo:block>
```

Notice that the higher-level `fo:block` has the attribute `margin-top=".5in"`. This is the source of the extra space above the paragraph. You need to find the attribute set this block uses and edit it to reduce or remove the top margin.

How do you find that attribute set? Well, you use a combination of logic and searching. You can search for the strings "margin-top" or ".5in" across all your attribute set files using an application like TextCrawler. That will probably turn up a lot of hits, so you can narrow the search using a little logic. You're dealing with title and paragraph formatting, both of which you know (or will learn) are in **commons-attr.xsl**, so you can safely limit your search to that file.

In this example, you find the extra `margin-top=".5in"` in the *conbody* attribute set. If you remove the attribute or set it to zero, you'll find the spacing between the title and the paragraph looks a lot better.

Hints on working with topic.fo:

- Always use a full-fledged XML editor to view **topic.fo**.

  Unlike a DITA topic, which you probably author in an XML editor and which, even when viewed in an application like Notepad++, usually retains its indents nicely, **topic.fo** is one big glob of text. Unless you view it in an XML editor that also includes a tree view or outline view, you'll never find anything.

- When trying to locate a specific section to look at or troubleshoot, search for a word that's likely to be used only in that location. For example, if you're looking at the formatting for a specific table and the first row of that table contains a fairly unusual word, search for that word. If there isn't a unique word nearby, then add one as a test and re-build. "Kumquat" usually works nicely. (Unless, of course, you're actually writing about kumquats.)

  It's usually not a good idea to search for words in a title. Titles are also found in the table of contents and bookmarks, and so will appear in several places other than the place you're looking for. Table and figure titles are also going to be in the List of Tables and List of Figures, if you've included those in your bookmap.

Exercise: Retain the topic.fo file for troubleshooting

Objective: Customize your ANT build file to retain **topic.fo**.

By default, the DITA Open Toolkit discards the **topic.fo** file after your PDF has been created. Because this file is such a valuable troubleshooting tool, it's a good idea to retain it along with your PDF.

While **topic.fo** is always available in the temp folder that is created during each build, it's not always convenient or possible to retrieve it from there. There used to be a handy parameter, *retain.topic.fo*, that you could simply add to your ANT build file to copy the topic.fo file to your output folder. That parameter has been deprecated—an unfortunate casualty of the effort to streamline the Open Toolkit and make the behavior of all the plugins more consistent. (No other plugin offered an option to retain a temporary file in the output folder.)

Instead, you now need to use an ANT copy task within your build file to do manually what *retain.topic.fo* used to do automatically.

1. Add the following property to your ANT build file after the line that defines *output.dir*:

```
<property name="dita.temp.dir"
    location="C:${file.separator}Users${file.separator}<username>${file.separator}
 ▶ Buildfiles${file.separator}temp"/>
```

As you might guess, this property tells the Open Toolkit exactly what folder to use as your temp folder. You have to specify this in order to copy a file out of it. This path assumes you created a Buildfiles folder in your users directory—the example offered in **Create an ANT build file** *(p. 40)*. If you want to create the `temp` folder elsewhere, add the appropriate path to this property.

2. Add the following property to your ANT build file after the line that defines the *transtype* property:

```
<property name="clean.temp" value="no"/>
```

This property tells the Open Toolkit not to empty the `temp` folder after the build is complete. The build will complete before the **copy-topic.fo** action takes place. If you then empty the `temp` folder, there's no **topic.fo** to copy.

3. Add the following just after the `</ant>` tag:

```
<copy todir="${output.dir}">
    <fileset dir="${dita.temp.dir}" includes="topic.fo"/>
</copy>
```

This task copies **topic.fo** file from the `temp` folder to the output folder.

4. Add the following just after the `</copy>` tag:

```
<delete includeEmptyDirs="true">
    <fileset dir="${dita.temp.dir}"/>
</delete>
```

This task deletes the `temp` folder after the copy action has taken place.

5. Run your ANT build file and confirm that **topic.fo** exists in the output folder.

Unless you specified another location, the `temp` folder should be in the same location as your build files.

Reality check

Your ANT build file should now be similar to this:

```
<?xml version="1.0" encoding="UTF-8" ?>
<!-- This file is part of the DITA Open Toolkit project hosted on
     Sourceforge.net. See the accompanying license.txt file for
     applicable licenses.-->
<!-- (c) Copyright IBM Corp. 2004, 2006 All Rights Reserved. -->

<project name="WidgetUserGuide" default="samples.pdf" basedir=".">

   <property name="dita.dir" location="C:${file.separator}DITA-OT"/>
   <property name="output.dir"
       location="C:${file.separator}Widget${file.separator}output${file.separator}pdf"/>
   <property name="dita.temp.dir"
       location="C:${file.separator}Users${file.separator}<username>${file.separator}
       ▶Buildfiles${file.separator}temp"/>
```

```
<target name="samples.pdf" description="build the samples as PDF"
    depends="clean.samples.pdf">
    <ant antfile="${dita.dir}/build.xml">
        <property name="args.input"
            location="C:${file.separator}Widget${file.separator}WidgetUserGuide.ditamap"/>
        <property name="transtype" value="custpdf"/>
        <property name="clean.temp" value="no"/>
    </ant>

    <copy todir="${output.dir}">
        <fileset dir="${dita.temp.dir}" includes="topic.fo"/>
    </copy>

    <delete includeEmptyDirs="true">
        <fileset dir="${dita.temp.dir}"/>
    </delete>

</target>

<target name="clean.samples.pdf" description="remove the sample PDF output">
    <delete dir="${output.dir}"/>
</target>

</project>
```

So there you have it. Four steps to do what *retain.topic.fo* used to do so easily.

Chapter 4

Attribute sets

This chapter explains what attribute sets are, how they work, how you can identify which attribute sets you need to edit, and how you edit them. It also explains how to create and call a new attribute set.

For a list of all the attribute sets in the DITA Open Toolkit and a description of their functions, see **Attribute set lists and descriptions** *(p. 403)*.

What are attribute set files?

Attribute set files control how specific elements look.

If you've done any Web design using Cascading Style Sheets (CSS), then you're already familiar with attribute sets. For example, you might have seen a line similar to the following, either in an HTML file or in a CSS file referenced by an HTML file:

```
p {font:12pt Arial; color:990000; font-weight:bold; line-height:14pt}
```

This line tells the browser to display text marked with the <p> tag in 12 point bold dark red Arial.

Notice I said, "tells the browser." CSS, in this form, is meant to format online output, such as Web pages, HTML Help, and so forth. For PDF output, you use a similar method to tell the FO formatter how to render the text in a PDF. Look at the following line:

```
<fo:block font-family="Arial" font-size="12pt"
        color="#990000" font-weight="bold" line-height="14pt">Here
        is some text.</fo:block>
```

This line is from an FO file produced by the DITA Open Toolkit. The PDF renderer turns instructions like this into formatted text. It's very similar to the line from the CSS; in fact, the properties (*color*, *font-weight*, *line-height*) are almost the same. In the CSS line, the font face and size were combined into a single font property, whereas in the FO line, they are separate *font-family* and *font-size* properties. (Although you can combine certain properties in FO as well.)

Attribute set files let you specify these properties for each element in your DITA content.

Attribute set files are found in the `fo/attrs` subfolder of the `cfg` folder. Take a look in that subfolder and you'll see there are a lot of XSL files, each with a name ending in "**-attr**".

Each of these files contains attribute sets for a group of elements. The files are mostly intuitively named. So, for example, the file **hi-domain-attr.xsl** contains attributes for all of the elements in the highlight domain (`<b>`, `<i>`, `<>u`, etc.).

If you open **hi-domain-attr.xsl**, you can see the following entry:

```
<xsl:attribute-set name="b">
  <xsl:attribute name="font-weight">bold</xsl:attribute>
</xsl:attribute-set>
```

This entry defines attributes for the `<b>` element. In this case, only one attribute (font-weight) is defined—text wrapped in the `<b>` tag should be bold. You can add or change the attributes for `<b>` by editing this attribute set. For example, here's an attribute set that reproduces the FO example above.

```
<xsl:attribute-set name="b">
  <xsl:attribute name="font-size">12pt</xsl:attribute>
  <xsl:attribute name="color">#990000</xsl:attribute>
  <xsl:attribute name="font-weight">bold</xsl:attribute>
  <xsl:attribute name="line-height">14pt</xsl:attribute>
</xsl:attribute-set>
```

Covering all of the many XSL-FO attributes is beyond the scope of this book. The best thing to do is to have a good XSL-FO reference handy, so that when you need to apply a certain kind of formatting to an element, you can find the right attribute.

How attribute set defaults work

Each attribute set contains certain attributes by default. To see the defaults for an attribute set, open any of the attribute set files in the `DITA-OT/plugins/org.dita.pdf2/cfg/fo/attrs/` folder. The attributes you see there are the defaults.

When you copy an attribute set file to your plugin and change the attribute sets in it, you can change the value of one or more attributes or add attributes, but you cannot delete any of the default attributes. If you do, the results will not be what you expect.

‖ **Note** Commenting out an attribute is the same as deleting it, so you can't do that either.   ‖

Here's why. When you create your plugin, you probably won't customize every single aspect of `org.dita.pdf2`. Therefore, your plugins are considered **as an addition to** the specifications in the standard plugin, and your custom attribute sets are merged with the default attribute sets of the same name. If an attribute appears in your custom attribute set and the default attribute set, your value for that attribute overrides the default value. When the default attribute set includes an attribute that your custom set doesn't include, your custom set can't override it, and the plugin uses the default attribute value.

The image below illustrates this principle: the default *topic.title* attribute set contains four attributes—A, B, C, D. You create a custom *topic.title* attribute set and delete attribute D (say D defines the border below the text and you don't want a border, so you simply delete that attribute).

When you build your output, the border is still there! That's because your custom attribute set doesn't have its own D attribute to override the default D attribute, so the default D attribute stuck."

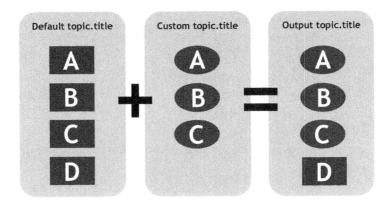

Figure 6: Attribute merge in attribute sets

What can you do? The default attribute set defines the title borders to be 3pt. You can simply change that value to 0pt, which creates the same effect as deleting the border altogether.

What can you do if you really need to remove an attribute from an attribute set?

If there is no way you can edit the value of an attribute to produce the same effect as deleting it altogether, you can create a new attribute set that does not include that attribute. You then need to replace all references to the original attribute set with references to your new attribute set.

This is a pretty rare situation, and in fact, no example of it springs to mind. The more common thing you'll need to do is simply set the value of the attribute in question to *inherit*. Most attributes accept the *inherit* value, but check your XSL-FO reference to be sure.

For example, the *userinput* attribute set formats the `<userinput>` element. The `<userinput>` element is an inline element, meaning you use it within a block element such as `<cmd>`:

```
<cmd>Select <userinput>Configure</userinput> from the drop-down list.</cmd>
```

By default, the *userinput* attribute set includes the following attribute:

```
<xsl:attribute name="font-family">monospace</xsl:attribute>
```

You don't always want `userinput` to be monospaced. Instead, you may want it to use the same font family as the element that contains it. You can't just delete the *font-family* attribute, and there seems to be no way to change the value to override it—any value you might use for *font-family* specifies, well, a font family.

In this case, set the *font-family* value to *inherit*. Doing so instructs the `userinput` element to use whatever font is specified for the block that contains `userinput`. Be aware, though, that the inheritance might not come from the immediate parent block. Given the very layered nature of XSL-FO, the inheritance might come from several blocks up the hierarchy.

Use-attribute-sets

Many attribute sets call other attribute sets using the *use-attribute-sets* attribute. *Use-attribute-sets* is another default attribute that you can't delete. If you delete it, the PDF plugin simply replaces the call with the one in the default attribute set.

So again, your best bet is to either use *inherit* or create a new attribute set that doesn't include *use-attribute sets* at all.

Special attributes for XEP, Antenna House, and FOP

In previous versions of the DITA Open Toolkit, stylesheets specifically for XEP, Antenna House, or FOP were all included in the `org.dita.pdf2` plugin, with suffixes added to the file names to distinguish them: the ones for XEP had an **_xep** suffix, the ones for Antenna House had an **_axf** suffix, and the ones for FOP had an **_fop** suffix.

These files still exist, but they have been moved into separate plugins: `org.dita.pdf2.axf`, `org.dita.pdf2.fop`, and `org.dita.pdf2.xep`.

Most of these files aren't independent. You're supposed to use them along with the generic file of the same name. For example, look at **commons-attr.xsl** and **commons-attr_xep.xsl**.

You can immediately see that **commons-attr.xsl** is a huge file with a gazillion attribute sets. On the other hand, **commons-attr_xep.xsl** only has one attribute set, *__fo__root*. The same attribute set exists in **commons-attr.xsl**. The attribute set in **commons-attr_xep.xsl** has a special attribute, *rx:link-back*, which takes advantage of additional functionality specific to XEP. Antenna House and FOP don't recognize *rx:link-back*, which would lead to a build error if you used this attribute with either of them. Because of this, and other similar reasons, there need to be renderer-specific attribute set files.

If you make changes to the generic file (such as **commons-attr.xsl**), copy the renderer-specific file for your renderer (such as **commons-attr_xep.xsl**) to your plugin as well. The reason is that the renderer-specific files override the generic files (just like your custom files override the generic files).

Because of this, there are cases where changing an attribute set in the generic file will not affect the PDF because that attribute set is also defined in the renderer-specific file. In those cases, you have to change the attribute set in the renderer-specific file. If you get in the habit of using renderer-specific files in your plugin, you'll get used to making changes there as well.

If you switch between different PDF renderers, you might need to create different plugins for each renderer.

> **Important:** The instructions in this book are for Apache FOP. The attributes and values described here work with FOP. Some of them give you different results with Antenna House or XEP. There's not a comprehensive list of what works where, so you will need to experiment. However, here are a few areas where you're likely to find differences:
>
> - image scaling
> - vertical spacing and alignment
> - justification and right-alignment
> - padding vs. margin, especially in headers and footers

Which attribute set do you customize?

If you can find the attribute set (or XSL template) you need to customize, you're 90% of the way there.

Customizing an attribute set is not hard; it's just a matter of knowing which attributes are available and what effect they have on the text. What can be difficult is knowing which attribute set you need to customize.

In some cases, it's pretty obvious. Say you want all your `<codeph>` text to appear in a monospaced font, dark gray. Well, if you know that the `<codeph>` element is part of the software (sw) domain, you can guess that you will find the attribute set you need to customize in the **sw-domain-attr.xsl** file. If you open that file, you will find an attribute set named *codeph* that is already set to use a monospaced font, so all you need to do is add the *color* attribute with a value that makes the text dark gray and you're all set.

But, alas, things are rarely so easy. More often, finding the right attribute set takes a combination of these three actions:

- logical guesswork
- searching the XSL files
- trial and error

Always start with logical guesswork. First, consider the element you want to customize and guess which attribute set file it's in. This book includes a list of all of the attribute sets and the files they are found in, and it also has descriptions of many of the attribute sets (see **Attribute set file list** *(p. 403)*).

Once you find a likely attribute set file, you should be able to narrow the possibilities to a handful of candidate attribute sets. Then, search the files in the xsl subfolder of the plugin to find where those attribute sets are called.

For example, say you want to change the formatting of the number that precedes steps. You want it to be dark red, bold, and followed by a parenthesis rather than a period. With your Sherlockian powers of deduction, you figure, correctly, that the attribute set you want to edit is in **task-elements-attr.xsl**, because steps are, after all, task elements. Opening **task-elements-attr.xsl**, you see some likely attribute sets:

- cmd
- steps
- steps.step
- steps.step__label
- steps.step__label__content
- steps.step__body
- steps.step__content

You can eliminate *steps* immediately because the number is part of an individual step, not the parent steps element. The *cmd* attribute set is possible, too, but then you notice *steps.step__label* and *steps.step__label__content*. When you consider that the number is not actually part of your content—you don't actually type the step number; rather, you let the stylesheet supply it—you begin to think that maybe those "label" attribute sets are the key.

How do you find out? First search through the XSLT files and see where this attribute set is called. You search for the string "steps.step__label" in all of the files in the xsl subfolder of the plugin and discover it's called in **task-elements.xsl**. Open that file and search for the string "steps.step__label".

Now things can get a bit tricky, especially if you aren't familiar with XSLT and XPath. Don't let the code freak you out. Take a deep breath and think about it logically. Here is the code that contains the string:

```
1  <xsl:template match="*[contains(@class, ' task/steps ')]/*[contains(@class, ' task/step ')]">
2      <!-- Switch to variable for the count rather than xsl:number, so that step
         ▶specializations are also counted -->
3      <xsl:variable name="actual-step-count" select="number(count(preceding-sibling::
         ▶*[contains(@class, ' task/step ')])+1)"/>
4      <fo:list-item xsl:use-attribute-sets="steps.step">
5          <fo:list-item-label xsl:use-attribute-sets="steps.step__label">
6              <fo:block xsl:use-attribute-sets="steps.step__label__content">
7                  <fo:inline>
8                      <xsl:call-template name="commonattributes"/>
9                  </fo:inline>
10                 <xsl:if test="preceding-sibling::*[contains(@class, ' task/step ')]
         ▶| following-sibling::*[contains(@class, ' task/step ')]">
11                     <xsl:call-template name="getVariable">
```

```
12                      <xsl:with-param name="id" select="''Ordered List Number'"/>
13                      <xsl:with-param name="params">
14                          <number>
15                              <xsl:value-of select="$actual-step-count"/>
16                          </number>
17                      </xsl:with-param>
18                  </xsl:call-template>
19              </xsl:if>
20          </fo:block>
21      </fo:list-item-label>
    ...
```

Figure 7: Use of step-related attribute sets

Line 4 calls the attribute set *steps.step*, line 5 calls *steps.step__label*, and Line 6 creates a text block that uses the *steps.step__label__content* attribute set. Line 12 calls the variable "Ordered List Number". Without knowing XSLT or XPath, you can make a pretty good guess that the step number gets generated there, and you'd be right. So it stands to reason that the attribute set that's called immediately before the number is generated (*steps.step__label__content*) probably affects the format of the number.

Although the process differs some for each situation, this example shows how to search systematically and make logical guesses.

When you get to the trial-and-error approach, it's a good idea to make an easily visible change to the attribute set to see if you're right. For example, if you think the attribute set *steps.step__label__content* is the one that formats step numbers, add the *color* attribute to make text formatted by this attribute set bright red. If you run a build and your step numbers are bright red, you're in the right place, and you can make your actual changes. If they are not red, you've got the wrong attribute set. If step numbers are bright red but so are other things, you need to find a more specific attribute set.

The closest attribute set wins

It's very common throughout the plugin XSLT stylesheets to have several layers of formatting applied to an element via multiple attribute sets. For example, an unordered list is rendered as multiple nested fo: elements, each of which has its own attribute set (shown in square brackets):

```
<fo:list-block> [ul]
    <fo:list-item> [ul.li]
        <fo:list-item-label> [ul.li__label]
            <fo:block> [ul.li__label__content]
            </fo:block>
        </fo:list-item-label>
        <fo:list-item-body> [ul.li__body]
            <fo:block> [ul.li__content]
            </fo:block>
        </fo:list-item-body>
    </fo:list-item>
</fo:list-block>
```

This doesn't need to make a lot of sense to you right now. The main thing to understand is that the bullet (that is, the list item label) is formatted first by attribute set *ul.li__label__content* and whatever formatting isn't specified there is picked up from *ul.li__label*. Whatever formatting isn't specified by *ul.li__label* is picked up from *ul.li* and finally from *ul*. So if you specify a color for *ul*, but not for *ul.li*, *ul.li__label* or *ul.li__label__content*, then the bullet color comes from *ul*. On the other hand, if you specify a color for *ul* and another for *ul.li__label__content*, then the bullet color comes from *ul.li__label__content*.

In all cases, the attribute set closest to what you're formatting will take precedence. Sometimes this can make it tricky to apply formatting to text that has several parts, like a list item with its bullet/number and text or a TOC entry with its label, title, leader, and page number. You might have to experiment to tease everything apart and discover exactly which attribute set is formatting that particular part.

Attribute sets that call other attribute sets

As you work in the attribute set files, you'll see a lot of attribute sets that call other attribute sets. For example, take a look at *topic.title* in **commons-attr.xsl**. Notice that it uses the attribute set *common.title*.

```
<xsl:attribute-set name="topic.title" use-attribute-sets="common.title">
```

This means that the attributes and values specified for *common.title* will be added to those specified for *topic.title*. Out of the box, *topic.title* specifies values for:

- border-bottom
- space-before
- space-after
- font-size
- font-weight
- padding-top
- keep-with-next.within-column

The *common.title* attribute set specifies a value for *font-family*.

By combining them, you get a more complete set of formatting attributes for level-1 topic titles. The thought behind creating the *common.title* attribute set and calling it within all the other topic-related attribute sets is that most topics have certain characteristics in common (pun unintended but inevitable). Rather than specify the same attributes over and over for each topic-related attribute set, you can just specify them once in *common.title* and be done.

> **Important:** The "closest attribute set wins" rule also applies here, so that an attribute directly defined in *topic.title* takes precedence over the same attribute defined in *common.title*.

For example, if you decide you want to use a serif font for level-3 titles, just add the *font-family* attribute to the *topic.topic.topic.title* attribute set and it overrides the specification in *common.title*.

> **Note** No, *topic.topic.topic.title* isn't a typo. The title-related attribute sets use this naming convention. The number of "topics" in the attribute set name corresponds to the hierarchical level of the topic that contains the title.

While this setup was always part of the PDF plugin, it's used much more heavily in recent versions of the DITA Open Toolkit. Maybe a little too heavily. There are several instances where attribute set A calls attribute set B, which calls attribute set C, and so forth. One example is the border-related attribute sets:

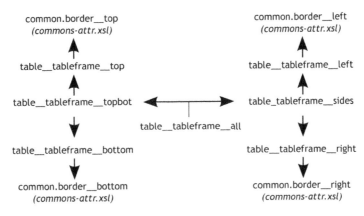

Figure 8: Interaction between attribute sets

None of the *table_tableframe* attribute sets actually contains any attributes by default. To find the properties that are being applied to a table where @frame="all", you have to trace through four separate paths of attribute sets that keep referring you to other attribute sets until you arrive at the *common.border* attribute sets, which are in a completely different file. Incidentally, the *common.border* attribute sets also control the formatting of borders above or below topic titles, so by making tables-related border changes, it's easy to make inadvertent (and incorrect) changes to topic title formatting.

The goal was to eliminate redundancy, but this is one of those cases where simpler is not easier, especially for people new to XSL and FO. Even though it reintroduces some redundancy, it's best for you to add the appropriate attribute to the attribute set that you're certain is closest to the element you're trying to format. It will override any formatting specified for any attribute sets further up the chain.

Don't try deleting the *use-attribute-set* attribute on an attribute set to eliminate its dependency on another attribute set and then add the attributes you need for the formatting you want. That won't work. Why? Because the attribute set in the default file still includes *use-attribute-set*. Remember that? If not, take a minute to review **How attribute set defaults work** *(p. 56)*.

Basic-settings variables in attribute sets

Many attributes in an attribute set can take either a hardcoded value, like this:

```
<xsl:attribute name="font-size>11pt</xsl:attribute>
```

or a variable, like this:

```
<xsl:attribute name="font-size">
    <xsl:value-of select="$default-font-size"/>
</xsl:attribute>
```

These variables live in **basic-settings.xsl**. By default, **basic-settings.xsl** contains only a few variables for things like page dimensions, page margins, and default font size. You can add as many more variables as you need and use them throughout your attribute sets.

For example, say your PDF uses the company color scheme, which consists of black, dark red and gray. The HTML hex values for these colors are #000000 (black); #990033 (dark red); and #8a8a8a (gray). You could use these hex values throughout your attribute sets, but what if your company decides to rebrand?

Now the colors are dark gray (#404040), dark green (#008000) and light green (#40ff40). Everything that was black is now dark gray, everything that was dark red is now dark green and everything that was gray is now light green. If you used the hex values throughout your attribute sets, you'll have to search and replace all the attribute set files in your PDF plugin. But if you used variables, you'd simply need to change the hex values for those three variables in **basic-settings.xsl**:

```
<xsl:variable name="text-color">#000000</xsl:variable>
<xsl:variable name="dark-accent-color">#990033</xsl:variable>
<xsl:variable name="light-accent-color">#8a8a8a</xsl:variable>
```

Other good uses of basic-settings variables include creating standard sets of font sizes, border types, or background images. It really only makes sense to use variables for attributes that can take any value. When an attribute already has a small fixed set of acceptable values, such as leader-pattern (dots, space, rule), it's not much of a time-saver to create a variable for it.

Syntax for variables in attribute values

There are two ways to express the use of a variable as an attribute value:

```
<xsl:attribute name="font-size">
    <xsl:value-of select="$default-font-size"/>
</xsl:attribute>
```

and

```
<xsl:attribute name="font-size" select="$default-font-size"/>
```

While the first method is used in the overwhelming majority of cases in the DITA Open Toolkit, the second method is more efficient. The examples in this book follow the convention used in the surrounding code, whichever that may be.

Other things you can do

Here are a few additional customizations you might want to make to your plugin. These customizations aren't part of the PDF specifications you're following, but they can be useful (and cool).

Exercise: Create a new attribute set

Objective: Create a custom attribute set to apply additional formatting in your PDF plugin.

Remember that when you customize an attribute set, you can't delete or comment out any of the attributes that are included by default. What you can do instead is create a new attribute set that doesn't include the attributes you don't want and then edit the appropriate stylesheet(s) in your plugin to call your new attribute set instead of the original.

You might also need a new attribute set if you have a special situation that needs different formatting. For example, you might want to add some boilerplate text to your title page that has characteristics different from other text blocks. The easiest way to create a new attribute set is to copy an existing one into your plugin and modify it. Ideally, copy one that's similar to the attribute set you want to create.

Let's create a new attribute set based on the existing *b* attribute set, to use for a special instance of bold text. This is an over-simplified example, but you'll use it as the basis for more interesting examples later.

1. If **hi-domain-attr.xsl** is not already in your plugin folder, create it following the instructions in **Create an attribute set file in your plugin** *(p. 31)*.

   Don't forget to add an xsl:import line to **custom.xsl** (attrs). Also add the **hi-domain.xsl** file to your plugin and an xsl:import line for it to **custom.xsl** (xsl). If you don't remember why, review **Add an attribute set to your plugin** *(p. 32)*.

2. In your copy of **hi-domain-attr.xsl**, copy the *b* attribute set.

3. Rename the copy *b_red*.

```
<xsl:attribute-set name="b_red">
  <xsl:attribute name="font-weight">bold</xsl:attribute>
</xsl:attribute-set>
```

4. Add the *color* attribute to the *b_red* attribute set with a value **red**.

```
<xsl:attribute-set name="b_red">
  <xsl:attribute name="font-weight">bold</xsl:attribute>
  <xsl:attribute name="color">red</xsl:attribute>
</xsl:attribute-set>
```

5. Save and close **hi-domain-attr.xsl**.

You're ready to edit the template that processes b to call your new *b_red* attribute set. Continue to **Call a new attribute set** *(p. 66)*.

Exercise: Call a new attribute set

Objective: Use the custom attribute set you created to apply custom formatting.

Before you start

Complete the exercise **Create a new attribute set** *(p. 65)*.

After you create a new attribute set to replace an existing one, you have to modify the corresponding template to call the new attribute set instead of the default one.

1. Search the files in DITA-OT/plugins/org.dita.pdf2/xsl/fo/ to find the template that uses the original attribute set.

You just created a new attribute set called *b_red* to replace the default *b* attribute set. Now you need to find all instances of the original attribute set and replace them with your new one.

Search for all instances of this text string: xsl:use-attribute-sets="b".

It turns out that one stylesheet calls this attribute set: **hi-domain.xsl** (no surprise there). In that stylesheet, the template that uses the *b* attribute set begins with

```
<xsl:template match="*[contains(@class,' hi-d/b ')]">
```

This is the template you need to modify to call your *b_red* attribute set instead.

2. If **hi-domain.xsl** is not already in your plugin folder, create it following the instructions in **Create an XSLT stylesheet in your plugin** *(p. 32)*.

Don't forget to add an `xsl:import` line to **custom.xsl** (xsl).

3. Copy this template into your copy of **hi-domain.xsl**:

```
<xsl:template match="*[contains(@class,' hi-d/b ')]">
    <fo:inline xsl:use-attribute-sets="b">
        <xsl:call-template name="commonattributes"/>
        <xsl:apply-templates/>
    </fo:inline>
</xsl:template>
```

This template matches all `<b>` elements and assigns the *b* attribute set to the contents of those elements. That strange-looking match attribute does the work, and you'll learn how later. For now, you just need to know that this is the template you need to edit.

4. Change the line `<fo:inline xsl:use-attribute-sets="b">` to

```
xsl:use-attribute-sets="b_red"
```

5. Save and close **hi-domain.xsl**.

6. Save changes, run a PDF build, and check your work.

Reality check

In your PDF, all text within `<b>` elements should now be bold and red.

Exercise: Conditionalize an attribute set

Objective: Set up an attribute set so that it formats an element one way in one context and a different way in another context.

Before you start

Complete the exercise **Call a new attribute set** *(p. 66)*.

In some cases, you might need to create a new attribute set for use only under certain conditions. For example, you might want text in a `<p>` that follows a `<ul>` to be formatted differently from other instances of `<p>`. This is conditionalizing based on context. In other cases, the context might not be predictable. When that is the case, you might want to use the *outputclass* attribute to call for alternate formatting.

For this example, let's use your new attribute set, *b_red*. After the previous exercise, all instances of b use this attribute set. Now you just want to apply it to instances of b where `@outputclass="red"`.

1. Open your copy of **hi-domain.xsl**.

2. Find the following code:

```
<xsl:template match="*[contains(@class,' hi-d/b ')]">
    <fo:inline xsl:use-attribute-sets="b_red">
        <xsl:call-template name="commonattributes"/>
        <xsl:apply-templates/>
    </fo:inline>
</xsl:template>
```

You want to test the *@outputclass* value of each occurrence of b and determine whether to use attribute set *b* or *b_red*.

There are three possible conditions:

- @outputclass="red"
- @outputclass=*[some other value]*
- no *outputclass* attribute at all

You're lucky in this example because the only time you want to apply your *b_red* attribute set is in the first case. You don't care about any other conditions; they will all use the original *b* attribute set. So, you can write a simple xsl:choose test.

3. Change the code to:

```
<xsl:template match="*[contains(@class,' hi-d/b ')]">
    <xsl:choose>
        <xsl:when test="./@outputclass='red'">
            <fo:inline xsl:use-attribute-sets="b_red" id="{@id}">
                <xsl:call-template name="commonattributes"/>
                <xsl:apply-templates/>
            </fo:inline>
        </xsl:when>
        <xsl:otherwise>
            <fo:inline xsl:use-attribute-sets="b" id="{@id}">
                <xsl:call-template name="commonattributes"/>
                <xsl:apply-templates/>
            </fo:inline>
        </xsl:otherwise>
    </xsl:choose>
</xsl:template>
```

4. Save and close **hi-domain.xsl**.

5. Somewhere in your test DITA files, include an instance of the <b> element with outputclass="red":

```
<b outputclass="red">This text should be bold and red.</b>
```

6. Save changes, run a PDF build, and check your work.

You might be asking yourself, "Do I really even need an attribute set here? Can't I just specify the color directly?"

```
<xsl:when test="./@outputclass='red'">
    <fo:inline id="{@id}" color="#ff0000">
    ...
```

That's a fair question and the answer is yes. It's your choice, really. Attribute sets are a convenient way to re-use a group of attributes and values multiple times without having to re-list them individually within multiple templates. When you have an ad hoc usage of a single attribute, it's probably just as fast (if not faster) to simply cite the attribute directly in the template.

Exercise: Conditionalize an attribute set based on a build parameter

Objective: Set up an attribute set so that it formats an element different ways depending on a parameter in your ANT build file.

Before you start

Complete the exercise **Call a new attribute set** *(p. 66)*.

You can conditionalize an attribute (or almost anything, really) based on attribute values in your content, such as *outputclass*. But suppose you want to conditionalize based not on anything within your content, but instead on some condition that comes from a higher, non-content-dependent level, such as your ANT build file.

As you probably know, you can specify the *args.draft* parameter in your ANT build file. By default, this triggers the Open Toolkit to include the content of `draft-comment` and `required-cleanup` elements in your PDF. In addition to those built-in behaviors, you might decide to use *args.draft* to mark the PDF in a way that makes it obvious that it's a draft copy and not for distribution.

> **Note** If you don't know about all the build parameters available to you, take a look at the topics in the `DITA-OT/doc/parameters` folder. You can get some serious mileage out of these parameters.

Suppose you've decided to add a magenta background to all topic titles. Here's how to do that.

1. Copy the attribute set *common.title* from `DITA-OT/plugins/org.dita.pdf2/xsl/fo/` **commons-attr.xsl** to your copy of **commons-attr.xsl**.

As you'll see in more detail later, this attribute set is used by all of the other title-related attribute sets. It's a one-stop shop for making formatting changes to titles.

2. In `DITA-OT/plugins/org.dita.pdf2/xsl/fo/`**commons.xsl**, search for the term "DRAFT" (all caps, no quotes).

You should find the first instance here:

```
<xsl:template match="*[contains(@class,' topic/titlealts ')]">
    <xsl:if test="$DRAFT='yes'">
        <xsl:if test="*">
            <fo:block xsl:use-attribute-sets="titlealts">
                <xsl:call-template name="commonattributes"/>
                <xsl:apply-templates/>
            </fo:block>
        </xsl:if>
    </xsl:if>
</xsl:template>
```

This template outputs alternate titles only if *args.draft* is set to "yes" in the ANT build files. This is exactly the same kind of test you want to perform to determine whether titles should have the magenta background or not.

3. In your copy of **commons-attr.xsl**, edit the *common.title* attribute set as follows:

```
<xsl:attribute-set name="common.title">
    <xsl:attribute name="font-family">sans-serif</xsl:attribute>
    <xsl:attribute name="background-color">
        <xsl:if test="$DRAFT='yes'">#FF00FF</xsl:if>
    </xsl:attribute>
</xsl:attribute-set>
```

> **Note** For simplicity, this code duplicates code found throughout the DITA OT. However, technically, a more flexible alternative is:
>
> ```
> <xsl:if test="$DRAFT=('yes')">#FF00FF</xsl:if>
> ```
>
> You could expand this syntax to:
>
> ```
> <xsl:if test="$DRAFT=('yes','on','true','1')">#FF00FF</xsl:if>
> ```
>
> In the event you weren't quite sure what the allowable values for the parameter were, you could cover all possibilities. For the same reason, you could also use `matches`:
>
> ```
> <xsl:if test="matches($DRAFT, 'yes|on|true|1')">#FF00FF</xsl:if>
> ```

4. Save changes, run a PDF build, and check your work.

Exercise: Select specific elements by context

Objective: Select a specific instance of an element based on its location within the structure.

Say you want to ensure that pages never break leaving a single list item stranded at the bottom of the page (a widow). You want to make sure that there are always at least two list items together at the bottom of a page, and if there is not sufficient space, the entire list flows to the following page.

To do this, you either need two attribute sets—one for the first `<li>` in an `<ol>`, `<ul>`, or `<sl>`—and another for the remaining `<li>` elements, or you need to conditionalize the existing attribute sets.

1. If you do not already have a **lists-attr.xsl** file in your plugin, create it.
2. In DITA-OT/plugins/org.dita.pdf2/cfg/fo/attrs/ **lists-attr.xsl**, find the *ul.li* attribute set.
3. Copy it and paste the copy in your **lists-attr.xsl** file.
4. Rename the copy *ul.li.first*.
5. Edit *ul.li.first* as follows:

```
<xsl:attribute-set name="ul.li.first">
    <xsl:attribute name="space-after">1.5pt</xsl:attribute>
    <xsl:attribute name="space-before">1.5pt</xsl:attribute>
    <xsl:attribute name="relative-align">baseline</xsl:attribute>
    <xsl:attribute name="keep-with-next.within-page">always</xsl:attribute>
</xsl:attribute-set>
```

> **Important:** This is a good place to say…be careful when using *keep-with*. If you have a number of elements in succession that are all set to *keep-with-next* or *keep-with-previous*, you could end up creating a very long block of content that has no opportunity to break at all. If that block is too long to fit on a single page, it cannot logically be rendered and will either cause missing content in the PDF or a failed build altogether.

6. Repeat steps 2-5 for the *ol.li* and the *sl.sli* attribute sets.
7. Save and close your **lists-attr.xsl** and the one in org.dita.pdf2.
8. If you do not already have a **lists.xsl** file in your plugin, create it.
9. In DITA-OT/plugins/org.dita.pdf2/xsl/fo/**lists.xsl**, find the template that begins with these two lines:

```
<xsl:template match="*[contains(@class, ' topic/ul ')]/*[contains(@class, ' topic/li ')]">
    <fo:list-item xsl:use-attribute-sets="ul.li">
```

This is the template that processes `<li>` elements in unordered lists (`<ul>`). Notice that it specifies that these `<li>` elements should use the *ul.li* attribute set.

10. Copy and paste this template in your **lists.xsl**.

You want to make this template use your new *ul.li.first* attribute set if the `<li>` is the first one in the unordered list and otherwise continue using the *ul.li* attribute set. To do that, you need to tell the template to choose between attribute sets based on the position of the `<li>` within the `<ul>`.

11. Just below the first line of the template, add an `<xsl:choose>` element as follows:

```
<xsl:template match="*[contains(@class, ' topic/ul ')]/*[contains(@class, ' topic/li ')]">
    <xsl:choose>
        <xsl:when>
        </xsl:when>
        <xsl:otherwise>
        </xsl:otherwise>
    </xsl:choose>
```

12. Select everything between `<fo:list-item` ... and `</fo:list-item>` and cut it (**Ctrl-X**).

13. Paste the cut content between the `<xsl:when>` and `</xsl:when>` tags:

```
<xsl:when>
  <fo:list-item xsl:use-attribute-sets="ul.li">
      <xsl:apply-templates select="*[contains(@class,' ditaot-d/ditaval-startprop ')]"
      ▶ mode="flag-attributes"/>
      <fo:list-item-label xsl:use-attribute-sets="ul.li__label">
          <fo:block xsl:use-attribute-sets="ul.li__label__content">
              <fo:inline>
                  <xsl:call-template name="commonattributes"/>
              </fo:inline>
              <xsl:call-template name="getVariable">
                  <xsl:with-param name="id" select="'Unordered List bullet'"/>
              </xsl:call-template>
          </fo:block>
      </fo:list-item-label>
      <fo:list-item-body xsl:use-attribute-sets="ul.li__body">
          <fo:block xsl:use-attribute-sets="ul.li__content">
              <xsl:apply-templates/>
          </fo:block>
      </fo:list-item-body>
  </fo:list-item>
</xsl:when>
```

14. Paste the cut content again between the `<xsl:otherwise>` and `</xsl:otherwise>` tags.

15. Edit the `<xsl:when>` as follows:

```
<xsl:when test="not(preceding-sibling::li)">
```

This template iterates through every `<li>` in a `<ul>`. As it processes each one, this `<xsl:when>` tests to see if the `<li>` that is currently being processed has no preceding `<li>` sibling. Only the first `<li>` in a `<ul>` satisfies this condition.

16. Within the `<xsl:when>`, change to call the *ul.li.first* attribute set:

```
<fo:list-item xsl:use-attribute-sets="ul.li.first">
```

The full template should now look like this:

```
<xsl:template match="*[contains(@class, ' topic/ul ')]/*[contains(@class, ' topic/li ')]">
   <xsl:choose>
      <xsl:when test="not(preceding-sibling::li)">
         <fo:list-item xsl:use-attribute-sets="ul.li.first">
            <xsl:apply-templates select="*[contains(@class,' ditaot-d/ditaval-startprop ')]"
            ▸ mode="flag-attributes"/>
            <fo:list-item-label xsl:use-attribute-sets="ul.li__label">
               <fo:block xsl:use-attribute-sets="ul.li__label__content">
                  <fo:inline>
                     <xsl:call-template name="commonattributes"/>
                  </fo:inline>
                  <xsl:call-template name="getVariable">
                     <xsl:with-param name="id" select="'Unordered List bullet'"/>
                  </xsl:call-template>
               </fo:block>
            </fo:list-item-label>
            <fo:list-item-body xsl:use-attribute-sets="ul.li__body">
               <fo:block xsl:use-attribute-sets="ul.li__content">
                  <xsl:apply-templates/>
               </fo:block>
            </fo:list-item-body>
         </fo:list-item>
      </xsl:when>
      <xsl:otherwise>
         <fo:list-item xsl:use-attribute-sets="ul.li">
            <xsl:apply-templates select="*[contains(@class,' ditaot-d/ditaval-startprop ')]"
            ▸ mode="flag-attributes"/>
            <fo:list-item-label xsl:use-attribute-sets="ul.li__label">
               <fo:block xsl:use-attribute-sets="ul.li__label__content">
                  <fo:inline>
                     <xsl:call-template name="commonattributes"/>
                  </fo:inline>
                  <xsl:call-template name="getVariable">
                     <xsl:with-param name="id" select="'Unordered List bullet'"/>
                  </xsl:call-template>
               </fo:block>
            </fo:list-item-label>
            <fo:list-item-body xsl:use-attribute-sets="ul.li__body">
               <fo:block xsl:use-attribute-sets="ul.li__content">
                  <xsl:apply-templates/>
               </fo:block>
            </fo:list-item-body>
         </fo:list-item>
      </xsl:otherwise>
   </xsl:choose>
</xsl:template>
```

17. Save and close your **lists.xsl** and the one in `org.dita.pdf2`.

18. Save changes, run a PDF build, and check your work.

You'll need to find or set up a list in a topic that normally breaks after the first `<li>` so that you can see the effect of the new attribute set and the *keep-with-next* attribute.

Repeat this process for `<li>` elements within ordered lists (`<ol>`) and simple lists (`<sl>`).

Chapter 5

Localization variables

What are localization variables?

If you plan to publish PDFs in multiple languages, you want to minimize the amount of translation required. One way to do this is to use language-specific variables or strings for boilerplate text, such as the prefix for admonitions (notes, cautions, warnings, etc.) or standard headings for the table of contents, index, related topics categories, etc. Often this boilerplate text does not even appear in the XML sent to the translation agency; it is added by the DITA Open Toolkit when your files are processed. That means you only pay to translate it once per language.

To ensure that the correct translation of boilerplate text appears in your PDFs, the Open Toolkit uses language-specific variables and strings files to insert the appropriate text.

The variables files are found in `DITA-OT/plugins/org.dita.pdf2/cfg/common/vars/`. The strings files are found in `DITA-OT/xsl/common/`.

Localization variable files are named using ISO language codes—for example, **en.xml**. Localization strings files are named using the same language codes plus locale codes—for example, **strings-en-us.xml**. These names usually reflect the name of the language and locale in that language. For example, **strings-de-de.xml** is the localization string file for German, because the name for the German language in German is *Deutsch* and the name for Germany in German is *Deutschland*.

‖ **Note** There is also a newer standard, ISO 639-2, which uses three-letter codes. ‖

How does the DITA Open Toolkit know which set of variables to use when creating the PDF? It reads the *xml:lang* attribute on the map (or topic or element) and uses the corresponding variables file. A map with the following *xml:lang* specification uses the **es.xml** variables file and the **strings-es-es.xml** strings file to translate boilerplate text to Spanish (es):

```
<map xml:lang="es">
```

By default, the Open Toolkit assumes that *xml:lang*="en" unless explicitly defined. This default is specified in DITA-OT/lib/**configuration.properties** using the *default.language* property. Any element, topic, or map that does not explicitly specify a value for *xml:lang* assumes "en". You can change the default value in **configuration.properties**, or you can pass in a different language as a parameter in your build.

The *xml:lang* code on maps and topics needs to exactly match the name of the localization variables file that should be used. For example, if you have a variables file named **fr.xml** but your map has an *xml:lang* value of "fr-fr" or "fr-ca", the Open Toolkit does not match the map and the **fr.xml** file; it falls back to the designated default language which out of the box is English and therefore it takes localization variable values from **en.xml**. So, if you use *xml:lang* codes such as "fr-fr" or "fr-ca", make sure that the localization variables files in your plugin are named accordingly.

For PDFs, the Open Toolkit uses the closest specified language as the default. If an element does not specify *xml:lang*, the *xml:lang* value of its parent topic is used. If the parent topic does not specify *xml:lang*, the *xml:lang* value of the root map is used. If the root map does not specify *xml:lang*, the value of *default.language* is used.

It's not necessary to specify *xml:lang* on every element, of course. You only need to do so if the language of that element is different from what is specified for the topic or map. For example, a topic might be in English (en) but perhaps it contains a word or phrase in Thai. If you wrap that word or phrase in <ph>, you only need to specify *xml:lang*="th" for that <ph> element. Everything else picks up the default "en".

You can, of course, create additional variables files for other languages. In addition to the variables that are included in these files out of the box, you can add other variables to an existing file as needed.

> **Important:** Never delete any variable from a localization variables file. If you delete a variable, you have to find all the places where it's called and delete those calls as well. It's likely you'll miss something and cause your builds to fail. If there is a localization variable whose value you don't want to output, simply delete the value but leave the variable.

Static variables vs. dynamic variables

The localization variables and strings are static. That is, the text they insert in your PDF is always the same, regardless of context. For example, the *Note* string always inserts the text "Note" (or the equivalent in the target language).

Many times, though, you need a dynamic variable whose value can change based on context. A good example of a dynamic variable is the product name in the page headers of a PDF. If you create deliverables for more than one product, you want the DITA Open Toolkit to pick up the product name from the map you're using as input. By default, the PDF plugin uses the variable *productName* in page headers. This variable, in turn, gets its value from the metadata in the input bookmap or from the product attribute on the input map.

You don't use localization variables for these dynamic variables. Usually, you maintain dynamic variables in the **root-processing.xsl** file, which you'll read about in **Where does header and footer information come from?** *(p. 121)*.

Literal characters and numeric character references

It's probably helpful to discuss encoding. At least, you should know that to include literal non-Latin characters such as Greek and Cyrillic in an XSL file, the file needs to use a Unicode encoding such as UTF-8 or UTF-16. If Greek and Cyrillic (or any other) literal characters show up in the PDF as ???????, boxes, gibberish, or nothing, your file is not using a Unicode encoding.

What is a literal character? A literal character is one that looks on the screen like it will look in print. For example, Σ and Щ are the literal Greek sigma and Cyrillic shcha characters, respectively. If you want to copy and paste characters from a Character Map application or type them directly from the keyboard, your file needs to be encoded as UTF-8, UTF-16, etc.

On the other hand, if you plan to use numeric character references, then you don't necessarily need UTF-8 files. A numeric character reference is a short sequence of characters that represents a single character. Numeric character references are used in languages such as HTML and XML.

There are often several valid numeric character references for a character. One numeric character reference for the Greek sigma is `&#x03A3;` (Σ), and one for the Cyrillic shcha is `&#x0429;` (Щ).

And you need to use a font that supports the characters you want to use. You have to use Unicode fonts, but just because a font is a Unicode font, it doesn't necessarily have glyphs for all Unicode characters (few do, other than Arial Unicode MS, GNU Unifon,t and few other special-purpose fonts). Be sure to research before you choose your font.

There are quite a few Unicode references online. Here are a few especially helpful ones:

- **https://en.wikibooks.org/wiki/Unicode/Character_reference/0000-0FFF**
- **http://www.ssec.wisc.edu/~tomw/java/unicode.html**
- **http://www.fileformat.info/info/unicode/**

In XML, numeric character references can be expressed either as a decimal (base 10) number or as a hexadecimal (base 16) number. For example, the decimal numeric character references for the Greek sigma are `&#931;` and `&#0931;`. The hexadecimal (hex) character references are `&#x3A3;`, `&#x03A3;`, and `&#x3a3;`.

Without going into a lot of detail, the decimal format uses digits 0-9. The hex format also uses digits 0-9 but adds letters A-F (or a-f) to complete the base-16 format. You've seen the hex format in the color specifications you've encountered so far in this book, such as "#ff0000" (red). If you're curious about the hex format, search online for " hexadecimal format." You'll find plenty of information.

The choice of which format to use is yours, but be aware that older versions of HTML do not allow the hex syntax. If you plan to generate HTML output from your content, be aware of the version of HTML you are targeting (and what version your users are using), so you can choose the most compatible format.

There is another kind of character reference called a character entity reference, which allows a character to be referred to by a name instead of a number. One common example is < (or less than, "<"). In general, using entities in your DITA content is not advisable, though the use of certain entities such as < in XSLT is common.

Exercises

There are no exercises for this chapter. However, you'll work with localization variables in many of the other exercises throughout this book.

Other things you can do

Here are a few additional customizations you might want to make to your plugin. These customizations aren't part of the PDF specifications you're following, but they can be useful (and cool).

Exercise: Create a localization variables file

Objective: Create a localization variables file for a language that is not in the Open Toolkit by default.

You can create a language-specific variables file for languages that aren't included in the DITA Open Toolkit. For example, let's create a variables file for Hungarian.

1. Copy one of the variables files (in DITA-OT/plugins/org.dita.pdf2/cfg/common/vars/) and rename it using the ISO 639-1 language code for Hungarian (**hu.xml**).
2. Move **hu.xml** to DITA-OT/plugins/com.company.pdf/cfg/common/vars/.
3. For each existing variable in the file, substitute the Hungarian text from your translated information. For example, let's look at the variable *Note*, which defines the label for note elements:

```
<!-- Text to use for 'Note' label generated from <note> element. -->
    <variable id="Note">Note</variable>
```

The Hungarian word for "Note" is "Megjegyzés". Edit the value of the *Note* variable accordingly:

```
<!-- Text to use for 'Note' label generated from <note> element. -->
    <variable id="Note">Megjegyzés</variable>
```

4. Save **hu.xml**.

Reality check

When you run a build using content that is marked `xml:lang="hu"`, the DITA Open Toolkit inserts appropriate boilerplate text from **hu.xml**.

Exercise: Create a localization strings file

Objective: Create a localization strings file for a language that is not in the Open Toolkit by default.

You can create a language-specific strings file for languages that aren't included in the DITA Open Toolkit.

1. Copy one of the strings files (in `DITA-OT/xsl/common/`) and rename it using the ISO 639-1 language code for Swahili and locale code for Kenya (**strings-sw-ke.xml**).

2. Move **strings-sw-ke.xml** to `DITA-OT/plugins/com.company.pdf/cfg/common/strings/`.

3. For each existing variable in the file, substitute the Swahili text from your translated information. For example, let's look at the variable *Note*, which defines the label for `note` elements:

```
<str name="Note">Note</str>
```

The Swahili word for "Note" is "Kumbuka". Edit the value of the *Note* variable accordingly:

```
<str name="Note">Kumbuka</str>
```

4. Save **strings-sw-ke.xml**.

5. Add the following lines to the first section of `DITA-OT/plugins/com.company.pdf/cfg/common/strings/` **strings.xml**.

```
<lang xml:lang="sw" filename="strings-sw-ke.xml"/>
<lang xml:lang="sw-ke" filename="strings-sw-ke.xml"/>
```

Reality check

When you run a build using content that is marked `xml:lang="sw"` or `xml:lang="sw-ke"`, the DITA Open Toolkit inserts appropriate boilerplate text from **strings-sw-ke.xml**.

Chapter 6

Fonts

Using the correct fonts is an integral part of any PDF customization. This chapter explains how to point your PDF renderer to the fonts you want to use and how to set up your plugin to use those fonts.

About fonts in the PDF plugin

Being able to use specific fonts in a PDF is an essential part of the design. Out of the box, the PDF plugin specifies the Helvetica, Arial Unicode MS, Tahoma, Times, Times New Roman, Courier, and Courier New fonts. To provide better support for Asian languages, the PDF plugin also specifies Batang, SimSun, AdobeSongStd-Light, KozMinProVI-Regular, and AdobeMyungjoStd-Medium. Zapf Dingbats is thrown in there as well. It's a pretty safe bet you have other fonts you'd rather use.

Configuring fonts is a two-part process:

1. Setting up your PDF renderer to find, recognize, and use those fonts:

   This part is different for each PDF renderer.

2. Telling your plugin which fonts to use:

   There are two ways to do this, depending on which PDF renderer you're using and how granular your font switching needs to be.

You also need to tell your build which PDF renderer you're using. If you haven't already set that property, see **Use a specific PDF renderer** *(p. 49)*.

Font mapping

In the past, the PDF plugin used a file called **font-mappings.xml** to map physical fonts found on your system to logical fonts named Sans, Serif, and Monospaced. For example, you might map the physical font Trebuchet MS to the logical font Sans. Then, when you wanted to use Trebuchet MS for a particular element, you would specify Sans in the attribute set associated with that element.

If this seems a little roundabout (why not just call Trebuchet MS directly?), it is. Font mapping was set up to get around FOP's inability to change the font for a single character if the font you were using didn't include a particular glyph. Unfortunately, FOP still doesn't support per-character font selection, so if you're using FOP, you might still need font mapping when your fonts don't contain all the necessary glyphs. For example, if you're using FOP and you need to display a word in Korean in the middle of a paragraph of English, and the font used by that paragraph doesn't include any Korean glyphs, you'll probably have to use font mapping.

The DITA Open Toolkit currently uses font mapping by default to maintain backwards compatibility, but it's an expensive process in terms of processing time and effort. This default is probably going to change in version 3.0 of the Open Toolkit. The font mapping process code will still be available, but it will not be enabled by default. Therefore, if you can move away from font mapping now and call physical fonts directly, you should. This chapter assumes that you aren't using font mapping, but in case you need to, it's covered in **Use font mapping in your plugin** *(p. 87)*.

That's the short explanation. You can read a little more about it at **elovirta.com**.

Font specifications

Here are the specifications for the fonts your PDF is going to use. Sizes and colors don't matter right now. At this point you just want to get the right fonts set up for your PDF.

General text

Chapter, appendix titles	28pt Trebuchet normal black, no borders. Autonumber is 16pt Trebuchet regular black.
Level 1 heading font	20pt Trebuchet normal black, no borders
Level 2 heading font	18pt Trebuchet normal dark red
Level 3 heading font	16pt Trebuchet normal dark red, .5in indent
Level 4+ heading font	14pt Trebuchet italic dark red, .5in indent
Section headings	12pt Trebuchet bold dark red
Regular body font	11pt Book Antiqua normal dark gray. Used for all body elements except where noted below.

Small body font	10pt Book Antiqua normal dark gray. Used for *note, info, stepxmp, stepresult, choice*.
Code samples, system messages	9pt Consolas normal black, light gray background
Line height (leading)	120%

Files you need

If you have completed other exercises, some of these files might already be present in your plugin. If not, create the following files in DITA-OT/plugins/com.company.pdf/cfg/fo/attrs/:

- **commons-attr.xsl**
- **hi-domain-attr.xsl**
- **markup-domain-attr.xsl**
- **pr-domain-attr.xsl**
- **sw-domain-attr.xsl**
- **ui-domain-attr.xsl**
- **xml-domain-attr.xsl**

Be sure to add the appropriate xsl:import statements to **custom.xsl** (attrs).

If you've determined you need to use font mapping, you need the **font-mappings.xml** file. That's explained in **Use font mapping in your plugin** *(p. 87)*.

If you're using Antenna House or XEP, you might also need to edit the configuration files specific to those PDF renderers.

Exercise: Specify fonts to use

Objective: Specify, in your custom PDF plugin, which fonts you're going to use.

The first step is to make sure your PDF renderer knows where all your fonts are (and, of course, that the fonts you want to use are actually in that location). FOP automatically looks for fonts in a default location, depending on your operating system. For Windows, this location is the Fonts folder. For Macintosh, there are two locations: the Library/Fonts folder and the Library/Fonts folder in your Users folder.

For Linux, the location varies depending on your Linux distribution (the most common location is /usr/share/fonts). Because this book's exercises use FOP, there's nothing more to do here.

The second step is to specify fonts for the various attribute sets. You don't have to specify fonts for every attribute set. Many designers use one font for titles and another font for most body text. There might be exceptions, but you can start with these two broad strokes. The PDF specifications for this book say to use Trebuchet MS for all titles and Book Antiqua for most body text. Start there.

1. Copy the *common.title*, *__fo__root* and *pre* attribute sets from DITA-OT/plugins/org.dita.pdf2/cfg/fo/attrs/**commons-attr.xsl** to your copy of **commons-attr.xsl**.

2. Edit the *font-family* attribute in *common.title* as follows:

```
<xsl:attribute-set name="common.title">
  <xsl:attribute name="font-family">Trebuchet MS, Arial Unicode MS, Helvetica
  ►</xsl:attribute>
</xsl:attribute-set>
```

By default, the *common.title* attribute set specifies sans-serif as its font-family. The three fonts you specified here are all sans-serif fonts. *common.title* is used by all of the other title attribute sets in **commons-attr.xsl**, so specifying the fonts here takes care of all six title levels as well as section titles, example titles, figure and table captions, the cover page title, and the Table of Contents header (some of which are defined in files other than **commons-attr.xsl**).

> **Note** The only attribute set other than *common.title* that specifies sans-serif as its font-family by default is *__toc__mini*, which is found in **toc-attr.xsl**.

Notice that this example actually specifies three fonts for *common.title*. The first is the main font to use. The others are fallbacks, to be used (in order) in case a particular system does not have the main font. So if a system does not have Trebuchet MS installed, it will use Arial Unicode MS instead. If Arial Unicode MS is not present, it will use Helvetica.

3. Edit the *font-family* attribute in *__fo__root* as follows:

```
<xsl:attribute-set name="__fo__root" use-attribute-sets="base-font">
  <xsl:attribute name="font-family">Book Antiqua, Times New Roman, Times</xsl:attribute>
  <xsl:attribute name="xml:lang" select="translate($locale, '_', '-')"/>
  <xsl:attribute name="writing-mode" select="$writing-mode"/>
</xsl:attribute-set>
```

By default, the *__fo__root* attribute set specifies serif as its font-family. The three fonts here are all serif fonts. The *__fo__root* attribute set isn't used by any other attribute sets but it defines the font for the entire PDF (at the "root" level). By defining fonts here, you're saying "Use these fonts throughout the PDF except when I say otherwise (by defining different fonts for specific attribute sets.)"

> **Note** *__fo__root* is the only attribute set that specifies serif as its font-family by default.

Based on your PDF specifications, you now need to specify Consolas as the font to use for code samples and system messages. Consolas is a monospaced font, meaning that you're going to need to edit all attribute sets that specify monospace as the font-family to instead specify Consolas. This step is going to be a bit more work because unlike sans-serif and serif, there are 23 attribute sets that use monospace. One of them, *pre*, is in **commons-attr.xsl**, and you've already copied it, so start there.

4. Edit the *font-family* attribute in *pre* as follows:

```
<xsl:attribute-set name="pre" use-attribute-sets="base-font common.block">
   <xsl:attribute name="white-space-treatment">preserve</xsl:attribute>
   <xsl:attribute name="white-space-collapse">false</xsl:attribute>
   <xsl:attribute name="linefeed-treatment">preserve</xsl:attribute>
   <xsl:attribute name="wrap-option">wrap</xsl:attribute>
   <xsl:attribute name="font-family">Consolas, Courier New, Courier</xsl:attribute>
   <xsl:attribute name="line-height">106%</xsl:attribute>
</xsl:attribute-set>
```

pre is used by the *codeblock* and *msgblock* attribute sets in **pr-domain-attr.xsl** and **sw-domain-attr.xsl**, respectively, so those two elements are taken care of with this one change. But there are others.

5. Copy the *tt* attribute set from `DITA-OT/plugins/org.dita.pdf2/cfg/fo/attrs/`**hi-domain-attr.xsl** to your copy of **hi-domain-attr.xsl**.

6. Edit the *font-family* attribute in *tt* as follows:

```
<xsl:attribute-set name="tt" use-attribute-sets="base-font">
   <xsl:attribute name="font-family">Consolas, Courier New, Courier</xsl:attribute>
</xsl:attribute-set>
```

7. Copy the *markupname* attribute set from `DITA-OT/plugins/org.dita.pdf2/cfg/fo/attrs/`**markup-domain-attr.xsl** to your copy of **markup-domain-attr.xsl**.

8. Edit the *font-family* attribute in *markupname* as follows:

```
<xsl:attribute-set name="markupname" use-attribute-sets="base-font">
   <xsl:attribute name="font-family">Consolas, Courier New, Courier</xsl:attribute>
</xsl:attribute-set>
```

I'll stop here. You get the idea. You now need to repeat these steps for all of the attribute sets in **pr-domain-attr.xsl**, **sw-domain-attr.xsl**, **ui-domain.xsl**, and **xml-domain-attr.xsl** that specify monospace as the font-family.

But do you really need to edit all of these attribute sets? Maybe, maybe not. For example, if you know that your content model never uses the screen element, you don't need to copy its attribute set into your plugin and edit it. Only do what you need to do. You can always go back later if the need arises.

9. Save changes, run a PDF build, and check your work.

Reality check

In earlier versions of the Open Toolkit and FOP, you had to edit the FOP configuration file, **fop.xconf**, which lives in `org.dita.pdf2/fop/conf`. You no longer have to do that, unless your fonts live in a location other than the default location for fonts on your system, and that's outside the scope of this exercise.

Although FOP and Antenna House automatically find fonts that are in the default location, XEP does not. Instead, the physical font name is itself an alias that XEP uses in its configuration files. If you're using XEP, check the documentation for details. However, here's a hint: XEP keeps font information in the **xep.xml** file, found in the XEP subfolder in the RenderX installation folder.

Other things you can do

Here are a few additional customizations you might want to make to your plugin. These customizations aren't part of the PDF specifications you're following, but they can be useful (and cool).

Exercise: Use a different font in special cases

Objective: Specify fonts by exception based on a condition.

You've seen how to assign specific fonts to various attribute sets. But what if you need to use a different font in the middle of an element? For example, say you're using Book Antiqua for body text (which you are in these exercises). In the middle of a paragraph (`<p>`), there is a phrase that needs to be in Korean. You've wrapped that phrase in a `<ph>` element and marked it `xml:lang="ko-kr"`.

Book Antiqua doesn't include any Korean glyphs, so for this phrase, you want to use Arial Unicode MS. Now what? How do you tell the PDF renderer to use? Easy. You set up a test for the language code of the phrase and specify Arial Unicode MS.

1. Open **commons-attr.xsl**.

2. Edit the *__fo__root* attribute set as follows:

```
<xsl:attribute-set name="__fo__root" use-attribute-sets="base-font">
    <xsl:attribute name="font-family">
        <xsl:choose>
            <xsl:when test="$locale.lang = 'kr'">Arial Unicode MS</xsl:when>
            <xsl:otherwise>Book Antiqua, Times New Roman, Times</xsl:otherwise>
        </xsl:choose>
    </xsl:attribute>
    <xsl:attribute name="xml:lang" select="translate($locale, '_', '-')"/>
    <xsl:attribute name="writing-mode" select="$writing-mode"/>
</xsl:attribute-set>
```

As you can see, the font-family attribute now includes a test for the locale. If it's "kr" (as in ko-kr), then the PDF renderer will use Arial Unicode MS. If it's anything else, the PDF renderer will use Book Antiqua (or Times New Roman or Times as fallbacks).

3. Save changes, run a PDF build, and check your work.

As a reminder, this procedure does not work if you're using FOP. For FOP, you'll have to include the **font-mappings.xml** file in your plugin (explained in **Use font mapping in your plugin** *(p. 87)*). For each of the three default logical fonts, there are special language cases already defined for Japanese, Chinese, and Korean. Ensure that these cases specify the fonts you want to use for those languages.

Exercise: Use font mapping in your plugin

Objective: Indicate which fonts you're going to use with a font-mappings file.

The first step in setting up your plugin to use font mapping is to map the three default logical fonts to corresponding physical fonts. FOP automatically looks for fonts in a default location, which depends on your operating system. For Windows, this location is the Fonts folder. For Macintosh, there are two locations: the Library/Font folder and the Library/Fonts folder in your Users folder. For Linux, the location varies depending on your Linux distribution (the most common location is /usr/share/fonts).

1. Copy **font-mappings.xml** from DITA-OT/plugins/org.dita.pdf2/cfg/fo/ to DITA-OT/plugins/com.company.pdf/cfg/fo/.

2. Add the following line to the **catalog.xml** file in com.company.pdf/cfg:

```
<uri name="cfg:fo/font-mappings.xml" uri="fo/font-mappings.xml"/>
```

3. Open **font-mappings.xml**.

This file, as the name suggests, maps physical fonts onto logical fonts. A physical font is an actual font in your Fonts folder, such as Times, Arial, Verdana, Courier, Garamond, and so forth. A logical font is a font type, such as sans-serif, serif, monospace, or others you might add yourself.

By default, the three logical fonts and the physical fonts they map to are: sans-serif, which maps to the physical fonts Helvetica and Arial Unicode; serif, which maps to the physical font Times New Roman and Times; and monospace, which maps to the physical fonts Courier New and Courier.

4. Scroll through **font-mappings.xml** and see how these mappings are set up.

You'll notice that the sans-serif font mapping is actually under the logical font name **Sans**. This is a legacy from the logical font names in previous versions of the DITA Open Toolkit. The

font-mappings.xml file uses aliases to map the new logical font names onto the legacy ones. You don't need to understand that, just know that the legacy name **Sans** refers to the sans-serif logical font, **Serif** refers to the serif logical font and **Monospaced** refers to the monospace logical font.

5. Add Trebuchet MS, or an alternative if you don't have that font, to the mapping for **Sans**.

```
<logical-font name="Sans">
    <physical-font char-set="default">
        <font-face>Trebuchet MS, Helvetica, Arial Unicode MS</font-face>
    </physical-font>
```

Notice that the *font-face* element lists the fonts that are mapped to the logical font. It's a good idea to leave the default fonts in the mapping, so that if anyone produces the PDF on a machine that happens not to have your specific font, there are predictable fallback fonts.

> **Important:** On a Windows machine, use the font names you see in the Fonts folder, not the font file names. This will make it easier for you to keep track of font names. On Macs or Linux machines, you should add the font file name to the font-face list.

6. Add Book Antiqua, or an alternative, to the mapping for **Serif**.

```
<logical-font name="Serif">
    <physical-font char-set="default">
        <font-face>Book Antiqua, Times New Roman, Times</font-face>
    </physical-font>
```

7. Add Consolas, or an alternative, to the mapping for **Monospaced**.

```
<logical-font name="Monospaced">
    <physical-font char-set="default">
        <font-face>Consolas, Courier New, Courier</font-face>
    </physical-font>
```

8. In each of the three logical font mappings, change the **SymbolsSuperscript** character set to the corresponding font as well, to ensure that superscripted characters use the same font as inline characters.

```
<physical-font char-set="SymbolsSuperscript">
    <font-face>Trebuchet MS, Helvetica, Arial Unicode MS</font-face>
    <baseline-shift>20%</baseline-shift>
    <override-size>smaller</override-size>
</physical-font>
```

Two or more physical fonts are specified for each logical font to accommodate different platforms. For example, Courier New is common on Windows machines while Courier is common on Macs or Linux machines. Specifying multiple fonts enables the PDF plugin to be portable between platforms.

9. In each of the three logical font mappings, you will see additional character sets defined: Simplified Chinese, Japanese, Korean, Symbols and SubmenuSymbols.

Many fonts include thousands of characters, such as Latin letters with numerous diacritics (accents, tildes, diaeresis, etc.) as well as the Greek, Cyrillic, Hebrew and Arabic alphabets. Some fonts include additional characters as well. However, when you need to render text in the Chinese, Japanese, or Korean scripts, you need to use fonts designed especially for those languages. That's why the character sets for those languages specify different fonts within each logical font mapping. If you publish in those languages, you will need to change these specifications to name the fonts you plan to use.

Likewise for the Symbols character sets. The Zapf Dingbats font is commonly used to render icons such as warning symbols or special list bullets. Its character set has to be handled separately.

For now, don't worry about these special character sets.

IO. Save **font-mappings.xml**.

II. In the **integrator.xml** file of your plugin, change the value of the *pdf2.i18n.skip* property to "false":

```
<property name="pdf2.i18n.skip" value="false"/>
```

Reality check

Most attribute sets that format text include the *font-family* attribute, usually with one of the logical fonts:

```
<xsl:attribute name="font-family">sans-serif</xsl:attribute>
<xsl:attribute name="font-family">serif</xsl:attribute>
<xsl:attribute name="font-family">monospace</xsl:attribute>
```

You can add this attribute to any attribute set that doesn't already have it, setting the logical font value for the *font-family* attribute. These attribute sets will pick up the physical fonts you've mapped to these logical font names.

Exercise: Use a custom font family with font mapping

Objective: Create an additional font family.

You've seen how to assign specific fonts to the serif, sans-serif and monospaced logical fonts. But what if you need to use a fourth font? For example, many companies use a custom font for their title and other text. How do you accommodate this? Easy, you add another logical font.

1. Open **font-mappings.xml**.

2. After the end of the last `<logical-font>` and before `</font-table>`, add the following:

```
<logical-font name="BigCorpFont">
    <physical-font char-set="default">
        <font-face>Big Corp Font</font-face>
    </physical-font>
</logical-font>
```

In this example, your custom font is "Big Corp Font." The logical font name, because it's an attribute, can't have spaces, so you need to edit it to something like "BigCorpFont." Now, let's say you want to use it for the main title on the front cover page.

3. Copy the file **front-matter-attr.xsl** from `DITA-OT/plugins/org.dita.pdf2/cfg/fo/attrs/` to your plugin and add the appropriate `<xsl:import>` statement to the **custom.xsl** file (attrs).

4. Add the following to the *__frontmatter__title* attribute set in your copy of **front-matter-attr.xsl**:

```
<xsl:attribute name="font-family">BigCorpFont</xsl:attribute>
```

Be sure to use the logical-font name, not the font-face. Also be sure that the font is installed in the default font location for your operating system.

5. Save changes, run a PDF build, and check your work.

Chapter 7

Page masters

One of the most basic things you do when creating a print publication is to design your pages—page size, margins, left and right facing pages, and so on. To be honest, the process is not simple in the DITA Open Toolkit's PDF plugin, but it's critical, so this chapter covers it in detail.

This chapter explains how to make some basic changes, such as setting up facing pages and margins. Then it explains page regions, which are an XSL-FO concept and very important to understand.

Finally, it explains how to fine-tune your pages, including tasks such as adding a watermark or background image, using multiple columns, and creating a landscape page.

For a list of attribute sets that affect headers and footers, see **Layout masters attribute sets** *(p. 429)*.

Page masters and regions in the PDF plugin

If you've used desktop publishing software like Adobe FrameMaker, InDesign, or Microsoft Word, then you're familiar with the concept of master pages. Most applications at least let you set up right and left master pages and often first and last master pages as well. The desktop publishing software automatically assigns these master pages to pages in the document. Some applications also let you set up additional special master pages that you can assign where you need them.

Typically, a master page specifies margins, header and footer areas, and maybe some boilerplate information that appears on the page, such as a logo or static information in the header or footer. You can usually set up headers and footers to include dynamic information as well, based on variables such as date/time, document title, nearest heading, page number, and so forth.

If you are familiar with the concept of master pages in desktop publishing software, then you are well on your way to understanding page masters in the DITA Open Toolkit. You define a page size, margins and other information for the page masters just as you do for master pages in desktop publishing software.

You define page masters in the DITA Open Toolkit using XSL-FO. XSL-FO defines four side regions and one body region:

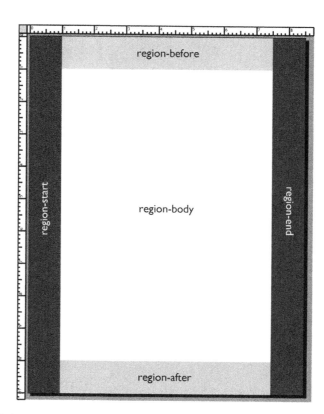

Figure 9: XSL-FO page regions

You can think of a region as a container for a *flow*, if you're used to that terminology. Be aware, though, that a region can contain more than one flow. Notice that even though region-before and region-after are at the top and bottom of the page, not the sides, they are still called side regions.

Side regions always start at the outer edge of the page and extend into the page by a defined amount. The amount they extend into the page is called the *extent*. By default, the start and end regions run the full length of the page. The before and after regions only extend to the inside edges of the start and end regions.

Region-before and region-after are analogous to the header and footer regions in other page layout applications. Region-body is the main flow of the page, where your content will flow.

In **Figure 9**, the regions are all contiguous, but they don't have to be. If you want, you can define region-body so that it doesn't touch the other regions.

This is a fairly simple explanation of page regions, but it's enough to get you started. Now that you have a basic understanding of page regions, you're ready to start designing the pages of your PDF.

> **Newsflash:** Page regions can be confusing. The way pages are defined in the DITA Open Toolkit isn't specific to the OT; it's a feature of the way XSL-FO works. After finishing this chapter, if you still don't understand (more or less) how page regions work, that would be a great time to turn to one of those XSL-FO references listed in the introduction to this book.

Default page masters, regions, and attribute sets

Here is a list of all the page masters that are available in the PDF plugin out of the box.

Page master	region-body attribute set
front-matter-first	region-body__frontmatter.odd
front-matter-even	region-body__frontmatter.even
front-matter-odd	region-body__frontmatter.odd
front-matter-last	region-body__frontmatter.even
back-cover-even	region-body__backcover.even
back-cover-odd	region-body__backcover.odd
back-cover-last	region-body__backcover.even
toc-first	region-body.odd
toc-even	region-body.even
toc-odd	region-body.odd
toc-last	region-body.even
body-first	region-body.odd
body-even	region-body.even
body-odd	region-body.odd

Page master	region-body attribute set
body-last	region-body.even
index-first	region-body__index.odd
index-even	region-body__index.even
index-odd	region-body__index.odd
glossary-first	region-body.odd
glossary-even	region-body.even
glossary-odd	region-body.odd

These page masters are found in **layout-masters.xsl**. Notice there are no index-last or glossary-last page masters. Why? In the past, the explanation would have been that glossaries were new in DITA, no one had yet implemented support for them, and, because FOP didn't do indexing, no one had implemented the full complement of index pages. Those explanations don't really hold anymore; glossaries are no longer all that new and FOP now does indexing. Probably, the best explanation now is that given the many updates needed to the Open Toolkit for DITA 1.3 features, index and glossary pages are a low priority.

The even and odd page masters are pretty self-explanatory, but how do the first and last page masters come into play? Well, let's say you have a double-sided PDF and you specify that chapters must end on an even page. If the chapter naturally ends on page 36, then page 36 is going to be an even page. But if the chapter naturally ends on page 35, then the plugin has to insert a blank even page, and that page is going to be a last page, not an even page. In other words, the last page is used only to force even pagination.

First pages are similar. By default, any topic that's a direct child of a map (and this includes chapters, appendices, and parts, because they are direct children of a bookmap) starts a new page sequence. And also by default, any new page sequence starts with the xxx-first page. So the TOC starts with toc-first, each body sequence starts with body-first, the index starts with index-first, and so on.

The following three regions are defined for each page master except front-matter-first, which has only region-body defined (because it is usually the title page, it does not need a header or footer):

- region-body (the main text flow)
- region-before (the header region)
- region-after (the footer region)

Region-start and region-end are not frequently customized, so they're not part of the default page master definitions. You might customize them if you want to set up tabs or have text in the left or right margins (not a sidehead—that's different. A sidehead is a heading that is to the left—in left-to-right languages—of the first paragraph of a topic rather than above it. Don't confuse it with a sidebar, which is usually a larger block of text.) Those are fairly rare edge cases, so this book doesn't talk about them.

By default, region-before on all page masters uses the attribute set region-before and region-after uses the attribute set region-after. Region-body uses different attribute sets (mostly region-body.odd and region-body.even) with a couple of exceptions for the frontmatter and index pages. For a full list, see **Layout masters attribute sets** *(p. 429)*. These attribute sets are found in **layout-masters-attr.xsl**.

You're not limited to what you see here. You can create new attribute sets and use them to describe existing page regions. You can also create new page masters and use them within PDFs as appropriate. You'll learn how to do both of these things as you work though the exercises in this section.

Page specifications

Let's review the template specifications for the page layout. The specifications include requirements for body pages and for the front and back cover pages.

Page layout

Here are the requirements for the body pages:

Page size	8.5in width by 11in height
Page layout	double-sided
Top margin	1in
Bottom margin	1in
Inside margin	1in
Outside margin	.75in
Header	text top-aligned .5in from top of page
Footer	text bottom-aligned .5in from bottom of page

The front and back cover pages have different requirements:

Top margin	1in
Bottom margin	1in
Inside margin	1in
Outside margin	1in
Header	none
Footer	none

These body page specifications produce a page that looks like **Figure 10**:

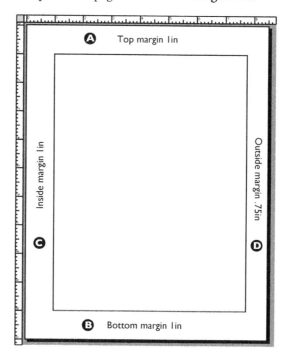

Figure 10: Body page diagram based on specifications

This is a right page. A left page would be a mirror image of this.

Files you need

If you have completed other exercises, some of these files might already be present in your plugin.

If not, create the following files in `DITA-OT/plugins/com.company.pdf/cfg/fo/attrs/`:

- **basic-settings.xsl**
- **commons-attr.xsl**
- **commons-attr_fop.xsl** (if you're using FOP)
- **commons-attr_xep.xsl** (if you're using XEP)
- **layout-masters-attr.xsl**
- **layout-masters-attr_xep.xsl** (if you're using XEP)

Be sure to add the appropriate `xsl:import` statement to **custom.xsl** (attrs).

You copied **layout-masters.xsl** from `DITA-OT/plugins/org.dita.pdf2/cfg/fo/` to `DITA-OT/plugins/com.company.pdf/cfg/fo/` when you first created your plugin. Now you need to add an import statement for it. Because **layout-masters.xsl** isn't in the `xsl` folder, but rather in the `fo` folder, the import statement needs to be different from the ones you've added so far. It needs to look up a level. Add the following to to **custom.xsl** (xsl):

```
<xsl:import href="../layout-masters.xsl"/>
```

(Notice this path is a URL and uses a forward slash (/) even on a Windows system.)

For instructions on creating attribute files and XSLT stylesheets in your PDF plugin, refer to **Create an attribute set file in your plugin** *(p. 31)* and **Create an XSLT stylesheet in your plugin** *(p. 32)*.

Do these exercises in order!

For best results, you should perform the exercises in this section in the order they're listed. Don't skip around and don't skip over exercises, even if you don't think you need to apply them in your work. To fully understand page masters, it's best to work through the entire process in order.

If you haven't already read the introductory material in this chapter, that would be a good idea too. There's a lot of background information there that will help you understand how page masters work.

- **Page masters and regions in the PDF plugin** *(p. 91)*
- **Default page masters, regions, and attribute sets** *(p. 93)*

Exercise: Set up double-sided pagination

Objective: Configure your custom PDF plugin to create left and right pages that are mirror images.

By default, the Open Toolkit produces a single-sided PDF, meaning that all pages are right, or odd, pages. The header content is always in the top right corner. It's easy to change this setting to produce double-sided pages, where the header content is in the top right corner on right/odd pages and in the top left corner on left/even pages.

1. Copy the following from `DITA-OT/plugins/org.dita.pdf2/cfg/fo/attrs/`**basic-settings.xsl** to your copy of **basic-settings.xsl**.

   ```
   <xsl:variable name="mirror-page-margins" select="false()"/>
   ```

   This variable controls whether the PDF plugin creates mirrored pages. The default is "false()".

2. Change the value of the variable to "true()";

   ```
   <xsl:variable name="mirror-page-margins" select="true()"/>
   ```

3. Save changes, run a PDF build, and check your work.

   You should find that the page header content is to the left on even-numbered pages and to the right on odd-numbered pages.

Exercise: Set page dimensions

Objective: Specify the size of the physical page your custom PDF plugin creates.

One of the first steps in setting up page masters is to specify the size of the page you're using. By default the PDF plugin sets the page size as American letter, vertically/portrait-oriented.

The measurements are given in millimeters (mm), which is a little odd—metric units paired with American letter-size page dimensions. If you prefer working with metric measurements, you can leave the values as they are. In fact, any of the English measurements you encounter while working on your PDF plugin can just as well be expressed in metric values. Just be sure to include the appropriate unit (mm, cm).

You can also use points (pt), picas (pc), inches (in), and several other units of measurement.

1. Copy the following from `DITA-OT/plugins/org.dita.pdf2/cfg/fo/attrs/`**basic-settings.xsl** to your copy of **basic-settings.xsl**:

```
<!-- The default of 215.9mm x 279.4mm is US Letter size (8.5x11in) -->
<xsl:variable name="page-width">215.9mm</xsl:variable>
<xsl:variable name="page-height">279.4mm</xsl:variable>
```

2. Change the value of *page-width* to "8.5in".
3. Change the value of *page-height* to "11in".
4. Save changes, run a PDF build, and check your work.

Exercise: Set page margins

Objective: Specify the top, bottom, inside, and outside margins of the pages in your custom PDF.

After defining page dimensions, the next logical step is to define the four outer page margins.

1. Copy the following from `DITA-OT/plugins/org.dita.pdf2/cfg/fo/attrs/`**basic-settings.xsl** to your copy of **basic-settings.xsl**:

```
<!-- Change these if your page has different margins on different sides. -->
<xsl:variable name="page-margin-inside" select="$page-margins"/>
<xsl:variable name="page-margin-outside" select="$page-margins"/>
<xsl:variable name="page-margin-top" select="$page-margins"/>
<xsl:variable name="page-margin-bottom" select="$page-margins"/>
```

2. Change the values as shown here:

```
<xsl:variable name="page-margin-inside">1in</xsl:variable>
<xsl:variable name="page-margin-outside">.75in</xsl:variable>
<xsl:variable name="page-margin-top">1in</xsl:variable>
<xsl:variable name="page-margin-bottom">1in</xsl:variable>
```

> **Note** You might be tempted to just change *$page-margins* to '.75in' or '1in' in the *select* attribute: `<xsl:variable name="page-margin-inside" select='1in'/>`. You can do so and, depending on your specifications, it might make as much sense to use a fixed value as a variable. If you use a fixed value, use single quotes, not double.

3. Save changes, run a PDF build, and check your work.

Figure 11 shows what the new margins look like, with the variables you just defined attached to the margins they describe:

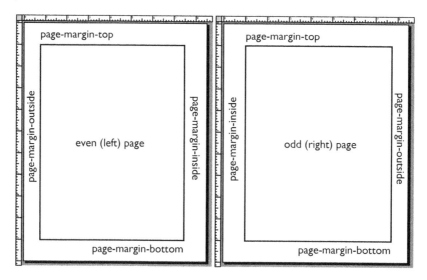

Figure 11: Page margin diagram

If you look at the rulers, you can see that the inside margin is 1in wide and the outside margin is .75in wide. The top and bottom margins are still 1in.

If all four of your page margins are the same and don't differ between left and right pages, then instead of using the four separate variables shown in this exercise, you can set them all at once using the *page-margins* variable:

```
<!-- This is the default, but you can set the margins individually below. -->
<xsl:variable name="page-margins">20mm</xsl:variable>
```

Exercise: Set up body regions

Objective: Determine the dimensions of the body region of your page masters.

In the previous exercise, you saw that the *page-margin-top* and *page-margin-bottom* variables define the top and bottom margins of the page. Both variables are set to 1in.

By default, the PDF plugin uses the same two variables to define the top and bottom margins of the body region of the page.

This makes the body region start where the top margin ends and end where the bottom margin begins. This gives you a page that looks like **Figure 12**:

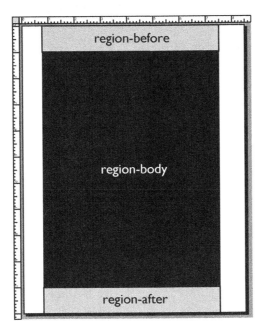

Figure 12: Body region without space above and below

But this isn't what you want now. You want .25in of white space between the top margin and the body and .25in of space between the body and the bottom margin.

In other words, you want a page like this:

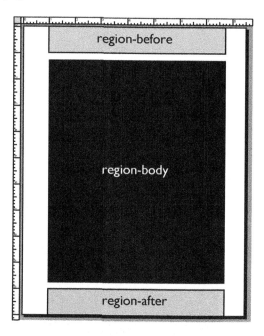

Figure 13: Body region with space above and below

To make this happen, you need to use variables other than *page-margin-top* and *page-margin-bottom* to define the top and bottom margin of the body region.

1. Open **layout-masters.xsl**.

2. Find the section that begins with `<!--BODY simple masters-->`.

   There are four body simple masters: body-first, body-even, body-odd and body-last. Notice that the region-body of the body-first and body-odd page masters is defined by the attribute set *region-body.odd*. Likewise, the region-body of the body-even and body-last page masters is defined by the attribute set *region-body.even*. Therefore, to change the dimensions of region-body, you need to edit the *region-body.odd* and *region-body.even* attribute sets.

3. Copy the *region-body.odd* and *region-body.even* attribute sets from `DITA-OT/plugins/org.dita.pdf2/cfg/fo/attrs/`**layout-masters-attr.xsl** to your copy of **layout-masters-attr.xsl**.

Both *region-body.odd* and *region-body.even* define *margin-top* to use the variable *page-margin-top* and *margin-bottom* to use the variable *page-margin-bottom*:

```
<xsl:attribute-set name="region-body.odd">
  <xsl:attribute name="margin-top">
    <xsl:value-of select="$page-margin-top"/>
  </xsl:attribute>
  <xsl:attribute name="margin-bottom">
    <xsl:value-of select="$page-margin-bottom"/>
  </xsl:attribute>
```

You need to define a new variable and call it instead of these two.

4. In **basic-settings.xsl**, just below the line `<xsl:variable name="page-margin-bottom"` `select="$page-margins"/>`, add the following line:

```
<xsl:variable name="body-margin">1.25in</xsl:variable>
```

Why 1.25in? Well, the top margin is 1in. Since you want the body region to start .25in below the bottom of the top margin, the body should start at 1.25in, right? The same goes for the relationship between the body region and the bottom margin.

5. In **layout-masters-attr.xsl**, in the *region-body.odd* attribute set, edit *margin-top* and *margin-bottom* to use the new variable:

```
<xsl:attribute-set name="region-body.odd">
  <xsl:attribute name="margin-top">
    <xsl:value-of select="$body-margin"/>
  </xsl:attribute>
  <xsl:attribute name="margin-bottom">
    <xsl:value-of select="$body-margin"/>
  </xsl:attribute>
```

6. Just below this section, add the following line:

```
<xsl:attribute name="margin-bottom">
  <xsl:value-of select="$body-margin"/>
</xsl:attribute>
<xsl:attribute name="background-color">#C0C0C0</xsl:attribute>
```

This attribute makes the body region light gray so you can more easily see its boundaries. It's just for testing; you can remove it later.

7. Make the same changes to the *region-body.even* attribute set.

8. Save changes, run a PDF build, and check your work.

The result should be similar to **Figure 14** (I set the headers and footers in dark gray here just to emphasize the space between them and the body; you won't see that in your PDF):

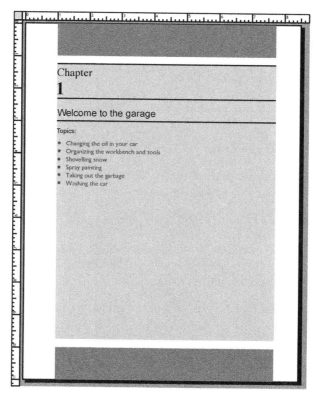

Figure 14: Body region setup complete

You now have all your margins set up. It's time to set up the areas that will hold the header and footer information.

Setting up header and footer regions

Objective: Understand how text is positioned in the header and footer regions.

Now that all the margins are set up, you might be wondering how to position text in the headers and footers. Headers go in the region-before and footers go in the region-after. From **Page masters and regions in the PDF plugin** *(p. 91)*, you know that the region-before starts at the top edge of the page and extends into the page a certain amount. Likewise, the region-after starts at the bottom edge and extends into the page a certain amount. The amount that these regions extend into the page is called the *extent*.

The region-before extent is defined using the *page-margin-top* variable. The region-after extent is defined using the *page-margin-bottom* variable. This is the default, and you've left it that way. And you've defined both *page-margin-top* and *page-margin-bottom* as "1in". So both regions have an extent of 1in.

> **Important:** Even though, in this case, the top margin is the same size as region-before, the two are completely different and don't have to be the same size. For example, you could define region-before to have an extent of .5in and define the top page margin to be 1in high.
>
> Don't confuse page margins with side regions. And bear in mind that, here, the term *page margins* really means *body region margins* because that's what the page margin variables set. XSL-FO distinguishes between margins on the page and margins for the body region.

If you built a PDF after completing the previous exercises, you saw that the header is placed exactly at the top edge of the page. There's no footer yet, but if there were one, it would be exactly at the bottom edge.

However, your template specifications say that the header should start .5in down from the top edge of the page and the footer should start .5in up from the bottom edge. How are you going to do that?

You can't do it by adjusting the size of the region-before and region-after. For example, you can't have region-before start .5in down from the top edge of the page; it **has** to start at the top edge.

You're going to have to adjust the position of the text blocks that contain the header and footer text within the region-before and region-after. You'll do that in the **Page headers and footers** *(p. 119)* chapter. For now, just appreciate the fact that you have your body pages set up.

Exercise: Set margins for the front cover page

Objective: Specify the top, bottom, inside, and outside margins for the front cover page.

By default, the margins for the front cover page are the same as for the body pages—in this case 1in inside and .75in outside—because odd/right frontmatter pages use the *region-body__frontmatter.odd* attribute set, which in turn uses the *region-body.odd* attribute set. Even/left frontmatter pages use the *region-body__frontmatter.even* attribute set, which in turn calls the *region-body.even* attribute set. Both *region-body.odd* and *region-body.even* use the *page-margin-inside* and *page-margin-outside* variables, which have the values "1in" and ".75in", respectively (if you made the changes in **Set page margins** *(p. 99)*).

But, you want the front cover page to have 1in margins all around. To do that you need to override the margins specified by *region-body.odd* and *region-body.even* by specifying margins directly within the *region-body__frontmatter.odd* and *region-body__frontmatter.even* attribute sets.

To be sure that your margins for the cover page don't collide with those for other pages, you can add new margin variables for the cover page.

1. Add the following new variables to your copy of **basic-settings.xsl**:

```
<xsl:variable name="page-margin-outside-front">1in</xsl:variable>
<xsl:variable name="page-margin-top-front">1in</xsl:variable>
<xsl:variable name="page-margin-bottom-front">1in</xsl:variable>
```

2. Copy the *region-body__frontmatter.odd* and *region-body__frontmatter.even* attribute sets from DITA-OT/plugins/org.dita.pdf2/cfg/fo/attrs/**layout-masters-attr.xsl** to your copy of **layout-masters-attr.xsl**.

3. Edit the *region-body__frontmatter.odd* attribute set as follows:

```
<xsl:attribute-set name="region-body__frontmatter.odd" use-attribute-sets="region-body.odd">
    <xsl:attribute name="margin-top">
        <xsl:value-of select="$page-margin-top-front"/>
    </xsl:attribute>
    <xsl:attribute name="margin-bottom">
        <xsl:value-of select="$page-margin-bottom-front"/>
    </xsl:attribute>
    <xsl:attribute name="{if ($writing-mode = 'lr') then 'margin-left' else 'margin-right'}">
        <xsl:value-of select="$page-margin-inside"/>
    </xsl:attribute>
    <xsl:attribute name="{if ($writing-mode = 'lr') then 'margin-right' else 'margin-left'}">
        <xsl:value-of select="$page-margin-outside-front"/>
    </xsl:attribute>
</xsl:attribute-set>
```

> **Tip:** Instead of typing all these lines, just copy them from the *region-body.odd* attribute set and make the changes as shown.

4. Find the *region-body__frontmatter.even* attribute set and edit it as follows:

```
<xsl:attribute-set name="region-body__frontmatter.even" use-attribute-sets=
▶"region-body.even">
    <xsl:attribute name="margin-top">
        <xsl:value-of select="$page-margin-top-front"/>
    </xsl:attribute>
    <xsl:attribute name="margin-bottom">
        <xsl:value-of select="$page-margin-bottom-front"/>
    </xsl:attribute>
    <xsl:attribute name="{if ($writing-mode = 'lr') then 'margin-left' else 'margin-right'}">
        <xsl:value-of select="$page-margin-outside-front"/>
    </xsl:attribute>
    <xsl:attribute name="{if ($writing-mode = 'lr') then 'margin-right' else 'margin-left'}">
        <xsl:value-of select="$page-margin-inside"/>
    </xsl:attribute>
</xsl:attribute-set>
```

5. To both the *region-body__frontmatter.odd* and *region-body__frontmatter.even* attribute sets, add the following line:

```
<xsl:attribute name="background-color">#ffc0ff</xsl:attribute>
```

This line makes the body region of the odd and even cover pages light pink, so you can see the margins more clearly. Again, you can remove this attribute later.

6. Save changes, run a PDF build, and check your work.

The front cover should look like **Figure 15** (this book is not in color so the body region in this example is black, not pink):

Figure 15: Front cover page margins diagram

Other things you can do

Here are a few additional customizations you might want to make to your plugin. These customizations aren't part of the PDF specifications you're following, but they can be useful (and cool).

Exercise: Eliminate blank last pages

Objective: Prevent your plugin from creating a blank last page when a chapter ends on an odd page.

By default, the DITA Open Toolkit ends all chapters on an even (or last) page, which is another nice thing not to have to worry about setting up. However, if that's not what you want, you can easily change it so that pages run continuously. Eliminating blank even pages is a good option for very short documents where page count is a consideration.

I. Copy the *__force__page__count* attribute set from
`DITA-OT/plugins/org.dita.pdf2/cfg/fo/attrs/`**commons-attr.xsl** to your copy of
commons-attr.xsl:

```
<xsl:attribute-set name="__force__page__count">
    <xsl:attribute name="force-page-count">
        <xsl:choose>
            <xsl:when test="name(/*) = 'bookmap'">
                <xsl:value-of select="'even'"/>
            </xsl:when>
            <xsl:otherwise>
                <xsl:value-of select="'auto'"/>
            </xsl:otherwise>
        </xsl:choose>
    </xsl:attribute>
</xsl:attribute-set>
```

This code says that if you're publishing from a bookmap (the `<xsl:when>` condition), then each chapter should end on an even page, even if that page has to be created as a blank page. These pages use the **last** simple page masters: body-last, toc-last, index-last, etc. If you're not publishing from a bookmap (the `<xsl:otherwise>` condition), pages should run continuously, with no forced even pages.

2. Within the `<xsl:when>` statement, change the value from "even" to "auto":

```
<xsl:when test="name(/*) = 'bookmap'">
    <xsl:value-of select="'auto'"/>
</xsl:when>
```

Reality check

Now pagination is the same whether you're publishing from a bookmap or not—pages run continuously, meaning that a chapter might start on the back of the last page of the preceding chapter and that the first page of the chapter will use an even or odd page master instead of the first page master.

Exercise: Design a separate layout for first pages

Objective: Create a separate design for the first page of chapters.

Let's say you want the first page of each chapter to have a 2in top margin and no header. First pages use the body-first page master. The body-first page master uses the *region-body.odd* attribute set to define the body region, which includes the top margin (along with the inside, outside, and bottom margins).

You don't want to change the margins on body pages, so you can't edit *region-body.odd*. What you need to do is define a new attribute set and use it to describe the body region on the body-first page master.

1. In **layout-masters.xsl**, find the definition for the body-first page master:

```
<fo:simple-page-master master-name="body-first" xsl:use-attribute-sets="simple-page-master">
    <fo:region-body xsl:use-attribute-sets="region-body.odd"/>
    <fo:region-before region-name="first-body-header" xsl:use-attribute-sets="region-before"/>
    <fo:region-after region-name="first-body-footer" xsl:use-attribute-sets="region-after"/>
</fo:simple-page-master>
```

2. Edit that definition so that the region-body uses a new *region-body.first* attribute set:

```
<fo:region-body xsl:use-attribute-sets="region-body.first"/>
```

Now you need to remove the header region from the body-first master. You might think you could just comment out the fo:region-before line to remove the header from the first page, like this:

```
<!--<fo:region-before region-name="first-body-header" xsl:use-attribute-sets="region-before"/>-->
```

However, this page master works like an attribute set. That is, if you comment out the region-before line in your copy of **layout-masters.xsl**, the Open Toolkit will use the body-first page master definition in the default copy of **layout-masters.xsl**. You'll have to take another tack.

3. Change the fo:region-before element to call a new attribute set:

```
<fo:region-before region-name="first-body-header" xsl:use-attribute-sets="region-before.first"/>
```

Now you need to create these two new attribute sets.

4. Open your copy of **layout-masters-attr.xsl**.

5. Create the new *region-body.first* attribute set as follows:

```
<xsl:attribute-set name="region-body.first">
    <xsl:attribute name="margin-top">
        <xsl:value-of select="$page-margin-top-first"/>
    </xsl:attribute>
    <xsl:attribute name="margin-bottom">
        <xsl:value-of select="$page-margin-bottom"/>
```

```
        </xsl:attribute>
        <xsl:attribute name="{if ($writing-mode = 'lr') then 'margin-left' else 'margin-right'}">
            <xsl:value-of select="$page-margin-inside"/>
        </xsl:attribute>
        <xsl:attribute name="{if ($writing-mode = 'lr') then 'margin-right' else 'margin-left'}">
            <xsl:value-of select="$page-margin-outside"/>
        </xsl:attribute>
        <xsl:attribute name="background-color">#c0ffc0</xsl:attribute>
    </xsl:attribute-set>
```

The *background-color* attribute here creates a pale green background.

> ‖ **Tip:** Copy and paste the *region-body.odd* attribute set and make changes.   ‖

This attribute set calls a new variable, *page-margin-top-first*. You need to add that to **basic-settings.xsl**.

6. Open **basic-settings.xsl** and add the new variable:

```
...
<xsl:variable name="page-margin-top-front" select="$page-margins"/>
<xsl:variable name="page-margin-bottom-front" select="$page-margins"/>
<xsl:variable name="page-margin-top-first">2in</xsl:variable>
```

Next, add the other new attribute set, *region-before.first*.

7. Return to **layout-masters-attr.xsl** and add the new attribute set:

```
<xsl:attribute-set name="region-before.first">
    <xsl:attribute name="extent">
        <xsl:value-of select="$header-extent-first"/>
    </xsl:attribute>
    <xsl:attribute name="display-align">before</xsl:attribute>
</xsl:attribute-set>
```

This attribute set uses a new variable, *header-extent-first*, so you'll want to add that to **basic-settings.xsl**.

8. Return to **basic-settings.xsl** and add the new variable:

```
...
<xsl:variable name="page-margin-bottom-front" select="$page-margins"/>
<xsl:variable name="page-margin-top-first">2in</xsl:variable>
<xsl:variable name="header-extent-first">0in</xsl:variable>
```

Setting the *header-extent-first* variable to 0in removes the header by giving it no space for content.

9. Save changes, run a PDF build, and check your work.

Reality check

The first body page should look like this (except the black area is actually yellow):

Figure 16: Example of first body page

Notice the body-region (the black area) has a 2in top margin as specified and the page has a footer (region-after; gray area) but no header (region-before).

Exercise: Use the bookmap page sequence for maps

Objective: Set up your custom PDF plugin to use the same pages when publishing from a map as when publishing from a bookmap.

By default, pages are assigned differently when you publish a PDF from a map as opposed to a bookmap. PDFs published from maps do not use the first and last pages in the body sequence, only the left and right. But, as with almost everything in the DITA Open Toolkit, you can change that.

> **Note** There are other differences between PDFs published from maps and those published from bookmaps. Most of these differences relate to pagination because maps do not have the chapter, appendix and part elements that help drive pagination in bookmap-published PDFs. While you can create the illusion of chapters using a map, your PDF won't have actual chapter numbering because that depends on the chapter element, which is available only in a bookmap. Other differences, including TOC and index generation, metadata, and a few other things, are discussed elsewhere in this book. Remember, though, that there are no inherent differences between maps and bookmaps that prevent doing any kind of processing you like.

1. In **layout-masters.xsl**, find this section (it's near the end of the file) in the `<!--Sequences-->` section:

```
<xsl:call-template name="generate-page-sequence-master">
    <xsl:with-param name="master-name" select="'ditamap-body-sequence'"/>
    <xsl:with-param name="master-reference" select="'body'"/>
    <xsl:with-param name="first" select="false()"/>
    <xsl:with-param name="last" select="false()"/>
</xsl:call-template>
```

This section describes the page masters to use for the odd and even pages in the body flow of a PDF published from a map. You can tell it refers to map-based publishing because the name of the page sequence masters is *ditamap-body-sequence*. Notice the first and last pages are set to "false()", meaning that only the odd and even pages will be used.

2. Change the two "false()" values to "true()":

```
<xsl:call-template name="generate-page-sequence-master">
    <xsl:with-param name="master-name" select="'ditamap-body-sequence'"/>
    <xsl:with-param name="master-reference" select="'body'"/>
    <xsl:with-param name="first" select="true()"/>
    <xsl:with-param name="last" select="true()"/>
</xsl:call-template>
```

This makes the body sequence pagination identical in PDFs published from both maps and bookmaps, but you still have one more thing to do.

3. In **commons-attr.xsl**, find this section:

```
<xsl:attribute-set name="__force__page__count">
    <xsl:attribute name="force-page-count">
        <xsl:choose>
            <xsl:when test="name(/*) = 'bookmap'">
                <xsl:value-of select="'even'"/>
            </xsl:when>
            <xsl:otherwise>
                <xsl:value-of select="'auto'"/>
            </xsl:otherwise>
        </xsl:choose>
    </xsl:attribute>
</xsl:attribute-set>
```

As explained in **Set up double-sided pagination** *(p. 98)*, this code sets up page sequencing when publishing from a bookmap (the `<xsl:when>` condition) or a map (the `<xsl:otherwise>` condition). You want the map sequence to match the bookmap sequence.

4. Change the "auto" value to "even" in the `<xsl:otherwise>` condition:

```
<xsl:otherwise>
    <xsl:value-of select="'even'"/>
</xsl:otherwise>
```

5. Save changes, run a PDF build, and check your work.

Exercise: Add a background image to a page

Objective: Place a background or watermark image in a specific location.

Background images are especially useful when you want to place a watermark or logo in a fixed position under text or other page content without interfering with the content's flow. In this exercise, you'll add a background image to the cover page.

1. Place the background image in `DITA-OT/plugins/com.company.pdf/cfg/common/artwork/`.

> **Tip:** The `artwork` folder is a good place to keep background images, logos, and header/footer images, because it keeps your plugin self-contained.

2. In **layout-masters-attr.xsl**, find the *region-body__frontmatter.odd* attribute set.

This exercise uses a default attribute set found in the out-of-the-box PDF plugin. In reality, you might create a separate first frontmatter page master and a *region-body__frontmatter.first* attribute set to format the body region of that page master. In that case, you'd add the background image to the *region-body__frontmatter.first* attribute set.

3. Add the following, highlighted, lines to the attribute set:

```
<xsl:attribute-set name="region-body__frontmatter.odd" use-attribute-sets="region-body.odd">
    <xsl:attribute name="margin-top">
      <xsl:value-of select="$page-margin-top"/>
    </xsl:attribute>
    <xsl:attribute name="margin-bottom">
      <xsl:value-of select="$page-margin-bottom"/>
    </xsl:attribute>
    ...
    <xsl:attribute name="background-image">url(Customization/OpenTopic/common/artwork/
    ▶ draft.svg)</xsl:attribute>
    <xsl:attribute name="background-repeat">no-repeat</xsl:attribute>
    <xsl:attribute name="background-position">100px 100px</xsl:attribute>
</xsl:attribute-set>
```

Of course, the URL for the background-image should name your file. The file name in this example is **draft.svg**, and it adds a "DRAFT" stamp to the page. The image in this example looks like this:

Figure 17: Example "DRAFT stamp" background image

> **Note** The "Customization/OpenTopic" part of the URL is a convention used within the PDF plugin. You can use this as a shorthand to any location in the current plugin folder.

The value "no-repeat" means that the image occurs only once on the page. You can use other values to specify that the image should repeat vertically, horizontally or in both directions as the page dimensions allow.

In this case, "100px 100px" represents x and y axes for positioning the image. You might have to experiment with different x and y values until the image is positioned exactly right. Take a look at your XSL-FO reference for more information on image positioning.

4. Save changes, run a PDF build, and check your work.

Reality check

The background image should look something like **Figure 18**.

Figure 18: Front cover page with background image

Exercise: Change the number of columns on a master page

Objective: Set up a page master to have more than one column.

By default, the DITA Open Toolkit renders pages in one column—except for index pages, which FOP, XEP, and Antenna House render in two columns by default. You can set up pages with any number of columns you want. For this exercise, you'll reformat the body-odd page master to have two columns.

The body-odd page master uses the *region-body.odd* attribute set to define the body region.

1. In **layout-masters-attr.xsl**, add the *column-count* attribute with a value of "2" to the *region-body.odd* attribute set (partially shown here)

```
<xsl:attribute-set name="region-body.odd">
    <xsl:attribute name="margin-top">
        <xsl:value-of select="$page-margin-top"/>
    </xsl:attribute>
    <xsl:attribute name="margin-bottom">
        <xsl:value-of select="$page-margin-bottom"/>
    </xsl:attribute>
    ...
    <xsl:attribute name="column-count">2</xsl:attribute>
</xsl:attribute-set>
```

2. Save changes, run a PDF build, and check your work.

3. To change the amount of space between columns, add the *column-gap* attribute with a value of "1in":

```
    ...
    <xsl:attribute name="column-count">2</xsl:attribute>
    <xsl:attribute name="column-gap">1in</xsl:attribute>
</xsl:attribute-set>
```

Reality check

The result is a page with two columns:

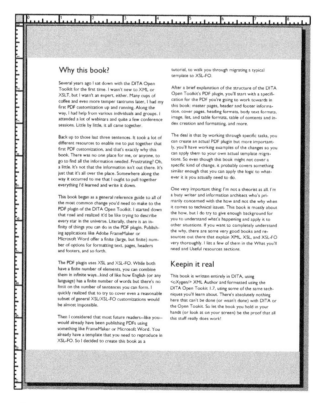

Figure 19: Two-column body page

Note Actually, if you scroll through the PDF, you might see that all the pages have two columns now, not just the odd ones and not just the body pages. This is because, out of the box, all of the odd page region attribute sets call *region-body.odd*. (Take a close look at **layout-masters-attr.xsl** to see.) If you haven't made any changes to the out-of-the-box page masters yet, you'll still see the default pagination.

To calculate the column width, the Open Toolkit takes the total width of the body region (in this example, 6.75in because your left and right margins total 1.75in and 8.5-1.75=6.75), subtracts the amount of space between columns (the default is .25in) and divides the remainder (6.5in) in half. The result in this example is two columns, each 3.25in wide.

Important: If your content includes items that are too large to fit within the column, such as tables with a fixed width or images with specified dimensions, you're likely to see some very strange results—overlapping text, only one column per page with a lot of empty space, and so forth.

Chapter 8

Page headers and footers

Most printed publications include page headers or footers or both. Setting these up in your PDF plugin is not straightforward, but if you take it step by step, it's not so bad.

This chapter explains each step of the process: where to get header and footer information, how to add that information to the appropriate page masters, and how to format your headers and footers.

For a list of attribute sets that affect headers and footers, see **Static content attribute sets** *(p. 439)*.

About headers and footers

Headers and footers are an integral part of any print publication. Because they repeat on each page, they rarely contain any manually-generated information. Instead, they include dynamically generated information such as the following:

- document title
- product name
- product version/release
- running header (nearest main topic title)
- current date
- document creation date
- publication date
- page number
- logo
- copyright information
- contact information
- author/owner
- filename and path

Most desktop publishing applications are limited in the kind of information you can include in headers and footers. Usually they allow only a fixed number of variables, or if they allow custom variables, they limit the information those variables can hold. Fortunately, DITA and the PDF plugin are a lot more flexible. If a piece of information exists in a map or can be generated, you can get that information into a header or footer. The trade-off for that flexibility is that the process is not as automatic as many desktop publishing applications make it.

PDF plugin defaults for headers and footers

By default, the PDF plugin puts the following information in headers and footers. The defaults are important to pay attention to because they can save you some work.

For example, if you refer to the header/footer specification in **Header and footer specifications** *(p. 129)*, you see that the Body odd header should include a running header. And the header you see in the Body odd header defaults listed below is the same thing. So that's one less thing you have to set up yourself.

Header/footer type	Default information
Body odd footer Body even footer Body first footer Preface odd footer Preface even footer Preface first footer TOC odd footer TOC even footer Index odd footer Index even footer Glossary odd footer Glossary even footer	Empty
Body first header Preface first header	Empty
Body odd header	product name | heading | page number
Body even header	page number | heading | product name
Preface odd header	product name | Introduction | page number

Header/footer type	Default information
Preface even header	page number | Introduction | product name
TOC odd header	product name | TOC | page number
TOC even header	page number | TOC | product name
Index odd header	product name | Index | page number
Index even header	page number | Index | product name
Glossary odd header	product name | Glossary | page number
Glossary even header	page number | Glossary | product name

Where does header and footer information come from?

Let's look at where information in headers and footers comes from. The information in the headers and footers in your PDF can be classified into four groups, shown here with examples:

- **External file:** product logo
- **Boilerplate text:** "Table of Contents," "Index"
- **Generated text:** running headers (chapter, appendix, or topic titles), page #
- **Map metadata:** book title, publication date, copyright info, product name, version

External files are things like logos or other image files. You call external files via a URI, which is basically a path to the file. (URI stands for Uniform Resource Indicator.)

Boilerplate text usually comes from a language-specific variables file such as **en.xml**. You could also hard code the text directly in the header or footer definition, but that makes translation difficult.

Generated text is created automatically by the PDF plugin.

Map metadata lives in various elements and attributes in your bookmap or map. You create variables to capture this metadata and then call those variables from within your header and footer definitions.

Metadata variables for headers and footers

Map metadata comes from elements and attributes in your map or bookmap. To get this information into a header or footer, you first have to put it into a variable, and most of the time, you have to create the variable yourself. When you create a variable, you give it a name and a value.

You can give a variable nearly any name you want, but to avoid possibly illegal values, I recommend you only use alphanumeric characters and avoid spaces and special characters. Do not choose a name that's already in use in the PDF plugin. A few names to avoid are: *count*, *topicNumber*, *topicTitle*, *topicType*, *locale*, and *value*.

> **Tip:** To find all variables used in the PDF plugin, use your text search application to search for instances of `xsl:variable`.

Even better, use a prefix for your variable names that is specific to the customization to make it easier to distinguish the custom variables and to ensure you don't conflict with existing variables. I use the prefix "bc." (for "BigCorp") in this chapter.

> **Note** You can also use namespaces for variable names instead of prefixes. This approach is not functionally different from using prefixes, but it gives you more control because the namespaces carry a degree of validation. If you put the wrong namespace on a variable (due to a typo, for example), you get a syntax error.
>
> To use a namespace, you must first declare it in the `<xsl:stylesheet>` element (note that `xsl:` is a namespace, too). Explaining namespaces is out of scope for this book, but you can find a good, quick overview at **http://www.w3schools.com/xml/xml_namespaces.asp**, and a more in-depth article at **http://www.lenzconsulting.com/namespaces/**.

You are going to define the five variables shown in the following table:

Header/footer info	Variable
book title	bc.bookTitle
publication date	bc.pubDate
copyright year	bc.copyYear
product name	bc.productName
product version	bc.productVersion

The value of each variable will be the contents of the corresponding `<topicmeta>` or `<bookmeta>` element. The next topic, **Models to use for map metadata** *(p. 123)*, maps each piece of header/footer information to the appropriate metadata element.

You create the variables that contain map metadata in **root-processing.xsl**. This file contains processing instructions and variable definitions that are executed early in the build process. You can't use a variable until you've given it a value (declared it), so you want to declare variables early. The best place to do that is in **root-processing.xsl**, which makes the declared variables available throughout the build process.

When you create new variables in **root-processing.xsl**, you need a way to define the value of that variable. This usually requires you to build a *path* to the elements and attributes that will supply the value. The language used within a stylesheet to do this is called *XPath*. While XPath is not hard to understand in theory, getting the paths right can sometimes be tricky. It's outside the scope of this book to teach XPath, but I will explain the paths for each variable you create.

> **Important:** Don't ever delete a variable from the plugin. If you delete a variable, you have to find all the places where it's called, remove every call, and possibly rewrite the XSL code that used that variable. If you miss a call, your build will fail. Instead, just leave the variable and delete the value.

Models to use for map metadata

Map metadata can come from a map, a bookmap or both, depending on which you are using. If you publish a bookmap that contains bookmeta, and the bookmap calls a map that contains topicmeta, the information in the bookmeta takes precedence. If you publish a bookmap that contains no bookmeta, and the bookmap calls a map that contains topicmeta, the information in the topicmeta takes precedence:

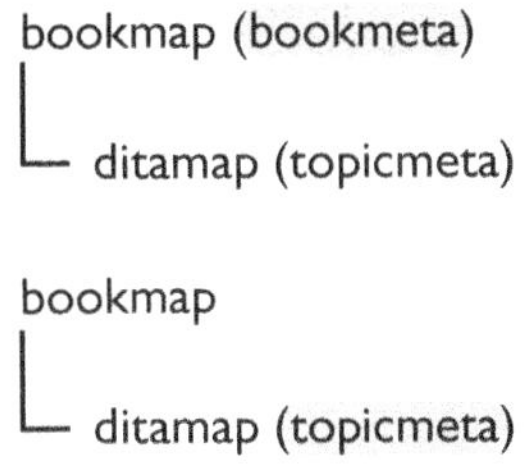

bookmap (bookmeta)
└─ ditamap (topicmeta)

bookmap
└─ ditamap (topicmeta)

Figure 20: Map metadata precedence

You need five pieces of information from the map or bookmap. In a map, most of this information is part of topicmeta. In a bookmap, it's part of bookmeta. Both topicmeta and bookmeta are fairly flexible. In many cases, there will be several elements you could logically use to hold these pieces of information. This book can't cover all the possibilities, so for the sake of this project, use the following:

	map	bookmap
book title	<title>	<mainbooktitle>
publication date	<revised golive> (topicmeta)	<revised golive> (bookmeta)
copyright year	<copyryear year> (topicmeta)	<copyrfirst><year> (bookmeta)
product name	<prodname> (topicmeta)	<prodname> (bookmeta)
product version	<vrm version> (topicmeta)	<vrm version> (bookmeta)

Ditamap topicmeta

If you are using a map for the exercises in this book, make sure it contains the topicmeta section as shown in this model, including the order the elements appear in. While the structure of topicmeta must be exactly like this model, you can substitute your own data.

```
<map xml:lang="en-us">
    <title>User Guide</title>
    <topicmeta>
        <copyright>
            <copyryear year="2016"/>
            <copyrholder>BigCorp, Inc.</copyrholder>
        </copyright>
        <critdates>
            <revised modified="" golive="10/01/2016"/>
        </critdates>
        <prodinfo>
            <prodname>WidgetPro</prodname>
            <vrmlist>
                <vrm version="2.0"/>
            </vrmlist>
        </prodinfo>
    </topicmeta>
    <mapref href="welcome.ditamap"/>
    <mapref href="about.ditamap"/>
</map>
```

Figure 21: Example of a map with topicmeta

Bookmap bookmeta

If you are using a bookmap for the exercises in this book, make sure it contains the bookmeta section as shown in this model.

```
<bookmap xml:lang="en-us">
    <booktitle>
        <mainbooktitle>User Guide</mainbooktitle>
    </booktitle>
    <bookmeta>
        <critdates>
            <revised modified="" golive="10/01/2016"/>
        </critdates>
        <prodinfo>
            <prodname>WidgetPro</prodname>
            <vrmlist>
                <vrm version="2.0"/>
            </vrmlist>
            <bookid>
                <booknumber>SC21-1234-00</booknumber>
            </bookid>
        </prodinfo>
        <bookrights>
            <copyrfirst>
                <year>2016</year>
            </copyrfirst>
            <bookowner>
                <organization>BigCorp, Inc.</organization>
            </bookowner>
        </bookrights>
    </bookmeta>
    <frontmatter>
        <booklists>
            <toc/>
        </booklists>
    </frontmatter>
    <chapter href="welcome.ditamap" format="ditamap"/>
    <chapter href="about.ditamap" format="ditamap"/>
    <backmatter>
        <booklists>
            <indexlist/>
        </booklists>
    </backmatter>
</bookmap>
```

Figure 22: Example of a bookmap with bookmeta

> **Note** As with many aspects of DITA, there may be multiple logical ways to represent this data, any of which would work just as well.

Notice that the elements that contain the publication date, product name and product version are the same for both topicmeta and bookmeta. Very convenient!

Now that your map contains the necessary metadata, you can create the variables you need to pull that metadata into headers and footers.

First, though, I'll explain the DITA *class* attribute, because it plays a critical part in these variable definitions.

DITA element classes

It's important to have a basic understanding of DITA element classes because you're going to use them to create variables that pull metadata from your map for use in headers and footers. You also use them anywhere else within your plugin where you need to identify a specific element.

If you've worked with XPath or XSL before, you might be used to selecting elements by name. DITA selects elements differently. It uses the *class* attribute rather than the element name. All DITA elements have a *class* attribute.

DITA *class* attributes have a particular structure. The *class* attribute always starts with a hyphen (-) if it is a structural element, or a plus sign (+) if it is a domain element.

> **Note** The easiest way to explain the difference is that domain elements are processed using the XSLT stylesheets that include **domain** in their names, such as **ui-domain.xsl** or **hi-domain.xsl**. Structural elements (usually block elements) are everything else.

After the - or + are one or more values. Each value has two parts: The first part is the short name of the vocabulary module the element is defined in, such as `topic`, `map`, `hi-d`, etc. The second part (after the slash) identifies an element type. For example, the class for the p element (which is a structural element) is `- topic/p`, meaning that the p element is defined in the topic module. The class for the `topicref` element is `- map/topicref`, meaning that the `topicref` element is used in maps.

> **Important:** Note there is a space between + or - and the value. This space is very important for matching. Don't leave it out.

The class for the b element (which is a domain element) is `+ topic/ph hi-d/b`. This one is a little more complicated to explain. Here goes.

DITA contains a small number of *base* elements. All other elements are specialized from these base elements. One example of a base element is ph. It's an inline element, meaning that you can use it to mark certain words or phrases within a block element like a paragraph (p). Many of the other inline elements, including a lot of the domain elements, are specialized from ph.

The b element is specialized from ph, and therefore, its class has to reflect that. That's why the class for b has two parts: `topic/ph`, which identifies the element b is specialized from and `hi-d/b`, which indicates the domain it's part of (the highlighting domain—think of **hi-domain.xsl**) and the element type. If you want to think of it as an analogy, you can: "b is to hi-d as ph is to topic".

If you don't completely understand DITA classes, don't worry—you don't really need to. All you need to understand is that rather than identifying elements by their name, you have to identify them by their class. Most XML editors hide elements' classes by default (both to unclutter the display and to keep you from inadvertently changing them), but you should be able to see the classes in your editor's attributes display.

Using classes, you can be as general or as specific as necessary when selecting elements for processing or for use as metadata. So, for example, consider the following statement:

```
<xsl:template match="*[contains(@class, ' topic/body ')]">
```

This statement will match body, conbody, taskbody, and refbody because all four of those elements contain topic/body in their *class* attributes (conbody, taskbody, and refbody are specialized from body, which is why all three have topic/body as part of their *class* attributes). On the other hand, the following statement will only match conbody because only conbody contains concept/conbody in its *class* attribute.

```
<xsl:template match="*[contains(@class, ' concept/conbody ')]">
```

But why?

You might wonder why not just match on the element name instead of the class. The reason boils down to portability of DITA content. Let's say that you specialized an element from ph and named it em (for emphasis). Its class would be topic/ph my-hi-d/em. Assume you've edited your DTDs and XSLT stylesheets correctly and everything works great.

Now suppose you share your content with another group that hasn't specialized an em element, and therefore has not edited its XSLT stylesheets to handle that element. If the DITA Open Toolkit was set up to call elements by name, that group's XSLT stylesheets would choke on em.

However, because the Open Toolkit calls elements by class, the other group's XSLT stylesheets will recognize topic/ph in the em element's class attribute and will process it without an error. Of course, the result will look the same as if the em element was actually a ph element, which might not be the look you want, but the build will not fail.

If you want to learn more about DITA element classes, Eliot Kimber's book *DITA for Practitioners, Volume I: Architecture and Technology* is a great place to start.

Why do I need to include the XSLT stylesheet in my plugin when I include the attribute set file?

In **Add an attribute set to your plugin** *(p. 32)*, you learned that if you include **commons.xsl** in your plugin, then when you add an attribute set file to your plugin, you must also copy the corresponding XSLT stylesheet, even if you don't need to change anything in it. Now that you understand that most DITA elements are specializations of base DITA elements, you can better understand why you have to do that.

Let's say that you want `wintitle` elements to be bold and italic, so you include **ui-domain-attr.xsl** in your plugin and add `font-style=italic` to the *wintitle* attribute set. You run a build and when you look at the PDF, `wintitle` elements are not italic. Why not? Well, because `wintitle` is a specialization of the `keyword` element (its class is `- topic/keyword ui-d/wintitle`). If you don't also include **ui-domain.xsl** in your plugin, you don't have a template that specifically matches the class - topic/keyword ui-d/wintitle. So the plugin looks for a template that more generally matches—that is, it looks for a template that matches on - topic/keyword. It finds just such a template—the one that processes the `keyword` element—in your **commons.xsl** file and therefore it processes `wintitle` just like `keyword`, which means that it uses the *keyword* attribute set because the keyword template calls the *keyword* attribute set.

Now let's complicate things a little more. Let's say you *have* included **ui-domain.xsl** in your plugin but `wintitle` elements are still not italic in the PDF output. What's wrong now? The first question is: does your plugin also include **commons.xsl**? It almost certainly does. So then the second question is: in **custom.xsl** (xsl) does the `xsl:import` statement for **commons.xsl** come after the `xsl:import` statement for **ui-domain.xsl**? Again, it almost certainly does. That's the issue, right there.

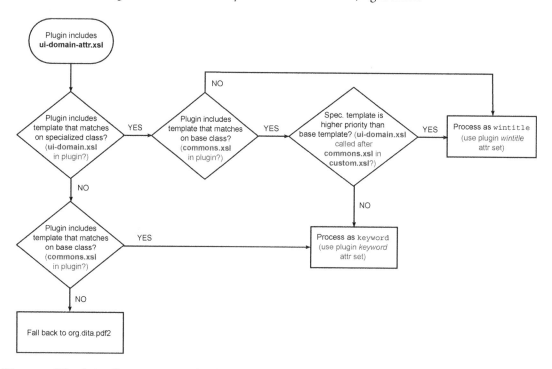

Figure 23: Attribute set priority processing

The order of xsl:import statements in **custom.xsl** (xsl) matters. The later in the list an XSLT stylesheet is called, the higher its priority. Any processing in a stylesheet higher in the list is superseded by identical processing in a stylesheet lower in the list. So processing for a specialized element in a XSLT stylesheet can be superseded by processing for its base element in an XSLT stylesheet called later (that is, lower in the list). In the current example, the processing for wintitle in **ui-domain.xsl** is superseded by the processing for keyword in **commons.xsl**. The solution, of course, is to call **commons.xsl** at or near the top of the **custom.xsl** (xsl) xsl:import list.

Why is priority an issue only if your plugin includes **commons.xsl**? First of all, because the base elements for most specialized elements are processed by templates in **commons.xsl**. So anytime your plugin does not include a specific template for a specialized element, odds are the fallback base template will be in **commons.xsl**. (The majority of those that are not in **commons.xsl** are in **links.xsl**, **lists.xsl**, and **tables.xsl**.)

If this explanation still isn't clear, **DITA element classes** *(p. 126)* shows it as a flow chart:

And if the explanation *still* is not clear, (and that is completely understandable!), then just follow this rule of thumb: if you include any attribute set files in your plugin, then be sure to include the corresponding XSLT stylesheets in your plugin as well. If you also include **commons.xsl** in your plugin, be sure to add it at or near the top of the **custom.xsl** (xsl) xsl:import list.

Header and footer specifications

Here are the template specifications for headers and footers.

Headers and footers

All footers are justified across the entire footer region width.

Header font	9pt Trebuchet normal black
Footer font	9pt Trebuchet normal black
Footer page number	9pt Trebuchet bold dark red
Header ruling	1pt solid black line below
Footer ruling	1pt solid black line above
Page number	includes chapter, appendix or part number and restarts at 1 in each chapter, appendix or part

Header - body first Header - toc first Header - index first	product logo
Header - body odd	running header | product logo
Header - body even Header - body last Header - toc even Header - toc last Header - index even Header - index last	book title
Header - toc odd	Table of Contents | product logo
Header - index odd	Index | product logo
Footer - body first Footer - body odd Footer - toc first Footer - toc odd Footer - index first Footer - index odd	publication date | Copyright ©copyright info | page #
Footer - body even Footer - body last Footer - toc even Footer - toc last Footer - index even Footer - index last	page # | product name version

These headers and footers include a mixture of information types. This combination of information types in a single header or footer is very common, which means that you'll often use several different files and methods to set up a header or footer.

You can (and should, if you haven't already) read all about the types of information in headers and footers in **Where does header and footer information come from?** *(p. 121)*.

Files you need

If you have completed other exercises, some of these files might already be present in your plugin.

If it is not already part of your plugin, create the following file:

- `DITA-OT/plugins/com.company.pdf/cfg/fo/attrs/`**static-content-attr.xsl**

Be sure to add the appropriate `xsl:import` statement to **custom.xsl** (attrs).

If they are not already part of your plugin, create the following files:

- `DITA-OT/plugins/com.company.pdf/cfg/fo/xsl/`**commons.xsl**
- `DITA-OT/plugins/com.company.pdf/cfg/fo/xsl/`**root-processing.xsl**

Be sure to add the appropriate `xsl:import` statement to **custom.xsl** (xsl).

For instructions on creating attribute files and XSLT stylesheets in your PDF plugin, refer to **Create an attribute set file in your plugin** *(p. 31)* and **Create an XSLT stylesheet in your plugin** *(p. 32)*.

You copied **layout-masters.xsl** from `DITA-OT/plugins/org.dita.pdf2/cfg/fo/` to `DITA-OT/plugins/com.company.pdf/cfg/fo/` when you first created your plugin. If you have not done so already, you need to add an import statement for it. Because **layout-masters.xsl** is in the `fo` folder, rather than the `xsl` folder, the import statement needs to be different; it needs to look up a level. Add the following to **custom.xsl** (xsl):

```
<xsl:import href="../layout-masters.xsl"/>
```

(Notice this path is a URL and uses a forward slash (/) even on a Windows system.)

If it isn't already in your plugin, copy `DITA-OT/plugins/org.dita.pdf2/xsl/fo/`**static-content.xsl** to `DITA-OT/plugins/com.company.pdf/cfg/fo/xsl/`. You'll probably end up changing almost all of the templates in this file, so it's easiest to have them all to begin with.

If it is not already part of your plugin, create the following file:

- `DITA-OT/plugins/com.company.pdf/cfg/common/vars/`**en.xml**

For instructions on creating variables files, refer to **Create a localization variables file** *(p. 78)*.

Be sure to add the appropriate language and locale mapping statements in **strings.xml**. You can find instructions in **Mapping strings files to xml:lang values in the strings.xml file** *(p. 37)*.

Complete these exercises in order!

For best results, you should perform the exercises in this section in order. Don't skip around and don't skip over exercises, even if you don't think they apply to your work. With headers and footers, it's a really good idea to work through the entire process to fully understand what's happening.

Several of these exercises assume that your page masters are already set up as described in the template specification. If you haven't worked through the **Page masters** *(p. 91)* chapter, consider doing that before you begin this chapter. If you opt not to, just be aware that some of the exercises might not give you the results you're expecting.

If you haven't already read the introductory material in this chapter, that would be a good idea too. There's background information there that will help you understand how headers and footers work.

Header setup

Exercise: Set up headers that include external files

Objective: Set up a header to include an external file such as a product logo.

Let's start simple. From your template specification, you know that the headers on the toc-first, body-first, and index-first pages should include just the product logo. This is an external file.

You also need the product logo in the headers on the first TOC and first index pages. By default, a PDF plugin doesn't include header and footer definitions for toc-first or index-first page masters. Later, you'll look at how to create new header and footer definitions, but for now, let's just look at how to add an external file to an existing header definition, body-first-header.

1. Create or select a small graphic, name it `logo.zzz` (where `.zzz` is the appropriate file extension), and copy it to `DITA-OT/plugins/com.company.pdf/cfg/common/artwork/`.

2. Find the following template in your copy of **static-content.xsl**:

```
<xsl:template name="insertBodyFirstHeader">
  <fo:static-content flow-name="first-body-header">
    <fo:block xsl:use-attribute-sets="__body__first__header">
      <xsl:call-template name="getVariable">
        <xsl:with-param name="id" select="'Body first header'"/>
        <xsl:with-param name="params">
          <heading>
            <fo:inline xsl:use-attribute-sets="__body__first__header__heading">
              <fo:retrieve-marker retrieve-class-name="current-header"/>
            </fo:inline>
          </heading>
          <pagenum>
            <fo:inline xsl:use-attribute-sets="__body__first__header__pagenum">
              <fo:page-number/>
            </fo:inline>
          </pagenum>
        </xsl:with-param>
      </xsl:call-template>
    </fo:block>
  </fo:static-content>
</xsl:template>
```

This is the template that creates the header on the first body pages. Even without understanding exactly how it works, you can spot the <heading> and <pagenum> sections (both highlighted), and you can probably assume that <heading> inserts some kind of running header and <pagenum> inserts the page number.

You don't want either one of these in your first body page headers.

3. Comment out the entire <xsl:call template> section, as shown:

```
<xsl:template name="insertBodyFirstHeader">
  <fo:static-content flow-name="first-body-header">
    <fo:block xsl:use-attribute-sets="__body__first__header">
      <!--
      <xsl:call-template name="getVariable">
        <xsl:with-param name="id" select="'Body first header'"/>
        <xsl:with-param name="params">
          <heading>
            <fo:inline xsl:use-attribute-sets="__body__first__header__heading">
              <fo:retrieve-marker retrieve-class-name="current-header"/>
            </fo:inline>
          </heading>
          <pagenum>
            <fo:inline xsl:use-attribute-sets="__body__first__header__pagenum">
              <fo:page-number/>
            </fo:inline>
          </pagenum>
        </xsl:with-param>
      </xsl:call-template>
      -->
    </fo:block>
  </fo:static-content>
</xsl:template>
```

Why are you commenting out the entire `<xsl:call-template>` section and not just the individual heading and pagenum parameters? Long story short, because the `<xsl:call-template>` section inserts information from other places into the header definition.

The first `<xsl:with-param>` statement within the `<xsl:call-template>` section inserts a localization variable into the header definition. You don't want that—you just want the external file.

The second `<xsl:with-param>` statement creates parameters that contain things like fo elements (such as `<fo:inline>` or `<fo:leader>`), metadata variables (as defined in **root-processing.xsl**), standard fo functions (such as `<fo:page-number>`), markers, or boilerplate text. You don't need any of those things, either. So because you don't need anything in the `<xsl:call-template>`, comment the whole thing out. Leaving parts of it in would insert things you don't want to insert.

4. Just after the section you commented out, add an `<fo:inline>` element as shown:

```
<xsl:template name="insertBodyFirstHeader">
  <fo:static-content flow-name="first-body-header">
    <fo:block xsl:use-attribute-sets="__body__first__header">
      <!--
      <xsl:call-template name="getVariable">
        <xsl:with-param name="id" select="'Body first header'"/>
        <xsl:with-param name="params">
          <heading>
            <fo:inline xsl:use-attribute-sets="__body__first__header__heading">
              <fo:retrieve-marker retrieve-class-name="current-header"/>
            </fo:inline>
          </heading>
          <pagenum>
            <fo:inline xsl:use-attribute-sets="__body__first__header__pagenum">
              <fo:page-number/>
            </fo:inline>
          </pagenum>
        </xsl:with-param>
      </xsl:call-template>
      -->
      <fo:inline>
        <fo:external-graphic src="url(Customization/OpenTopic/common/artwork/logo.zzz)"/>
      </fo:inline>
    </fo:block>
  </fo:static-content>
</xsl:template>
```

This code calls the logo file you created.

‖ **Important:** Be sure to change the file extension of the image file name.   ‖

5. Save changes, run a PDF build, and check your work.

Exercise: Set up headers that include boilerplate text

Objective: Set up a header to include boilerplate text, such as "Table of Contents" or "Index."

From your template specification, you know that the headers on the toc-odd and index-odd pages should include boilerplate text along with the external file.

1. Find the following template in your copy of **static-content.xsl**.

```
<xsl:template name="insertTocOddHeader">
  <fo:static-content flow-name="odd-toc-header">
    <fo:block xsl:use-attribute-sets="__toc__odd__header">
      <xsl:call-template name="getVariable">
        <xsl:with-param name="id" select="'Toc odd header'"/>
        <xsl:with-param name="params">
          <prodname>
            <xsl:value-of select="$productName"/>
          </prodname>
          <heading>
            <fo:inline xsl:use-attribute-sets="__body__odd__header__heading">
              <fo:retrieve-marker retrieve-class-name="current-header"/>
            </fo:inline>
          </heading>
          <pagenum>
            <fo:inline xsl:use-attribute-sets="__toc__odd__header__pagenum">
              <fo:page-number/>
            </fo:inline>
          </pagenum>
        </xsl:with-param>
      </xsl:call-template>
    </fo:block>
  </fo:static-content>
</xsl:template>
```

This template creates the header on the odd TOC pages. Even without understanding how it works, you can spot the `<prodname>`, `<heading>`, and `<pagenum>` sections (highlighted). You can safely assume that `<heading>` inserts some kind of running header and `<pagenum>` inserts the page number. You can also assume that `<prodname>` inserts the product name.

You don't want any of these in your odd TOC page headers. But notice the line

```
<xsl:with-param name="id" select="'Toc odd header'"/>
```

This line inserts a variable named *Toc odd header* into the odd TOC page header. This variable is found in a localization variables file like **en.xml**. You'll come back to this in a minute.

2. The first step in ensuring you don't get any of this information in your page header is to comment out the second `<xsl:with-param>` section shown in the code below (the second step follows shortly):

```
<xsl:template name="insertTocOddHeader">
  <fo:static-content flow-name="odd-toc-header">
    <fo:block xsl:use-attribute-sets="__toc__odd__header">
      <xsl:call-template name="getVariable">
        <xsl:with-param name="id" select="'Toc odd header'"/>
        <!--
        <xsl:with-param name="params">
          <prodname>
            <xsl:value-of select="$productName"/>
          </prodname>
          <heading>
            <fo:inline xsl:use-attribute-sets="__body__odd__header__heading">
              <fo:retrieve-marker retrieve-class-name="current-header"/>
            </fo:inline>
          </heading>
          <pagenum>
            <fo:inline xsl:use-attribute-sets="__toc__odd__header__pagenum">
              <fo:page-number/>
            </fo:inline>
          </pagenum>
        </xsl:with-param>
        -->
      </xsl:call-template>
    </fo:block>
  </fo:static-content>
</xsl:template>
```

3. Just after the section you commented out, add the `<fo:inline>` element that calls your logo image, making sure to change the file extension:

```
<xsl:template name="insertTocOddHeader">
  <fo:static-content flow-name="odd-toc-header">
    <fo:block xsl:use-attribute-sets="__toc__odd__header">
      <xsl:call-template name="getVariable">
        <xsl:with-param name="id" select="'Toc odd header'"/>
        <!--
        <xsl:with-param name="params">
          <prodname>
            <xsl:value-of select="$productName"/>
          </prodname>
          <heading>
            <fo:inline xsl:use-attribute-sets="__body__odd__header__heading">
              <fo:retrieve-marker retrieve-class-name="current-header"/>
            </fo:inline>
          </heading>
          <pagenum>
            <fo:inline xsl:use-attribute-sets="__toc__odd__header__pagenum">
              <fo:page-number/>
            </fo:inline>
          </pagenum>
        </xsl:with-param>
        -->
      </xsl:call-template>
      <fo:inline>
        <fo:external-graphic src="url(Customization/OpenTopic/common/artwork/logo.zzz)"/>
      </fo:inline>
    </fo:block>
  </fo:static-content>
</xsl:template>
```

4. Find the template that begins with `<xsl:template name="insertIndexOddHeader">`.

5. Make the same changes to this template that you made to the ***insertTocOddHeader*** template.

6. Copy the following variables from `DITA-OT/plugins/org.dita.pdf2/cfg/common/vars/`**en.xml** to your copy of **en.xml**:

```
<!-- The header that appears on the odd-numbered table of contents pages. -->
<variable id="Toc odd header"><param ref-name="prodname"/> | <param ref-name="heading"/> | <param
 ref-name="pagenum"/></variable>
```

```
<!-- The header that appears on the odd-numbered index pages. -->
    <variable id="Index odd header"><param ref-name="prodname"/> | <param
ref-name="heading"/> | <param ref-name="pagenum"/></variable>
```

These are the localization variables called by the ***insertTocOddHeader*** and ***insertIndexOddHeader*** templates. Notice how the three parameter names in the variable match the three sections that you commented out of the ***insertTocOddHeader*** and ***insertIndexOddHeader*** templates (*prodname, heading, pagenum*). That's not a coincidence. These three sections in the ***insertTocOddHeader*** and ***insertIndexOddHeader*** templates create parameters of those names and specify what information those parameters contain. The corresponding localization variables place those parameters into the headers. This is the second part of the two-part process mentioned in step 2.

By default, the *Toc odd header* variable sets up the odd TOC page headers to include the product name, a running header, and the page number. But you don't want this information in the headers, so you need to remove it from this definition and add the "Table of Contents" boilerplate text.

7. Edit the *Toc odd header* and *Index odd header* variables as shown:

```
<variable id="Toc odd header">Table of Contents</variable>
```

```
<variable id="Index odd header">Index</variable>
```

The change to *Toc odd header* leaves nothing but the boilerplate text, without the logo. The changes you made in **static-content.xsl** already added the logo.

8. Save changes, run a PDF build, and check your work.

Reality check

The result are headers that look like this (with your logo in place of the XML Press logo, of course):

There's no space between the text and the logo. That's okay for now. The important thing is that you have all the pieces in place for the header. You'll format everything later.

> **Tip:** If you plan to align the bottoms of the text and the image, make sure the image itself is flush with the bottom of the canvas. That will avoid unnecessary effort later on.

Exercise: Set up headers that include generated text

Objective: Set up a header to include generated text such as a running header or page number.

From your specification, you know the odd body page includes a running header and the product logo.

1. Find the following template in your copy of **static-content.xsl**:

```
<xsl:template name="insertBodyOddHeader">
  <fo:static-content flow-name="odd-body-header">
    <fo:block xsl:use-attribute-sets="__body__odd__header">
      <xsl:call-template name="getVariable">
        <xsl:with-param name="id" select="'Body odd header'"/>
        <xsl:with-param name="params">
          <prodname>
            <xsl:value-of select="$productName"/>
          </prodname>
          <heading>
            <fo:inline xsl:use-attribute-sets="__body__odd__header__heading">
                <fo:retrieve-marker retrieve-class-name="current-header"/>
            </fo:inline>
          </heading>
          <pagenum>
            <fo:inline xsl:use-attribute-sets="__body__odd__header__pagenum">
                <fo:page-number/>
            </fo:inline>
          </pagenum>
        </xsl:with-param>
      </xsl:call-template>
    </fo:block>
  </fo:static-content>
</xsl:template>
```

This is the template that creates the header on the odd body pages. You already know more or less what the `<prodname>`, `<heading>`, and `<pagenum>` sections do.

You don't want the product name or page number in your odd body page headers, but you do want the running header. So you need to get rid of `<prodname>` and `<pagenum>`.

2. Comment out the `<prodname>` and `<pagenum>` elements as shown:

```
<xsl:template name="insertBodyOddHeader">
  <fo:static-content flow-name="odd-body-header">
    <fo:block xsl:use-attribute-sets="__body__odd__header">
      <xsl:call-template name="getVariable">
        <xsl:with-param name="id" select="'Body odd header'"/>
        <xsl:with-param name="params">
          <!--
          <prodname>
            <xsl:value-of select="$productName"/>
          </prodname>
          -->
```

```
        <heading>
          <fo:inline xsl:use-attribute-sets="__body__odd__header__heading">
              <fo:retrieve-marker retrieve-class-name="current-header"/>
          </fo:inline>
        </heading>
        <!--
        <pagenum>
          <fo:inline xsl:use-attribute-sets="__body__odd__header__pagenum">
            <fo:page-number/>
          </fo:inline>
        </pagenum>
        -->
      </xsl:with-param>
    </xsl:call-template>
  </fo:block>
 </fo:static-content>
</xsl:template>
```

3. Add the `<fo:inline>` element that calls your logo file:

```
<xsl:template name="insertBodyOddHeader">
  <fo:static-content flow-name="odd-body-header">
    <fo:block xsl:use-attribute-sets="__body__odd__header">
      <xsl:call-template name="getVariable">
        <xsl:with-param name="id" select="'Body odd header'"/>
        <xsl:with-param name="params">
          <!--
          <prodname>
            <xsl:value-of select="$productName"/>
          </prodname>
          -->
          <heading>
            <fo:inline xsl:use-attribute-sets="__body__odd__header__heading">
                <fo:retrieve-marker retrieve-class-name="current-header"/>
            </fo:inline>
          </heading>
          <!--
          <pagenum>
            <fo:inline xsl:use-attribute-sets="__body__odd__header__pagenum">
              <fo:page-number/>
            </fo:inline>
          </pagenum>
          -->
        </xsl:with-param>
      </xsl:call-template>
      <fo:inline>
        <fo:external-graphic src="url(Customization/OpenTopic/common/artwork/logo.zzz)"/>
      </fo:inline>
    </fo:block>
  </fo:static-content>
</xsl:template>
```

‖ **Important:** Be sure to change the file extension of the image file name.   ‖

This code places the graphic into the header just after (in this case, to the right of) the running header created by the `<heading>` element. Notice that the graphic is inside an `<fo:inline>` element, which means that it does not appear on a separate line. It's part of the `<fo:block>` element that contains the running header.

If you completed the **Set up headers that include boilerplate text** *(p. 135)* exercise, then you can probably guess that you now need to edit a localization variable in **en.xml** to insert a running header. By looking at the code in this step, you can also guess that the variable you need to edit is *Body odd header.*

4. Copy the following from `DITA-OT/plugins/org.dita.pdf2/cfg/common/vars/`**en.xml** to your copy of **en.xml**:

```
<!-- The header that appears on odd-numbered pages. -->
<variable id="Body odd header"><param ref-name="prodname"/> | <param ref-name="heading"/> | <param
 ref-name="pagenum"/></variable>
```

By default, this variable sets up the odd body page headers to include the product name, a running header, and the page number. You don't want the product name or page number and you just commented them out of the *insertBodyOddHeader* template in **static-content.xsl**. You need to remove them from this variable definition as well.

5. Edit the variable as follows:

```
<variable id="Body odd header"><param ref-name="heading"/></variable>
```

The odd body page header also includes the logo, but that's not part of this variable definition.

6. Save changes, run a PDF build, and check your work.

Reality check

The result is a header that looks something like this:

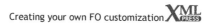

Again, you'll format everything later.

Adding information from the map to headers

Exercise: Create metadata variables for headers and footers

Objective: Create variables that pull in map metadata for use in headers and footers.

Before you get started with this exercise, be sure you've read **DITA element classes** *(p. 126)*.

Let's look at how to pull metadata from a map and put it into a variable.

You'll use **root-processing.xsl** for this and several other exercises in this chapter. **root-processing.xsl** contains, among other things, variables that you can use throughout the PDF generation process. These variables are defined at the beginning of the build process so they are available the whole way through.

I. Add the following variable declaration to your copy of **root-processing.xsl**:

```
<xsl:variable name="bc.productName">
    <xsl:variable name="mapProdname" select="(/*/opentopic:map//*
    ▸[contains(@class, ' topic/prodname ')])[1]"/>
    <xsl:choose>
        <xsl:when test="$mapProdname">
            <xsl:value-of select="$mapProdname"/>
        </xsl:when>
        <xsl:otherwise>
            <xsl:call-template name="getVariable">
                <xsl:with-param name="id" select="'Product Name'"/>
            </xsl:call-template>
        </xsl:otherwise>
    </xsl:choose>
</xsl:variable>
```

This defines the *bc.productVersion* variable. As its value, the `select` path chooses the content of the `prodname` element within the topicmeta or bookmeta. You can keep this definition simple because both topicmeta and bookmeta use the same element in the same structure.

> **Important:**   Notice there is a space between the quotes and the class name: `' topic/prodname '`. These spaces are **very** important. Don't leave them out! They ensure nothing is undermatched or overmatched.

Using `/*/opentopic:map//*` in this variable definition (or `//` in some other definitions) is overkill because it requires the processing of more of the merged map than is really necessary. It increases the time it takes to build the PDF (although you probably won't notice the difference in processing time unless you have an extremely large map to process).

Because the existing variables in **root-processing.xsl** use `/*/opentopic:map//*` and `//`, and the goal of the exercises in this book is to help you make changes with the least amount of background XSL knowledge, you'll follow the examples of the existing variables. When and if you learn more about XSL, you might want to return to these variables and make the paths more specific to reduce the processing time.

2. Add the following variable declaration to your copy of **root-processing.xsl**:

```
<xsl:variable name="bc.productVersion"
  ► select="(/*/opentopic:map//*[contains(@class,' topic/vrm')])/@version)[1]"/>
```

This defines the *bc.productVersion* variable. As its value, the `select` path chooses the content of the *version* attribute on the `vrm` element within the topicmeta or bookmeta. You can keep this definition simple as well because both topicmeta and bookmeta use the same element in the same structure.

3. Below that variable declaration, add the following:

```
<xsl:variable name="bc.pubDate"
  ► select="(/*/opentopic:map//*[contains(@class,' topic/revised ')])/@golive)[1]"/>
```

This defines the *bc.pubDate* variable. As its value, the `select` path chooses the content of the *golive* attribute on the `<revised>` element within the topicmeta or bookmeta. Again, both topicmeta and bookmeta use the same element in the same structure.

> **Note** The [1] tacked on to the end of these variable definitions is important. If you don't include it, you'll get multiple instances of the variable value in your PDF.

4. Below the *pubDate* variable, add the following variable definition:

```
<xsl:variable name="bc.bookTitle">
  <xsl:choose>
    <xsl:when test="exists($map/*[contains(@class, ' bookmap/booktitle ')]
      ► /*[contains(@class,' bookmap/mainbooktitle ')])">
      <xsl:value-of>
        <xsl:apply-templates select="$map/*[contains(@class, ' bookmap/booktitle ')]
          ► /*[contains(@class,' bookmap/mainbooktitle ')][1]" mode="dita-ot:text-only"/>
      </xsl:value-of>
    </xsl:when>
    <xsl:when test="exists($map/*[contains(@class,' topic/title ')])">
      <xsl:value-of>
        <xsl:apply-templates select="$map/*[contains(@class,' topic/title ')][1]"
          ► mode="dita-ot:text-only"/>
      </xsl:value-of>
    </xsl:when>
    <xsl:when test="exists(//*[contains(@class, ' map/map ')]/@title)">
      <xsl:value-of select="//*[contains(@class, ' map/map ')]/@title"/>
    </xsl:when>
    <xsl:otherwise>
      <xsl:value-of>
        <xsl:apply-templates select="descendant::*[contains(@class, ' topic/topic ')][1]
          ► /*[contains(@class, ' topic/title ')]" mode="dita-ot:text-only"/>
      </xsl:value-of>
    </xsl:otherwise>
  </xsl:choose>
</xsl:variable>
```

This code retrieves the title to use on the cover page. It includes three `<xsl:when>` tests and an `<xsl:otherwise>` fallback. The first `<xsl:when>` tests to see if the map includes a `<mainbooktitle>` element (that is, you are publishing from a bookmap), and if so, its content is selected as the variable value. The second `<xsl:when>` tests to see if the map includes a `<title`

element (that is, you are publishing from a map), and if so, its content is selected as the variable value. The third `<xsl:when>` tests to see if the map includes a *title* attribute, and if so, its content is selected as the variable value.

If none of these three conditions proves true, the `<xsl:otherwise>` fallback uses the title of the first topic in the map.

It's also worth noting that the `dita-ot:text-only` mode ensures that only the title text is used. This is important in cases where the title might also contain markup that you don't want to output.

5. Below the definition for *bc.bookTitle*, add the following:

```
<xsl:variable name="bc.copyYear">
   <xsl:choose>
      <xsl:when test="$map//*[contains(@class,' topic/data bookmap/copyrfirst ')]
      ▸ //*[contains(@class, ' topic/ph bookmap/year ')]">
         <xsl:apply-templates select="$map//*[contains(@class,' topic/data
         ▸ bookmap/copyrfirst ')]//*[contains(@class, ' topic/ph bookmap/year ')]"/>
      </xsl:when>
      <xsl:when test="$map//*[contains(@class,' topic/copyryear ')]/@year">
         <xsl:apply-templates select="$map//*[contains(@class,' topic/copyryear ')]/@year"/>
      </xsl:when>
      <xsl:otherwise>
         <xsl:call-template name="getVariable">
            <xsl:with-param name="id" select="'Copy Year'"/>
         </xsl:call-template>
      </xsl:otherwise>
   </xsl:choose>
</xsl:variable>
```

This defines the *bc.copyYear* variable. Like the definition for *bookTitle*, this variable can choose different content. It chooses between the content of a `<bookmeta>` element (when publishing from a bookmap) and a `<topicmeta>` element (when publishing from a map).

The paths are a little more complicated because those elements are nested farther down in the bookmeta/topicmeta hierarchy than `<title>`. The `<xsl:otherwise>` fallback requires a variable from **en.xml**: *Copy Year*, which you need to create.

6. Save your changes.

You now have all the variables you need to pull metadata from your map or bookmap. Next, you need to add that fallback variable for *bc.copyYear*.

Can you add `<xsl:otherwise>` fallback conditions to *bc.productVersion* and *bc.pubDate*? Of course you can. If you want to try, use the last three variables you added as a guide. Hint: you'll have only one `<xsl:when>` test.

Exercise: Add fallback header and footer variables to the localization variables file

Objective: Create variables to use as fallbacks when a map doesn't contain the necessary metadata for headers and footers.

When you created the *bc.productName*, *bc.bookTitle*, *bc.bookOwner*, and *bc.copyYear* variables in **root-processing.xsl**, you created fallback variables to be used in case the other possible values fail. Those fallback variables are named *Product Name*, *Book Title*, *Book Owner*, and *Copy Year*.

The variable *Product Name* is already present in **en.xml** by default, but you need to add the other three.

1. Add the following variable to your copy of **en.xml**:

```
<variable id="Copy Year">_____</variable>
```

> **Tip:** It's a good idea to add all custom localization variables to the end of the file, to make them easy to find and edit later.

2. In place of the _____, enter the value you want to appear if the variable is used.
3. Copy the following variable from `DITA-OT/plugins/org.dita.pdf2/cfg/common/vars/`**en.xml** to your copy of **en.xml**:

```
<variable id="Product Name"/>
```

4. Change the variable declaration to `<variable id="Product Name">_____</variable>` and enter the text you want to appear if this variable is used, in place of the _____.
5. Save your changes.

   You now have all the variables you need for your headers and footers.

Exercise: Set up headers that include map metadata

Objective: Set up a header to include metadata from your map or bookmap.

From your template specification, you know that the header on the body even, body last, toc even, toc last, index even, and index last page masters includes the book title. You also need this information on the last TOC and index pages, but by default, the PDF plugin doesn't include header and footer definitions for toc last or index last page masters.

Later, you'll learn how to create new header and footer definitions, but for now, let's just look at how to add map metadata to existing header definitions.

1. Find the following template in your copy of **static-content.xsl**:

```
<xsl:template name="insertBodyEvenHeader">
  <fo:static-content flow-name="even-body-header">
    <fo:block xsl:use-attribute-sets="__body__even__header">
      <xsl:call-template name="getVariable">
        <xsl:with-param name="id" select="'Body even header'"/>
        <xsl:with-param name="params">
          <prodname>
            <xsl:value-of select="$productName"/>
          </prodname>
          <heading>
            <fo:inline xsl:use-attribute-sets="__body__even__header__heading">
              <fo:retrieve-marker retrieve-class-name="current-header"/>
            </fo:inline>
          </heading>
          <pagenum>
            <fo:inline xsl:use-attribute-sets="__body__even__header__pagenum">
              <fo:page-number/>
            </fo:inline>
          </pagenum>
        </xsl:with-param>
      </xsl:call-template>
    </fo:block>
  </fo:static-content>
</xsl:template>
```

This template creates the header on even body pages. You already know what the `<prodname>`, `<heading>`, and `<pagenum>` sections do.

You don't want any of this information in your even body page headers. Instead, you want the book title. Remember that you created a variable in **root-processing.xsl** named *bc.bookTitle* that contains either the `<title>` of the map or the `<mainbooktitle>` of the bookmap. If you need to refresh your memory, refer to **Create metadata variables for headers and footers** *(p. 141)*.

2. Comment out the `<prodname>`, `<heading>`, and `<pagenum>` elements and add a `<title>` element:

```
<xsl:template name="insertBodyEvenHeader">
  <fo:static-content flow-name="even-body-header">
    <fo:block xsl:use-attribute-sets="__body__even__header">
      <xsl:call-template name="getVariable">
        <xsl:with-param name="id" select="'Body even header'"/>
        <xsl:with-param name="params">
          <title>
            <xsl:value-of select="$bc.bookTitle"/>
          </title>
          <!--
          <prodname>
            <xsl:value-of select="$productName"/>
          </prodname>
          <heading>
            <fo:inline xsl:use-attribute-sets="__body__even__header__heading">
              <fo:retrieve-marker retrieve-class-name="current-header"/>
            </fo:inline>
          </heading>
          <pagenum>
            <fo:inline xsl:use-attribute-sets="__body__even__header__pagenum">
              <fo:page-number/>
            </fo:inline>
```

```
        </pagenum>
        -->
      </xsl:with-param>
    </xsl:call-template>
  </fo:block>
 </fo:static-content>
</xsl:template>
```

3. Following the example of the *insertBodyEvenHeader* template, make the same changes to these templates:

- *insertTocEvenHeader*
- *insertIndexEvenHeader*

As you did in previous exercises, the next step is to edit the corresponding localization variables.

4. Copy the *Body even header* variable from DITA-OT/plugins/org.dita.pdf2/cfg/common/vars/**en.xml** to your copy of **en.xml**:

```
<!-- The header that appears on even-numbered pages. -->
<variable id="Body even header"><param ref-name="pagenum"/> | <param ref-name="heading"/> | <param
 ref-name="prodname"/></variable>
```

By default, this variable sets up the even body page headers to include the page number, a running header, and the product name. You don't want any of this.

5. Edit the *Body even header* variable as follows:

```
<variable id="Body even header"><param ref-name="title"/></variable>
```

The odd body page header also includes the logo, but that's not part of this variable definition.

6. Following the example of *Body even header* variable, copy the following variables to your copy of **en.xml** and make similar changes to them:

- *Toc even header*
- *Index even header*

Why aren't you making any changes to a variable named *Body last header*? Because there isn't one in **en.xml**. You'll add it later.

7. Save changes, run a PDF build, and check your work.

Footer setup

Exercise: Set up odd footers

Objective: Set up footers that include generated text and map metadata.

So far, you've set up headers with external files, boilerplate text, generated text, and map metadata. Now let's look at some footer definitions. The footers in your specification are a bit more complex than the headers. All of the footers in your PDF template include both map metadata and generated text.

1. Find the following template in your copy of **static-content.xsl**:

```
<xsl:template name="insertBodyOddFooter">
  <fo:static-content flow-name="odd-body-footer">
    <fo:block xsl:use-attribute-sets="__body__odd__footer">
      <xsl:call-template name="getVariable">
        <xsl:with-param name="id" select="'Body odd footer'"/>
        <xsl:with-param name="params">
          <heading>
            <fo:inline xsl:use-attribute-sets="__body__odd__footer__heading">
              <fo:retrieve-marker retrieve-class-name="current-header"/>
            </fo:inline>
          </heading>
          <pagenum>
            <fo:inline xsl:use-attribute-sets="__body__odd__footer__pagenum">
              <fo:page-number/>
            </fo:inline>
          </pagenum>
        </xsl:with-param>
      </xsl:call-template>
    </fo:block>
  </fo:static-content>
</xsl:template>
```

This is the template that creates the footer on the odd body pages. You already know what the `<heading>` and `<pagenum>` elements do.

You don't want a running header in your odd body page footers, but you do want the page number.

You also want the publication date and copyright information. Remember that you created a variable in **root-processing.xsl** named *bc.pubDate* that contains the content of the *golive* attribute on the `<revised>` element within the topicmeta or bookmeta.

If you need to refresh your memory on these metadata variables, refer to **Create metadata variables for headers and footers** *(p. 141)*.

2. Comment out the `<heading>` section as shown:

```
<xsl:template name="insertBodyOddFooter">
  <fo:static-content flow-name="odd-body-footer">
    <fo:block xsl:use-attribute-sets="__body__odd__footer">
      <xsl:call-template name="getVariable">
        <xsl:with-param name="id" select="'Body odd footer'"/>
        <xsl:with-param name="params">
          <!--
          <heading>
            <fo:inline xsl:use-attribute-sets="__body__odd__footer__heading">
              <fo:retrieve-marker retrieve-class-name="current-header"/>
            </fo:inline>
          </heading>
          -->
          <pagenum>
            <fo:inline xsl:use-attribute-sets="__body__odd__footer__pagenum">
              <fo:page-number/>
            </fo:inline>
          </pagenum>
        </xsl:with-param>
      </xsl:call-template>
    </fo:block>
  </fo:static-content>
</xsl:template>
```

3. Below `<pagenum>`, add blocks for `<pubdate>` and `<copyright>` as shown below.

```
<xsl:template name="insertBodyOddFooter">
  <fo:static-content flow-name="odd-body-footer">
    <fo:block xsl:use-attribute-sets="__body__odd__footer">
      <xsl:call-template name="getVariable">
        <xsl:with-param name="id" select="'Body odd footer'"/>
        <xsl:with-param name="params">
          <!--
          <heading>
            <fo:inline xsl:use-attribute-sets="__body__odd__footer__heading">
              <fo:retrieve-marker retrieve-class-name="current-header"/>
            </fo:inline>
          </heading>
          -->
          <pagenum>
            <fo:inline xsl:use-attribute-sets="__body__odd__footer__pagenum">
              <fo:page-number/>
            </fo:inline>
          </pagenum>
          <pubdate>
            <xsl:value-of select="$bc.pubDate"/>
          </pubdate>
          <copyright>
            <xsl:value-of select="$bc.copyYear"/>
          </copyright>
        </xsl:with-param>
      </xsl:call-template>
    </fo:block>
  </fo:static-content>
</xsl:template>
```

Until now, each parameter in a header or footer definition has contained only one variable; however, a parameter can contain more than one variable and other things as well.

You also want the word "Copyright" and the copyright symbol (©) in the footer. As with any other boilerplate text, it's best to put this information in the localization variables file so that it can more easily be translated.

Also notice that although you want the order of the information in the footer to be publication date, copyright, and page number, the corresponding blocks are not in that order here. That's okay. This is not where the order of the information is determined. In **static-content.xsl**, you only determine what you want in the headers and footers. As you might have noticed by now, you determine the order of the information in the corresponding localization variable in **en.xml**.

4. Following the example of the ***insertBodyOddFooter*** template, find the following templates in your copy of **static-content.xsl** and make the same changes:

- ***insertTocOddFooter***
- ***insertIndexOddFooter***

Now it's time to change the corresponding localization variables.

5. Copy the following variable from `DITA-OT/plugins/org.dita.pdf2/cfg/common/vars/`**en.xml** to your copy of **en.xml**:

```
<variable id="Body odd footer"/>
```

By default, there is no content in the odd body page footers. So even though the ***insertBodyOddFooter*** template includes the *header* and *pagenum* parameters by default, they are not used in the out-of-the-box PDF.

> **Important:** This is a one-way street. You can include parameters in a header/footer definition template that you don't include in the corresponding localization variable, but you cannot include parameters in a localization variable that you don't include in the corresponding header/footer definition template. If you do, your builds will fail.

6. Edit the variable as shown:

```
<variable id="Body odd footer"><param ref-name="pubdate"/>Copyright &#169;<param
ref-name="copyright"/>
<param ref-name="pagenum"/></variable>
```

Notice how the order of the `<param>` elements in this variable definition reflect the actual order you want the information to appear in the footer. Also notice that the variable includes the boilerplate "Copyright" text and the copyright symbol (`&#169;`).

7. Following the example of the *Body odd footer* variable, copy the following variables into your copy of **en.xml** and make the same changes.

- *Body first footer*
- *Toc odd footer*
- *Index odd footer*

8. Save changes, run a PDF build, and check your work.

Exercise: Set up even footers

Objective: Set up footers that include generated text and map metadata.

You've set up the odd footers. Now set up the even ones.

I. Find the following template in your copy of **static-content.xsl**:

```
<xsl:template name="insertBodyEvenFooter">
    <fo:static-content flow-name="even-body-footer">
        <fo:block xsl:use-attribute-sets="__body__even__footer">
            <xsl:call-template name="getVariable">
                <xsl:with-param name="id" select="'Body even footer'"/>
                <xsl:with-param name="params">
                    <heading>
                        <fo:inline xsl:use-attribute-sets="__body__even__footer__heading">
                            <fo:retrieve-marker retrieve-class-name="current-header"/>
                        </fo:inline>
                    </heading>
                    <pagenum>
                        <fo:inline xsl:use-attribute-sets="__body__even__footer__pagenum">
                            <fo:page-number/>
                        </fo:inline>
                    </pagenum>
                </xsl:with-param>
            </xsl:call-template>
        </fo:block>
    </fo:static-content>
</xsl:template>
```

This is the template that creates the footer on the even body pages. You know what the <heading> and <pagenum> elements do.

You don't want a running header in your even body page footers, but you do want the page number.

You also want the product name and product version. Remember that you created a variable in **root-processing.xsl** named *bc.productName* that contains the content of the <prodname> element within the topicmeta or bookmeta. You created another variable named *bc.productVersion* that contains the value of the *version* attribute of the <vrm> element within the topicmeta or bookmeta.

If you need to refresh your memory on these metadata variables, refer to **Create metadata variables for headers and footers** *(p. 141)*.

2. Comment out the `<heading>` section.

3. Below `<pagenum>`, add blocks for `<prodname>` and `<version>` as shown below.

```
<xsl:template name="insertBodyEvenFooter">
    <fo:static-content flow-name="even-body-footer">
        <fo:block xsl:use-attribute-sets="__body__even__footer">
            <xsl:call-template name="getVariable">
                <xsl:with-param name="id" select="'Body even footer'"/>
                <xsl:with-param name="params">
                    <!--<heading>
                        <fo:inline xsl:use-attribute-sets="__body__even__footer__heading">
                            <fo:retrieve-marker retrieve-class-name="current-header"/>
                        </fo:inline>
                    </heading>-->
                    <pagenum>
                        <fo:inline xsl:use-attribute-sets="__body__even__footer__pagenum">
                            <fo:page-number/>
                        </fo:inline>
                    </pagenum>
                    <prodname>
                        <xsl:value-of select="$bc.productName"/>
                    </prodname>
                    <version>
                        <xsl:value-of select="$bc.productVersion"/>
                    </version>
                </xsl:with-param>
            </xsl:call-template>
        </fo:block>
    </fo:static-content>
</xsl:template>
```

4. Following the example of the ***insertBodyEvenFooter*** template, find the following templates in your copy of **static-content.xsl** and make the same changes:

- ***insertTocEvenFooter***
- ***insertIndexEvenFooter***

Now change the corresponding localization variables.

5. Copy the following variables from `DITA-OT/plugins/org.dita.pdf2/cfg/common/vars/`**en.xml** to your copy of **en.xml**:

```
<variable id="Body even footer"/>
```

By default, there is no content in the even body page footers. So even though the ***insertBodyOddFooter*** template includes the *prodname* and *version* parameters by default, they are not used in the out-of-the-box PDF.

6. Edit the variable as shown:

```
<variable id="Body even footer"><param ref-name="pagenum"/> <param ref-name="prodname"/> <param
ref-name="version"/></variable>
```

7. Following the example of the *Body even footer* variable, copy the following variables into your copy of **en.xml** and make the same changes.

- *Toc even footer*
- *Index even footer*

8. Save changes, run a PDF build, and check your work.

Exercise: Create new header and footer definitions

Objective: Create definitions for headers and footers that don't exist in the PDF plugin by default.

So far, you've edited existing header definition templates in **static-content.xsl**. But there are a few header definitions that your PDF specifications call for that do not exist by default in **static-content.xsl**: TOC first, index first, TOC last, and index last. There are some footer definitions that also don't exist by default: TOC first footer, index first footer, TOC last footer, and index last footer.

What do I mean by "do not exist by default"? I mean that if you open **static-content.xsl** and look at all the `<xsl:call-template>` statements at the beginning of the file, you won't see anything like `<xsl:call-template name="insertTocFirstHeader"/>` or `<xsl:call-template name="insertTocFirstFooter"/>`.

You need to create these templates yourself and then use them on the corresponding page masters.

1. Find the following template in your copy of **static-content.xsl**:

```
<xsl:template name="insertTocStaticContents">
    <xsl:call-template name="insertTocOddFooter"/>
        <xsl:if test="$mirror-page-margins">
            <xsl:call-template name="insertTocEvenFooter"/>
        </xsl:if>
    <xsl:call-template name="insertTocOddHeader"/>
        <xsl:if test="$mirror-page-margins">
            <xsl:call-template name="insertTocEvenHeader"/>
        </xsl:if>
</xsl:template>
```

2. Below the second `</xsl:if>` (highlighted), add this line

```
<xsl:call-template name="insertTocFirstHeader"/>
```

You are adding a call to a new template, which you now need to create.

3. At the end of **static-content.xsl**, just above the line `</xsl:stylesheet>`, add the following template:

```
<xsl:template name="insertTocFirstHeader">
  <fo:static-content flow-name="first-toc-header">
    <fo:block xsl:use-attribute-sets="__toc__odd__header">
      <fo:inline>
        <fo:external-graphic src="url(Customization/OpenTopic/common/artwork/logo.zzz)"/>
      </fo:inline>
    </fo:block>
  </fo:static-content>
</xsl:template>
```

Be sure to change the file extension of the image file name.

Notice the name of this template corresponds to the name of the template in the `<call-template>` statement you added. This is critical—they must match. Also notice that unlike the body header definitions, there are no sections like `<heading>` or `<title>`. This is because you are not pulling in any information from the topicmeta or bookmeta. You are only including the product logo.

4. Save **static-content.xsl**.

5. Find the following in your copy of **layout-masters.xsl**:

```
<fo:simple-page-master master-name="toc-first" xsl:use-attribute-sets="simple-page-master">
    <fo:region-body xsl:use-attribute-sets="region-body.odd"/>
    <fo:region-before region-name="odd-toc-header" xsl:use-attribute-sets="region-before"/>
    <fo:region-after region-name="odd-toc-footer" xsl:use-attribute-sets="region-after"/>
</fo:simple-page-master>
```

6. Change the region-before name (highlighted) from "odd-toc-header" to "first-toc-header" as shown:

```
<fo:region-before region-name="first-toc-header" xsl:use-attribute-sets="region-before"/>
```

Notice that the region-name is the same as the flow-name in the *insertTocFirstHeader* template that you added to **static-content.xsl**. This is also critical—they must match.

7. Save **layout-masters.xsl**.

Reality check

Now you know how to create a new header template in **static-content.xsl** and how to place the contents of that template into the header or footer of a specific page master. You can use what you've learned here to create templates for index first header, TOC last header, index last header, TOC first footer, TOC last footer, index first footer, and index last footer as well, and you can expand that to set up unique headers and footers for any other page master.

Exercise: Add chapter, appendix, or part numbers to page numbers (Antenna House, XEP)

Objective: Use a custom marker to add the chapter, appendix, or part number to page numbers in the footer.

Out of the box, the PDF plugin doesn't add chapter, part, or appendix numbers to page numbers. Fortunately, it's easy to add them yourself. You create a marker that contains the current chapter, part, or appendix number and add that marker to page numbers in your headers or footers.

Unfortunately, this procedure does not work with FOP. Testing with FOP failed, producing a number of `Element "fo:marker" is missing required property "marker-class-name"`! errors, which appear to be false even though the build failed.

A marker retrieves information from some other place. For example, in **commons.xsl**, the template that processes topic titles includes a marker named `current-header` that essentially says, "Okay, whatever topic title you're processing now, mark that as the current header." Then in **static-content.xsl**, the template that creates the odd body header includes a line that retrieves whatever the value of the `current-header` marker happens to be at that point in the processing and puts it in the header.

A marker simply says, "Note this piece of information because I'm going to use it elsewhere."

> **Note** This procedure only creates chapter, appendix, and part numbers inside chapters, appendices, and parts. It doesn't create anything for pages in the frontmatter, preface, TOC, index, or glossary. For those, you might choose to have no page number prefix or you might use something fixed like **TOC-5** or **I-3**. If you want to use a fixed prefix in those sections, refer to **Add a prefix to non-body page numbers** *(p. 171)* for instructions.

This approach only adds the number to the header or footer after the page number is generated. You're not actually affecting the generated page numbers created by `<fo:page-number-citation>`. If you want to add a chapter, appendix, or part number, that is best done with folios on the page sequences, which is not covered in this book.

1. Copy the following template from `DITA-OT/plugins/org.dita.pdf2/xsl/fo/`**commons.xsl** to your copy of **commons.xsl**:

```
<xsl:template match="*[contains(@class, ' bookmap/chapter ')] |
            opentopic:map/*[contains(@class, ' map/topicref ')]" mode="topicTitleNumber"
        ▸ priority="-1">
   <xsl:variable name="chapters">
      <xsl:document>
         <xsl:for-each select="$map/descendant::*[contains(@class, ' bookmap/chapter ')]">
            <xsl:sequence select="."/>
         </xsl:for-each>
      </xsl:document>
   </xsl:variable>
```

```
<xsl:for-each select="$chapters/*[current()/@id = @id]">
   <xsl:number format="1" count="*[contains(@class, ' bookmap/chapter ')]"/>
</xsl:for-each>
</xsl:template>
```

In this template, the `<xsl:number>` line generates the current chapter number. You want the marker value to be the same as the current chapter number, so it makes sense to use this same line to create the marker value, right?

2. Add a marker, like this:

```
<xsl:template match="*[contains(@class, ' bookmap/chapter ')] |
               opentopic:map/*[contains(@class, ' map/topicref ')]" mode="topicTitleNumber"
               ▶ priority="-1">
   <xsl:variable name="chapters">
      <xsl:document>
         <xsl:for-each select="$map/descendant::*[contains(@class, ' bookmap/chapter ')]">
            <xsl:sequence select="."/>
         </xsl:for-each>
      </xsl:document>
   </xsl:variable>
   <xsl:for-each select="$chapters/*[current()/@id = @id]">
      <fo:inline>
         <fo:marker marker-class-name="current-chapter-number">
            <xsl:number format="1" count="*[contains(@class, ' bookmap/chapter ')]"/>
         </fo:marker>
      </fo:inline>
      <xsl:number format="1" count="*[contains(@class, ' bookmap/chapter ')]"/>
   </xsl:for-each>
</xsl:template>
```

3. Next, copy the following template from `DITA-OT/plugins/org.dita.pdf2/xsl/fo/`**commons.xsl** to your copy of **commons.xsl**:

```
<xsl:template match="*[contains(@class, ' bookmap/appendix ')]" mode="topicTitleNumber">
   <xsl:number format="A" count="*[contains(@class, ' bookmap/appendix ')]"/>
</xsl:template>
```

This template generates the current appendix number. It works just like the chapter number section (although it's a little simpler), so you will make a similar change to it.

4. Add a marker, like this:

```
<xsl:template match="*[contains(@class, ' bookmap/appendix ')]" mode="topicTitleNumber">
   <fo:inline>
      <fo:marker marker-class-name="current-chapter-number">
         <xsl:number format="A" count="*[contains(@class, ' bookmap/appendix ')]"/>
      </fo:marker>
   </fo:inline>
   <xsl:number format="A" count="*[contains(@class, ' bookmap/appendix ')]"/>
</xsl:template>
```

5. Finally, copy the following template from
`DITA-OT/plugins/org.dita.pdf2/xsl/fo/`**commons.xsl** to your copy of **commons.xsl**:

```
<xsl:template match="*[contains(@class, ' bookmap/part ')]" mode="topicTitleNumber">
    <xsl:number format="I" count="*[contains(@class, ' bookmap/part ')]"/>
</xsl:template>
```

This template generates the current part number. It works just like the chapter number section, so again, you will make a similar change to it.

6. Add a marker, like this:

```
<xsl:template match="*[contains(@class, ' bookmap/part ')]" mode="topicTitleNumber">
    <fo:inline>
        <fo:marker marker-class-name="current-chapter-number">
            <xsl:number format="I" count="*[contains(@class, ' bookmap/part ')]"/>
        </fo:marker>
    </fo:inline>
    <xsl:number format="I" count="*[contains(@class, ' bookmap/part ')]"/>
</xsl:template>
```

Notice that regardless of whether you are picking up a chapter, appendix, or part number, you always put the value inside the same `current-chapter-number` marker. This means that you don't have to distinguish between the three in other code. The marker, no matter where you use it, always contains the chapter, appendix, or part number needed at that location. Now that you've created the marker, you can use it.

7. Open your copy of **static-content.xsl**.

8. Find the first template that contains a `<pagenum>` that you haven't commented out.

As an example, I use the *insertBodyOddFooter* template, which you customized to include the page number (if you're doing these exercises in order; if not, you may need to edit a different template).

By default, `<pagenum>` within *insertBodyOddFooter* looks like this:

```
<pagenum>
    <fo:inline xsl:use-attribute-sets="__body__odd__footer__pagenum">
        <fo:page-number/>
    </fo:inline>
</pagenum>
```

9. Add the following code as shown here.

```
<pagenum>
    <fo:inline xsl:use-attribute-sets="__body__odd__footer__pagenum">
        <fo:retrieve-marker retrieve-class-name="current-chapter-number"/>
        <xsl:text>-</xsl:text>
        <fo:page-number/>
    </fo:inline>
</pagenum>
```

The `<fo:retrieve-marker>` element retrieves the value of the `current-chapter-number` marker you created in **commons.xsl**, which is the current chapter number. The code shown also adds a hyphen between the chapter number and the page number. You can substitute a space or any other character for the hyphen.

10. Repeat steps 9 and 10 to add the chapter number to other instances of `<pagenum>` in *body* templates in **static-content.xsl**

Don't bother doing the TOC, index, frontmatter, preface, back cover, or glossary-related templates. They are not part of chapters, appendices, or parts and so will not have a `current-chapter-number` marker value.

> **Tip:** If you haven't finished setting up all your headers and footers, you might want to return to this exercise after you've done so and add the chapter/appendix/part page number prefixes.

11. Save changes, run a PDF build, and check your work.

Because this procedure doesn't work with FOP but does work with the other two PDF renderers, consider placing the `<fo:inline>` inside of an `<xsl:if>` test that tests for the PDF formatter, like this:

```
<xsl:if test="$pdfFormatter != 'fop'">
    <fo:inline>
        <fo:marker marker-class-name="current-chapter-number">
            <xsl:number format="1" count="*[contains(@class, ' bookmap/chapter ')]"/>
        </fo:marker>
    </fo:inline>
</xsl:if>
```

(The `!=` means "not equal to".) So the marker will be created if the PDF renderer is Antenna House or XEP but not if it's FOP. Poor FOP.

Exercise: Restart page numbering in each chapter, appendix, or part

Objective: Restart page numbering at 1 for each chapter, appendix, or part.

By default, the Open Toolkit generates PDFs that start at page one and use continuous page numbering. Your template specs say that page numbering should restart with each chapter. Here's how to set that up.

1. Copy the following template from `DITA-OT/plugins/org.dita.pdf2/xsl/fo/`**commons.xsl** to your copy of **commons.xsl**:

```
<xsl:template name="startPageNumbering" as="attribute()*">
    <!--BS: uncomment if you need reset page numbering at first chapter-->
    <!--
    <xsl:variable name="id" select="ancestor-or-self::*[contains(@class, ' topic/topic ')]
    ►[1]/@id"/>
    <xsl:variable name="mapTopic" select="key('map-id', $id)"/>
    <xsl:if test="not(($mapTopic/preceding::*[contains(@class, ' bookmap/chapter ') or
    ► contains(@class, ' bookmap/part ')])
        or ($mapTopic/ancestor::*[contains(@class, ' bookmap/chapter ') or contains
        ►(@class, ' bookmap/part ')]))">
        <xsl:attribute name="initial-page-number">1</xsl:attribute>
    </xsl:if>
    -->
</xsl:template>
```

Notice the contents of this template are commented out.

2. Uncomment the contents.

3. In the `<xsl:if>` statement, comment out the `<xsl:if>` element and the `</xsl:if>` statement, leaving only the `<xsl:attribute>` element:

```
<xsl:template name="startPageNumbering" as="attribute()*">
    <!--BS: uncomment if you need reset page numbering at first chapter-->
    <xsl:variable name="id" select="ancestor-or-self::*[contains(@class, ' topic/topic ')]
    ►[1]/@id"/>
    <xsl:variable name="mapTopic" select="key('map-id', $id)"/>
    <!--<xsl:if test="not(($mapTopic/preceding::*[contains(@class, ' bookmap/chapter ') or
    ► contains(@class, ' bookmap/part ')])
        or ($mapTopic/ancestor::*[contains(@class, ' bookmap/chapter ') or contains
        ►(@class, ' bookmap/part ')]))">-->
        <xsl:attribute name="initial-page-number">1</xsl:attribute>
    <!--</xsl:if>-->
</xsl:template>
```

4. Save changes, run a PDF build, and check your work.

Header and footer formatting

Exercise: Set top and bottom margins for headers and footers

Objective: Format the top and bottom margins for headers and footers to add appropriate space above and below the text.

While you were working through the exercises to set up header and footer regions, you probably noticed that the header text is too close to the top edge of the page and the footer text is too close to the bottom. When you set up header and footer regions on your page masters (**Setting up header and footer regions** *(p. 104)*), you saw that you can't adjust the dimensions of those regions to control where text appears. Instead, you have to adjust the text blocks that contain the header/footer text. Here's how to do that.

1. Copy the *odd__header*, *even__header*, *odd__footer*, and *even__footer* attribute sets from `DITA-OT/plugins/org.dita.pdf2/cfg/fo/attrs/`**static-content-attr.xsl** to your copy of **static-content-attr.xsl**.

   The specifications call for .5in of space above the header text and .5in of space below the footer text:

Figure 24: Header and footer vertical alignment

You use the *padding-top* and *padding-bottom* attributes to create this space. Why not use *margin-top* and *margin-bottom*? You certainly could, but padding is probably a better choice in this case, because any background color or rules will be inside the margin but outside the padding. Depending on your design, the distinction could be important. You can always experiment to see how each choice works.

2. Add the *padding-top* and *padding-bottom* attributes to the *odd__header* attribute set, both with a value of ".25in", as shown here:

```
<xsl:attribute-set name="odd__header">
    <xsl:attribute name="text-align">end</xsl:attribute>
    <xsl:attribute name="end-indent">10pt</xsl:attribute>
    <xsl:attribute name="space-before">10pt</xsl:attribute>
    <xsl:attribute name="space-before.conditionality">retain</xsl:attribute>
    <xsl:attribute name="padding-top">.25in</xsl:attribute>
    <xsl:attribute name="padding-bottom">.25in</xsl:attribute>
</xsl:attribute-set>
```

3. Make the same changes to the *even__header*, *odd__footer*, and *even__footer* attribute sets.

4. Save changes, run a PDF build, and check your work.

> **Note** The pages that include an image in the header might look a little off, depending on the size of the image. If you're including images in headers or footers, you might need to do some extra adjusting and positioning of the image. You'll learn about that in the **Add space around a header or footer image** *(p. 165)* exercise.

Exercise: Set left and right margins for headers and footers

Objective: Format the left and right margins for headers and footers to create the correct alignment for left and right page masters.

When you were running your test builds, you probably also noticed that the left and right header and footer margins are way off—the text is aligned to the far left and the far right of the page. You need to add separate left and right margin attribute sets for even and odd headers and footers.

1. If not already present, copy the *odd__header* and *even__header* attribute sets from `DITA-OT/plugins/org.dita.pdf2/cfg/fo/attrs/`**static-content-attr.xsl** to your copy of **static-content-attr.xsl**.

2. Add the *margin-left* and *margin-right* attributes to the *odd__header* attribute set, as shown:

```
<xsl:attribute-set name="odd__header">
    <xsl:attribute name="text-align">end</xsl:attribute>
    <xsl:attribute name="end-indent">10pt</xsl:attribute>
    <xsl:attribute name="space-before">10pt</xsl:attribute>
    <xsl:attribute name="space-before.conditionality">retain</xsl:attribute>
    <xsl:attribute name="margin-top">.5in</xsl:attribute>
    <xsl:attribute name="margin-left">
        <xsl:value-of select="$page-margin-inside"/>
    </xsl:attribute>
    <xsl:attribute name="margin-right">
        <xsl:value-of select="$page-margin-outside"/>
    </xsl:attribute>
</xsl:attribute-set>
```

These attributes use as their values the same variables that set the page's left and right margins. therefore, if you ever need to adjust the page margins, and you change these variables' values in **basic-settings.xsl**, you'll adjust the header margins as well.

3. In the *even__header* attribute set, add the same attributes, as shown:

```
<xsl:attribute-set name="even__header">
    <xsl:attribute name="start-indent">10pt</xsl:attribute>
    <xsl:attribute name="space-before">10pt</xsl:attribute>
    <xsl:attribute name="space-before.conditionality">retain</xsl:attribute>
    <xsl:attribute name="margin-top">.5in</xsl:attribute>
    <xsl:attribute name="margin-left">
        <xsl:value-of select="$page-margin-outside"/>
    </xsl:attribute>
    <xsl:attribute name="margin-right">
        <xsl:value-of select="$page-margin-inside"/>
    </xsl:attribute>
</xsl:attribute-set>
```

Notice the attributes use the opposite variables as in the *odd__header* attribute set, which makes sense because the page margins are the opposite on odd and even pages (that is, the outside margin is the left margin on even pages and the right margin on odd pages, and vice versa).

4. Add the same attributes to the *odd__footer* and *even__footer* attribute sets being sure to use the same values as the corresponding *odd__header* and *odd__footer* attribute sets.

5. Save changes, run a PDF build, and check your work.

Exercise: Format the appearance of headers and footers

Objective: Format additional aspects of headers and footers such as size, color, weight, and borders.

Your header and footer text is now in the right place on the page. But there's more work to do. The template specifications say that header text should be 9pt Trebuchet regular black and that there should be a 1pt solid black line below headers and above footers.

Of course, while you're setting up headers and footers, you can set up other criteria such as point size, weight, and color.

Technically, the headers and footers for each page master have their own attribute sets. For example, the odd body header uses attribute set *__body__odd_header*, and the odd body footer uses attribute set *__body__odd__footer*. But each of these specific attribute sets also calls the corresponding *odd__header*, *even__header*, *odd__footer*, or *even__footer* attribute set.

If any formatting is common to all headers or all footers (or all even headers and footers or all odd headers and footers), you can specify that formatting using the common attributes rather than the specific attributes.

In your template, all headers are formatted the same and all footers are formatted the same, so in the following procedure, you will make the same changes in even and odd attribute sets for both headers and footers.

1. If not already present, copy the *__odd__header*, *even__header*, *odd__footer*, *even__footer*, and *pagenum* attribute sets from DITA-OT/plugins/org.dita.pdf2/cfg/fo/attrs/**static-content-attr.xsl** to your copy of **static-content-attr.xsl**.

2. Add the following attributes to the *odd__header* and *even__header* attribute sets:

```
<xsl:attribute-set name="odd__header">
    <xsl:attribute name="text-align">end</xsl:attribute>
    <xsl:attribute name="end-indent">10pt</xsl:attribute>
    <xsl:attribute name="space-before">10pt</xsl:attribute>
    <xsl:attribute name="space-before.conditionality">retain</xsl:attribute>
    <xsl:attribute name="margin-top">.5in</xsl:attribute>
    <xsl:attribute name="margin-left">
        <xsl:value-of select="$page-margin-inside"/>
    </xsl:attribute>
    <xsl:attribute name="margin-right">
        <xsl:value-of select="$page-margin-outside"/>
    </xsl:attribute>
    <xsl:attribute name="font-family">Trebuchet MS, Arial Unicode MS, Helvetica
    ▶</xsl:attribute>
    <xsl:attribute name="font-size">9pt</xsl:attribute>
    <xsl:attribute name="font-weight">regular</xsl:attribute>
    <xsl:attribute name="color">#000000</xsl:attribute>
    <xsl:attribute name="border-bottom">1pt solid black</xsl:attribute>
</xsl:attribute-set>
```

3. Add the following attributes to the *odd__footer* and *even__footer* attribute sets:

```
<xsl:attribute-set name="odd__footer">
    <xsl:attribute name="text-align">end</xsl:attribute>
    <xsl:attribute name="end-indent">10pt</xsl:attribute>
    <xsl:attribute name="space-after">10pt</xsl:attribute>
    <xsl:attribute name="space-after.conditionality">retain</xsl:attribute>
    <xsl:attribute name="margin-bottom">.5in</xsl:attribute>
    <xsl:attribute name="margin-left">
        <xsl:value-of select="$page-margin-inside"/>
    </xsl:attribute>
    <xsl:attribute name="margin-right">
        <xsl:value-of select="$page-margin-outside"/>
    </xsl:attribute>
    <xsl:attribute name="font-family">Trebuchet MS, Arial Unicode MS, Helvetica
    ▶</xsl:attribute>
    <xsl:attribute name="font-size">9pt</xsl:attribute>
    <xsl:attribute name="font-weight">regular</xsl:attribute>
    <xsl:attribute name="color">#000000</xsl:attribute>
    <xsl:attribute name="border-top">1pt solid black</xsl:attribute>
</xsl:attribute-set>
```

The only difference between these changes and the ones in step 2 is that you define *border-top* in the footer attribute sets and *border-bottom* in the header attribute sets.

The template specifications also say that page numbers in footers should be bold and dark red. The odd body footers use the attribute set *__body__odd__footer__pagenum*, which in turn calls the attribute set *pagenum*. So again, if all your page numbers have the same formatting, you can just use the *pagenum* attribute set to format them all at once.

4. Edit the *pagenum* attribute set as follows:

```
<xsl:attribute-set name="pagenum">
    <xsl:attribute name="font-weight">bold</xsl:attribute>
    <xsl:attribute name="color">#990033</xsl:attribute>
</xsl:attribute-set>
```

5. Save your changes, run a PDF build, and check your work.

Reality check

You now understand how to identify what attribute set a particular header or footer is using and how to make basic changes to it. You also now know how to define margins, add rules, and set font properties. Use what you've learned here to format headers and footers for the other page masters.

There are many other attributes you can use to further control the appearance of your page headers and footers; you've seen just a few common ones here. Take a look at an XSL-FO reference for a complete list and description.

Exercise: Justify elements of a header or footer

Objective: Use leaders to justify headers or footers all the way across the page.

So far, header and footer information has been all together, like this:

06/01/2017Copyright ©20173-7

But what if you want to spread it across the entire width of the header or footer area, like this?

You're going to need to insert a leader between the elements of the header or footer as well as change the alignment of the header or footer.

1. Find the ***insertBodyOddFooter*** template in your copy of **static-content.xsl**.

2. Make changes to the template as shown:

```
<fo:static-content flow-name="odd-body-footer">
  <fo:block xsl:use-attribute-sets="__body__odd__footer">
    <xsl:call-template name="getVariable">
      <xsl:with-param name="id" select="'Body odd footer'"/>
      <xsl:with-param name="params">
        <!--<heading>
            <fo:inline xsl:use-attribute-sets="__body__odd__footer__heading">
              <fo:retrieve-marker retrieve-class-name="current-header"/>
            </fo:inline>
        </heading>-->
        <pagenum>
            <fo:inline xsl:use-attribute-sets="__body__odd__footer__pagenum">
              <fo:page-number/>
            </fo:inline>
        </pagenum>
        <pubdate>
            <xsl:value-of select="$bc.pubDate"/>
          <fo:leader xsl:use-attribute-sets="__hdrftr__leader"/>
        </pubdate>
        <copyright>
            <xsl:value-of select="$bc.copyYear"/>
          <fo:leader xsl:use-attribute-sets="__hdrftr__leader"/>
        </copyright>
      </xsl:with-param>
    </xsl:call-template>
  </fo:block>
</fo:static-content>
```

The <fo:leader> element creates a leader wherever you use it. Notice that even though you want the leader between the publication date and the copyright information, you added it inside <pubdate> rather than between <pubdate> and <copyright>.

The simplest explanation is that the <xsl:with-param> statement only uses items inside elements, such as <pubdate> and <copyright>, and ignores content outside. If you put the leader between <pubdate> and <copyright>, nothing would happen. For the same reason, the leader must be inside <pagenum> instead of between <pagenum> and <pubdate>.

> **Tip:** Remember that the order of the parameters in the output is determined by the localization variable in **en.xml** and not the order they are in here. Keep the actual parameter order in mind when determining where to add the leader. I recommend keeping the parameters in the same order in both places.

Notice also that you assigned the leader the attribute set *__hdrftr__leader*. If you didn't assign an attribute set, the leader would be rendered according to its default, which is dots. You probably don't want a line of dots between each element of your footer, right?

By default, there isn't an attribute set named *__hdrftr__leader*. Add it now.

3. At the end of your copy of **static-content-attr.xsl**, just above the line `</xsl:stylesheet>`, add the following:

```
<xsl:attribute-set name="__hdrftr__leader">
    <xsl:attribute name="leader-pattern">space</xsl:attribute>
</xsl:attribute-set>
```

The *leader-pattern* attribute says to use a space instead of dots. You're all set, right? Wrong.

The attribute set that controls the formatting of odd body page footers (*__body__odd__footer*) is still set to align *right*, meaning that all the footer content is pushed to the right side of the page, making leaders meaningless. If you want leaders, you have to force the footer content to spread out; the alignment that makes that happen is *justify*.

The *__body__odd__footer* attribute set uses the *odd__footer* attribute set, so that is where you need to make your changes.

4. Edit the attribute set as follows:

```
<xsl:attribute-set name="odd__footer">
    <xsl:attribute name="text-align">justify</xsl:attribute>
    <xsl:attribute name="text-align-last">justify</xsl:attribute>
    <xsl:attribute name="end-indent">0pt</xsl:attribute>
    <xsl:attribute name="space-after">0pt</xsl:attribute>
    <xsl:attribute name="space-after.conditionality">retain</xsl:attribute>
    <xsl:attribute name="font-family">sans-serif</xsl:attribute>
    <xsl:attribute name="font-size">9pt</xsl:attribute>
    <xsl:attribute name="font-weight">regular</xsl:attribute>
    <xsl:attribute name="color">#000000</xsl:attribute>
    <xsl:attribute name="border-top">1pt solid black</xsl:attribute>
</xsl:attribute-set>
```

The attribute set *text-align-last* is often needed with *text-align* to truly force justification.

5. Save changes, run a PDF build, and check your work.

Exercise: Add space around a header or footer image

Objective: Add space between the header or footer text and the image.

Although you've formatted the text in your headers and footers, the header image is still flush against the header text. You need to add white space around that image to make it look attractive.

1. In the *insertBodyOddHeader* template in your copy of **static-content.xsl**, find the `<fo:inline>` element that adds the image to the header.

2. Edit that element as shown:

```
<fo:static-content flow-name="odd-body-header">
  <fo:block xsl:use-attribute-sets="__body__odd__header">
    <xsl:call-template name="getVariable">
      <xsl:with-param name="id" select="'Body odd header'"/>
      <xsl:with-param name="params">
        <!--<prodname>
          <xsl:value-of select="$productName"/>
        </prodname>-->
        <heading>
          <fo:inline xsl:use-attribute-sets="__body__odd__header__heading">
            <fo:retrieve-marker retrieve-class-name="current-header"/>
          </fo:inline>
        </heading>
        <!--<pagenum>
          <fo:inline xsl:use-attribute-sets="__body__odd__header__pagenum">
            <fo:page-number/>
          </fo:inline>
        </pagenum>-->
      </xsl:with-param>
    </xsl:call-template>
    <fo:inline xsl:use-attribute-sets="__header__image">
      <fo:external-graphic src="url(Customization/OpenTopic/common/artwork/logo.zzz)"/>
    </fo:inline>
  </fo:block>
</fo:static-content>
```

This change assigns a new attribute set named *__header__image*, which you now need to create.

3. At the end of your copy of **static-content-attr.xsl**, just above the line `</xsl:stylesheet>`, add the following:

```
<xsl:attribute-set name="__header__image">
  <xsl:attribute name="padding-left">10pt</xsl:attribute>
</xsl:attribute-set>
```

4. Save changes, run a PDF build, and check your work.

You can use other margin- and padding-related attributes to further position the image.

Other things you can do

Here are a few additional customizations you might want to make to your plugin. These customizations aren't part of the PDF specifications you're following, but they can be useful (and cool).

Exercise: Put header or footer content on multiple lines

Objective: Split lengthy header or footer information into multiple lines.

So far, all the header and footer information has been on a single line. What if you want to split the information across multiple lines, like so:

```
Widget Pro 1.1.1
36
```

Figure 25: Page footer with multiple lines

To do this you need more than one `<fo:block>` element. Each `<fo:block>` element represents, in the very simplest terms, a hard return.

1. If it is not already present, copy the *insertBodyEvenFooter* template from `DITA-OT/plugins/org.dita.pdf2/xsl/fo/`**static-content.xsl** to your copy of **static-content.xsl**.

2. Edit the template to match this example:

```
<xsl:template name="insertBodyEvenFooter">
  <fo:static-content flow-name="even-body-footer">
    <fo:block xsl:use-attribute-sets="__body__even__footer">
      <xsl:call-template name="getVariable">
        <xsl:with-param name="id" select="'Body even footer'"/>
        <xsl:with-param name="params">
          <!--<heading>
            <fo:inline xsl:use-attribute-sets="__body__even__footer__heading">
              <fo:retrieve-marker retrieve-class-name="current-header"/>
            </fo:inline>
          </heading>-->
          <prodname>
            <xsl:value-of select="$bc.productName"/>
          </prodname>
          <version>
            <xsl:value-of select="$bc.productVersion"/>
          </version>
        </xsl:with-param>
      </xsl:call-template>
    </fo:block>
    <fo:block xsl:use-attribute-sets="__body__even__footer">
      <xsl:call-template name="getVariable">
        <xsl:with-param name="id" select="'Body even footer'"/>
        <xsl:with-param name="params">
          <pagenum>
            <fo:inline xsl:use-attribute-sets="__body__even__footer__pagenum">
              <fo:page-number/>
            </fo:inline>
          </pagenum>
        </xsl:with-param>
      </xsl:call-template>
    </fo:block>
  </fo:static-content>
</xsl:template>
```

You're creating a new `<fo:block>` element and moving the `<prodname>` and `<version>` sections into it. The first `<fo:block>` creates a line with the product name and version in it. The second `<fo:block>` creates a line with the page number in it.

Now you need to edit the *Body even footer* variable in **en.xml**.

3. Copy the *Body even footer* variable from `DITA-OT/plugins/org.dita.pdf2/cfg/common/vars/`**en.xml** to your copy of **en.xml**.

4. Edit this variable definition to match the example, if it does not already.

```
<variable id="Body even footer"><param ref-name="pagenum"/><param ref-name="prodname"/> <param
ref-name="version"/></variable>
```

5. Save your changes, run a build, and check your work.

You'll notice that there is an extra border, and the space between the first and second lines is too wide. This is because the two `<fo:block>` elements use the same attribute set, *__body__even__footer*, which in turn uses the *even__footer* attribute set, which has a .5in bottom margin and a border above. You want the border above on the first line but you don't want the bottom margin. You want the bottom margin on the second line but not the border above.

To do that you need different attribute sets for the first and second lines of the footer.

6. Copy the *__body__even__footer* attribute set from `DITA-OT/plugins/org.dita.pdf2/cfg/fo/attrs/`**static-content-attr.xsl** to your copy of **static-content-attr.xsl** (*even__footer* should already be present).

7. In your copy of **static-content-attr.xsl**, make a copy of the *__body__even__footer* attribute set and rename the copy and rename the copy *__body__even__footer__first*:

```
<xsl:attribute-set name="__body__even__footer__first" use-attribute-sets="even__footer">
</xsl:attribute-set>
```

Because *__body__even__footer* and *__body__even__footer__first* both use *even__footer*, by default they look exactly alike, and they should in most ways. The only two things that should be different are the border and the padding. Rather than specifying these attributes in *even__footer*, you should specify them in *__body__even__footer* and *__body__even__footer__first*, as appropriate.

8. Cut the following lines from *even__footer* and paste them into the *__body__even__footer__first* attribute set:

```
<xsl:attribute name="border-top">1pt solid black</xsl:attribute>
<xsl:attribute name="padding-top">.25in</xsl:attribute>
```

9. Cut the following line from *even__footer* and paste it into the *__body__even__footer* attribute set:

```
<xsl:attribute name="padding-bottom">.25in</xsl:attribute>
```

Next, you need to tell the ***insertBodyEvenFooter*** template to use the correct attribute sets.

10. In your copy of **static-content.xsl**, edit the first `<fo:block>` element in the *insertBodyEvenFooter* template to use your new *__body__even__footer__first* attribute set, as follows:

```
<fo:block xsl:use-attribute-sets="__body__even__footer__first">
```

11. Save changes, run a PDF build, and check your work.

Exercise: Add the current date to headers and footers

Objective: Automatically add the current date (the system date on the computer used for generating PDF output) to headers or footers.

Many popular publishing applications let you add the current date to headers or footers using a pre-defined variable. You can do the same thing with the PDF plugin.

1. Open your copy of **root-processing.xsl**.

2. Add a variable definition like the following:

```
<xsl:variable name="bc.currentDate"
      select="format-date(current-date(), '[M01]/[D01]/[Y0001]')"/>
```

Edit this definition to get the date format you want. Here are some sample format strings:

[M01]/[D01]/[Y0001] = 01/01/2011

[M01]-[D01]-[Y0001] = 01-01-2011

[Y0001]/[M01]/[D01] = 2011/01/01

[M01]/[D01]/[Y01] = 01/01/11

[M1]/[D1]/[Y01] = 1/1/11

[h1]:[m01] [P] on [D] [MNn] = 11:53 a.m. on 1 October

[H01]:[m01] [z] = 11:53 GMT-6

After you create this variable, you can use it anywhere in the PDF, not just in headers and footers.

3. In your copy of **static-content.xsl**, add the variable to header or footer definitions as needed.

For example:

```
<xsl:template name="insertBodyEvenHeader">
  <fo:static-content flow-name="even-body-header">
    <fo:block xsl:use-attribute-sets="__body__even__header">
      <xsl:call-template name="getVariable">
        <xsl:with-param name="id" select="'Body even header'"/>
        <xsl:with-param name="params">
          <title>
            <xsl:value-of select="$bookTitle"/>
          </title>
          <currentdate>
            <xsl:text>(</xsl:text>
            <xsl:value-of select="$bc.currentDate"/>
            <xsl:text>)</xsl:text>
          </currentdate>
          <!--<prodname>
            <xsl:value-of select="$productName"/>
          </prodname>
          <heading>
            <fo:inline xsl:use-attribute-sets="__body__even__header__heading">
              <fo:retrieve-marker retrieve-class-name="current-header"/>
            </fo:inline>
          </heading>
          <pagenum>
            <fo:inline xsl:use-attribute-sets="__body__even__header__pagenum">
              <fo:page-number/>
            </fo:inline>
          </pagenum>-->
        </xsl:with-param>
      </xsl:call-template>
    </fo:block>
  </fo:static-content>
</xsl:template>
```

This code adds the current date, enclosed in parentheses, after the book name, like this:

Getting started with Widget Pro (01/25/2013)

If you prefer, you can omit the parentheses, use another character, or use nothing—it's your choice.

> **Tip:** Instead of using `<xsl:text>` to insert parentheses, you can also use `concat`:
>
> ```
> <xsl:value-of select="concat('(', $bc.currentDate, ')')"/>
> ```

4. In **en.xml**, edit the *Body even header* variable to include the new currentdate parameter:

```
<!-- The header that appears on even-numbered pages. -->
<variable id="Body even header"><param ref-name="title"/> <param
ref-name="currentdate"/></variable>
```

5. Save changes, run a PDF build, and check your work.

Exercise: Add a prefix to non-body page numbers

Objective: Add a prefix to page numbers on non-body pages, such as adding the prefix TOC to pages within the Table of Contents.

Earlier in this section, in **Add chapter, appendix, or part numbers to page numbers (Antenna House, XEP)** *(p. 154)*, you learned how to add chapter, appendix, and part numbers as page number prefixes on body pages. You might also want to have prefixes in non-body sections such as the frontmatter, TOC, index, or glossary. Here's how.

These steps explain how to add a prefix for TOC header/footer page numbers, but they apply equally well to the index or glossary or any other non-body section.

> **Important:** This exercise assumes you have already set up the odd TOC page footers to match the odd body page footers—that is, to include the publication date, copyright info, and page number.

Before you look at this topic, take a look at one way not to do it. You should not hardcode the value in **static-content.xsl** like this:

```
<pagenum>
    <fo:inline xsl:use-attribute-sets="__toc__odd__footer__pagenum">
        <xsl:text>TOC-</xsl:text>
        <fo:page-number/>
    </fo:inline>
</pagenum>
```

If you publish in languages other than English, or plan to in the future, this approach leaves you stuck with the **TOC-** prefix in every language. Obviously, **TOC** only stands for "Table of Contents" in English. It might mean something else entirely in another language; you could end up unintentionally amusing or offending someone.

Now that you've seen the wrong way, here's the right way.

1. In your copy of **en.xml**, edit the *Toc odd footer* variable as shown:

```
<variable id="Toc odd footer">
    <param ref-name="pubdate"/><param ref-name="copyright"/>TOC-<param ref-name="pagenum"/>
</variable>
```

You'll notice that TOC- is not formatted like the actual page number (bold red). That's because the pagenum attribute set only affects things that are within the pagenum parameter. If you really need the prefix formatted just like the page number, you can take advantage of **The closest attribute set wins** *(p. 61)* principle. By default, the *__toc__odd__footer* attribute set applies to the entire odd TOC page footer. Set it up to apply the formatting you want to the prefix. Then set up additional attribute sets to apply to each parameter within the footer as necessary to override the formatting in *__toc__odd__footer*. (Use `<fo:inline>`.)

If you set up a TOC prefix and then decide not to include a TOC in your PDF, nothing bad happens. The variable never gets used, and the PDF builds as usual.

2. Save changes, run a PDF build, and check your work.

Exercise: Use a specific title level in the running header

Objective: Determine the title level (using the map hierarchy) you want to use in running page headers.

Based on the template specifications, you set page headers on body odd pages to include a running header and a logo. By default, the PDF plugin uses the level-1 title as the running header in page headers. This running header is created using the *heading* parameter, and the *heading* parameter by default picks up text marked with the `current-header` marker:

```
<heading>
    <fo:inline xsl:use-attribute-sets="__body__odd__header__heading">
        <fo:retrieve-marker retrieve-class-name="current-header"/>
    </fo:inline>
</heading>
```

A search through the plugin files for "current-header" shows that the following titles have this marker and are used as running headers:

- chapter titles
- appendix titles
- part titles
- notices titles
- glossary title
- table of contents title
- index title
- list of tables title
- list of figures title
- other level-1 titles

Say you want to use a different title level as the running header, for example the closest topic title regardless of level. Here's how.

1. Open **commons.xsl**. (You don't need to change anything in this file; you're just going to look at some code to understand what's happening.)

2. Find the template that begins with this line:

```
<xsl:template match="*" mode="processTopicTitle">
```

3. Within that template, find the following code:

```
<xsl:if test="$level = 1">
    <xsl:apply-templates select="." mode="insertTopicHeaderMarker"/>
</xsl:if>
<xsl:if test="$level = 2">
    <xsl:apply-templates select="." mode="insertTopicHeaderMarker">
        <xsl:with-param name="marker-class-name" as="xs:string">current-h2</xsl:with-param>
    </xsl:apply-templates>
</xsl:if>
```

Notice that the first `<xsl:if>` tests for level-1 titles. It then applies this template, also in **commons.xsl**:

```
<xsl:template match="*" mode="insertTopicHeaderMarker">
    <xsl:param name="marker-class-name" as="xs:string">current-header</xsl:param>
    <fo:marker marker-class-name="{$marker-class-name}">
        <xsl:apply-templates select="." mode="insertTopicHeaderMarkerContents"/>
    </fo:marker>
</xsl:template>
```

That template, of course, grabs whatever title has the `current-header` marker.

The second `<xsl:if>` tests for level-2 titles and applies the `current-h2` marker to them. Because this marker is already there, use it.

> **Note** You can see how easy it would be to edit this code to add additional checks for level-3 titles, level-4 titles, and so on.

4. Open your copy of **static-content.xsl**.

5. In the *insertBodyOddHeader* template, change the `<heading>` section to retrieve the `current-h2` marker:

```
<heading>
    <fo:inline xsl:use-attribute-sets="__body__odd__header__heading">
        <fo:retrieve-marker retrieve-class-name="current-h2"/>
    </fo:inline>
</heading>
```

6. Save changes, run a PDF build, and check your work.

Chapter 9

Cover pages

Most PDFs have a front cover, or title, page. Some also have a back cover page. The DITA Open Toolkit creates a front cover page by default and can also create back cover. This chapter explains how to customize (or eliminate) the front cover and how to create and customize a back cover. For a list of attributes that affect cover pages, refer to **Frontmatter attribute sets** *(p. 426)*. Most of the exercises in this chapter depend on variables that you created in Chapter 8. If you have not worked through that chapter, complete at least the **Create metadata variables for headers and footers** *(p. 141)* exercise before continuing.

Cover specifications

By default, the PDF plugin produces a front cover with the book title centered and no back cover.

Front cover page

The front cover page includes:

Specification	Details
Product name	14pt Trebuchet black left-aligned
Product version	14pt Trebuchet black left-aligned
Book title	36pt Trebuchet bold dark red left-aligned
Product logo	left-aligned
Link to company website	12pt Trebuchet black left-aligned, aligned to bottom page margin

Table 1: Front cover page

The book title is already on the front cover by default. If you did the exercises in the chapter titled **Page headers and footers** *(p. 119)*, you've already seen how to place the product name, product version, and product logo in headers and footers, and it will be straightforward to include the same information on the title page using the variables you already created. Here are those variables:

metadata	variable
book title	bookTitle
product name	productName
product version	productVersion

Table 2: Metadata variables for the cover

If you didn't work through the headers and footers chapter, it's a good idea to do so now. Even if you don't need to set up headers and footers, you still need to set up the variables. Work through the exercises in the section titled **Metadata variables for headers and footers** *(p. 122)*.

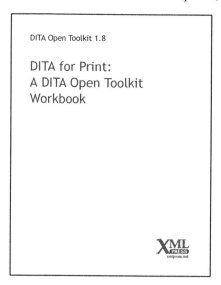

Figure 26: Front cover example

After the variables are set up, you just need to call them on the front cover page. The only new thing you're going to do is create the link. The front cover should look like **Figure 26**:

Back cover page

The back cover page includes:

Product logo	Support contact info
Book item number	Legal disclaimer
Publication date	Copyright info

The back cover page is more of a challenge. You need to make a few changes to your plugin to create one. But, as with the front cover, you already have variables set up to insert most of the information you need:

metadata	variable
publication date	pubDate
book owner	bookOwner
copyright year	copyYear

You'll create one new variable to insert the book item number and insert some boilerplate text for the support information and legal disclaimer.

Files you need

If you have completed other exercises, some of these files might already be present in your plugin.

If not already part of your plugin, create the following files in `DITA-OT/plugins/com.company.pdf/cfg/fo/attrs/`:

- **basic-settings.xsl**
- **front-matter-attr.xsl**

Be sure to add the appropriate `xsl:import` statement to **custom.xsl** (attrs).

If not already part of your plugin, create the following files in
`DITA-OT/plugins/com.company.pdf/cfg/fo/xsl/`:

- **commons.xsl**
- **front-matter.xsl**
- **root-processing.xsl**

For instructions on adding attribute files and XSLT stylesheets to your PDF plugin, refer to **Add an attribute set to your plugin** *(p. 32)* and **Add a template to your plugin** *(p. 34)*.

You copied **layout-masters.xsl** from `DITA-OT/plugins/org.dita.pdf2/cfg/fo/` to `DITA-OT/plugins/com.company.pdf/cfg/fo/` when you first created your plugin. If you have not done so already, you need to add an import statement for it. Because **layout-masters.xsl** is in the `fo` folder, rather than the `xsl` folder, the import statement needs to be different from the ones you've added so far. It needs to look up a level. Add the following to **custom.xsl** (xsl):

```
<xsl:import href="../layout-masters.xsl"/>
```

(Notice this path is an URL and uses a forward slash (/) even on a Windows system.)

If it is not already part of your plugin, create the following file:

- `DITA-OT/plugins/com.company.pdf/cfg/common/vars/`**en.xml**

For instructions on creating variables and strings files, refer to **Create a localization variables file** *(p. 78)* and **Create a localization strings file** *(p. 79)*.

Be sure to add the appropriate language and locale mapping statements in **strings.xml**. You can find instructions in **Mapping strings files to xml:lang values in the strings.xml file** *(p. 37)*.

Front cover customization

Exercise: Add bookmap information to the front cover

Objective: Pull information from the metadata in your bookmap and add it to the front cover page.

Before you start

Complete the exercise **Create metadata variables for headers and footers** *(p. 141)*.

The book title is present on the front cover page by default. You need to add the product name, product version, product logo, and a link. Here's how.

1. Copy the following templates from `DITA-OT/plugins/org.dita.pdf2/xsl/fo/`**front-matter.xsl**
to your copy of **front-matter.xsl**:

```
<xsl:template name="createFrontMatter">
    <xsl:if test="$generate-front-cover">
        <fo:page-sequence master-reference="front-matter" xsl:use-attribute-sets=
        ▶"page-sequence.cover">
            <xsl:call-template name="insertFrontMatterStaticContents"/>
            <fo:flow flow-name="xsl-region-body">
                <fo:block-container xsl:use-attribute-sets="__frontmatter">
                    <xsl:call-template name="createFrontCoverContents"/>
                </fo:block-container>
            </fo:flow>
        </fo:page-sequence>
    </xsl:if>
</xsl:template>
```

```
<xsl:template name="createFrontCoverContents">
<!-- set the title -->
    <fo:block xsl:use-attribute-sets="__frontmatter__title">
        <xsl:choose>
            <xsl:when test="$map/*[contains(@class,' topic/title ')][1]">
                <xsl:apply-templates select="$map/*[contains(@class,' topic/title ')][1]"/>
            </xsl:when>
            <xsl:when test="$map//*[contains(@class,' bookmap/mainbooktitle ')][1]">
                <xsl:apply-templates select="$map//*[contains(@class,'
                ▶ bookmap/mainbooktitle ')][1]"/>
            </xsl:when>
            <xsl:when test="//*[contains(@class, ' map/map ')]/@title">
                <xsl:value-of select="//*[contains(@class, ' map/map ')]/@title"/>
            </xsl:when>
            <xsl:otherwise>
                <xsl:value-of select="/descendant::*[contains(@class, ' topic/topic ')][1]
                ▶/*[contains(@class, ' topic/title ')]"/>
            </xsl:otherwise>
        </xsl:choose>
    </fo:block>
    <!-- set the subtitle -->
    <xsl:apply-templates select="$map//*[contains(@class,' bookmap/booktitlealt ')]"/>
    <fo:block xsl:use-attribute-sets="__frontmatter__owner">
        <xsl:apply-templates select="$map//*[contains(@class,' bookmap/bookmeta ')]"/>
    </fo:block>
</xsl:template>
```

2. In the first template, notice the highlighted lines

Everything you want on the front cover page must go within this `<fo:block-container>`. This
block-container in turn calls the second template, which includes two `<fo:block>` elements. The
first contains the main map or bookmap title and the second contains the subtitle, if there is one.

The first thing to do is delete the default code that adds the title and instead use your custom variable.
You don't need to re-create the title from scratch now that your have the *bc.bookTitle* variable.

3. Comment out the entire `<xsl:choose>` section as shown:

```
<xsl:template name="createFrontCoverContents">
<!-- set the title -->
    <fo:block xsl:use-attribute-sets="__frontmatter__title">
        <!--<xsl:choose>
            <xsl:when test="$map/*[contains(@class,' topic/title ')][1]">
                <xsl:apply-templates select="$map/*[contains(@class,' topic/title ')][1]"/>
            </xsl:when>
            <xsl:when test="$map//*[contains(@class,' bookmap/mainbooktitle ')][1]">
                <xsl:apply-templates select="$map//*[contains(@class,
                ▶' bookmap/mainbooktitle ')][1]"/>
            </xsl:when>
            <xsl:when test="//*[contains(@class, ' map/map ')]/@title">
                <xsl:value-of select="//*[contains(@class, ' map/map ')]/@title"/>
            </xsl:when>
            <xsl:otherwise>
                <xsl:value-of select="/descendant::*[contains(@class,
                ▶' topic/topic ')][1]/*
                ▶[contains(@class, ' topic/title ')]"/>
            </xsl:otherwise>
        </xsl:choose>-->
    </fo:block>
    <!-- set the subtitle -->
    <xsl:apply-templates select="$map//*[contains(@class,' bookmap/booktitlealt ')]"/>
    <fo:block xsl:use-attribute-sets="__frontmatter__owner">
        <xsl:apply-templates select="$map//*[contains(@class,' bookmap/bookmeta ')]"/>
    </fo:block>
</xsl:template>
```

4. Just below the commented-out `<xsl:choose>`, add the highlighted line:

```
    ...
    </xsl:choose>
    -->
    <xsl:value-of select="$bc.bookTitle"/>
</fo:block>
```

5. Just before the book title `<fo:block>`, add the following highlighted code, which inserts the product name, a space, and the product version:

```
<xsl:template name="createFrontCoverContents">
    <fo:block>
        <xsl:value-of select="$bc.productName"/><xsl:text> </xsl:text>
        <xsl:value-of select="$bc.productVersion"/>
    </fo:block>
<!-- set the title -->
```

6. Just after the book title `<fo:block>`, add the following highlighted code, which inserts the same product logo that's used in the page headers:

```
    ...
    <xsl:value-of select="$bc.bookTitle"/>
</fo:block>
<fo:block>
    <fo:external-graphic src="url(Customization/OpenTopic/common/artwork/logo.xxx)"/>
</fo:block>
<!-- set the subtitle -->
```

Remember to change the extension of the logo file.

7. Just after the `<fo:block>` that inserts the logo, add the following:

```
<fo:block>
    <fo:basic-link external-destination="http://xmlpress.net/">http://xmlpress.net/</fo:basic-link>
</fo:block>
```

This code inserts the URL for this book's publisher, XML Press. Notice this `<fo:block>` does not call a variable like the other blocks. Instead, you're using the `<basic-link>` element with a hardcoded URL. You could create the URL as a variable, but because it's not going to change from publication to publication, you might as well hardcode it.

8. Save changes, run a PDF build, and check your work.

All of the information is present on the cover page, but it's not formatted the way you want. The product name and version are centered at the top of the page and are the wrong font and size. The logo and link are centered immediately below the title. Now you need to format all this information.

Exercise: Format information on the cover page

Objective: Determine the appearance and position of text and images on the front cover page.

Before you start

Complete the exercise **Add bookmap information to the front cover** *(p. 178)*.

After you get the information you want on the cover page, you need to format it. First you assign attribute sets to the `<fo:block>` elements that insert the information, then you create those attribute sets.

1. Open your copy of **front-matter.xsl**.

The `<fo:block>` you created to insert the book title already uses the attribute set *__frontmatter__title*, so you don't need to change it (you'll edit *__frontmatter__title* in a later step).

2. Make the following change to the `<fo:block>` that inserts the product name and version:

```
<fo:block xsl:use-attribute-sets="__frontmatter__product">
    <xsl:value-of select="$bc.productName"/>
    <xsl:text> </xsl:text>
    <xsl:value-of select="$bc.productVersion"/>
</fo:block>
```

3. Make the following change to the `<fo:block>` that inserts the logo:

```
<fo:block xsl:use-attribute-sets="__frontmatter__logo">
    <fo:external-graphic src="url(Customization/OpenTopic/common/artwork/logo.png)"/>
</fo:block>
```

4. Make the following change to the `<fo:block>` that inserts the link:

```
<fo:block xsl:use-attribute-sets="__frontmatter__link">
    <fo:basic-link external-destination="http://xmlpress.net/">http://xmlpress.net/</fo:basic-link>
</fo:block>
```

5. Save **front-matter.xsl**.

6. Copy the *__frontmatter__title* attribute set from `DITA-OT/plugins/org.dita.pdf2/cfg/fo/attrs/`**front-matter-attr.xsl** to your copy of **front-matter-attr.xsl**.

7. Edit the *__frontmatter__title* attribute set as shown:

```
<xsl:attribute-set name="__frontmatter__title" use-attribute-sets="common.title">
    <xsl:attribute name="space-before">24pt</xsl:attribute>
    <xsl:attribute name="space-before.conditionality">retain</xsl:attribute>
    <xsl:attribute name="font-size">36pt</xsl:attribute>
    <xsl:attribute name="font-weight">bold</xsl:attribute>
    <xsl:attribute name="line-height">140%</xsl:attribute>
    <xsl:attribute name="text-align">left</xsl:attribute>
    <xsl:attribute name="color">#990033</xsl:attribute>
</xsl:attribute-set>
```

8. Immediately below *__frontmatter__title*, add the following attribute sets:

```
<xsl:attribute-set name="__frontmatter__product">
    <xsl:attribute name="font-size">14pt</xsl:attribute>
    <xsl:attribute name="text-align">left</xsl:attribute>
    <xsl:attribute name="font-family">Trebuchet MS, Arial Unicode MS, Helvetica</xsl:attribute>
</xsl:attribute-set>
```

```
<xsl:attribute-set name="__frontmatter__logo">
    <xsl:attribute name="text-align">right</xsl:attribute>
</xsl:attribute-set>
```

```
<xsl:attribute-set name="__frontmatter__link">
    <xsl:attribute name="font-size">12pt</xsl:attribute>
    <xsl:attribute name="font-weight">bold</xsl:attribute>
    <xsl:attribute name="text-align">right</xsl:attribute>
    <xsl:attribute name="font-family">Trebuchet MS, Arial Unicode MS, Helvetica</xsl:attribute>
</xsl:attribute-set>
```

9. Save changes, run a PDF build, and check your work.

The product information and the book title look good. The logo and link are right-aligned, but they're too high on the page. You want to align them with the bottom margin of the page.

Exercise: Place cover page information in a specific location

Objective: Position text or images in an absolute location on the page.

Before you start

Complete the exercise **Format information on the cover page** *(p. 181)*.

The product information and the book title on the front cover look good. The logo and link are correctly right-aligned, but they're too high on the page. You want to align them with the bottom margin.

One thing you could do, of course, is to add space below the title using either *margin-bottom* or *padding-bottom* in the *__frontmatter__title* attribute set. However, that could get you into trouble. Spacing that would work for a one-line title would be too much for a two-line title, forcing the logo and link onto the back side of the cover page.

Normally, each <fo:block> is placed relative to the preceding <fo:block>. If a block moves, grows, or shrinks, the block below it will also move on the page. This is called *relative positioning*. However, you want the logo and link to be in the same position regardless of the position or size of the preceding <fo:block>. This is called *absolute positioning*. You can't assign an absolute position to an <fo:block>. Instead, you need to place it inside an <fo:block-container>.

1. Open your copy of **front-matter.xsl** and find the ***createFrontCoverContents*** template.

2. Enclose the *__frontmatter__logo* and *__frontmatter__link* blocks in an <fo:block-container> element, and have it use the *__frontmatter__logo__container* attribute set:

```
<fo:block-container xsl:use-attribute-sets="__frontmatter__logo__container">
    <fo:block xsl:use-attribute-sets="__frontmatter__logo">
        <fo:external-graphic src="url(Customization/OpenTopic/common/artwork/logo.png)"/>
    </fo:block>

    <fo:block xsl:use-attribute-sets="__frontmatter__link">
        <fo:basic-link external-destination="http://xmlpress.net/">http://xmlpress.net/
        ▸ </fo:basic-link>
    </fo:block>
</fo:block-container>
```

3. Open your copy of **front-matter-attr.xsl** and just below the *__frontmatter__link* attribute set, add the following:

```
<xsl:attribute-set name="__frontmatter__logo__container">
    <xsl:attribute name="position">absolute</xsl:attribute>
    <xsl:attribute name="top">8.35in</xsl:attribute>
</xsl:attribute-set>
```

This places the `<fo:block-container>` 8.35in from the top of the page. This value was calculated assuming your template specification of 1in top and bottom margins for the front cover page. You may need to change this value to match your format (logo, page size, etc.).

4. Save changes, run a PDF build, and check your work.

Back cover creation and customization

Exercise: Create a back cover

By default, the PDF plugin doesn't create a back cover. However, it has the code in place to add one; you just need to activate it. If you used older versions of the PDF plugin, you'll recognize this as a very nice new feature!

1. Copy the following variable from `DITA-OT/plugins/org.dita.pdf2/xsl/fo/`**basic-settings.xsl** to your copy of **basic-settings.xsl**:

```
<xsl:variable name="generate-back-cover" select="false()"/>
```

2. Change the *select* value to "true()".

3. Save changes, run a PDF build, and check your work.

Reality check

You now have a back cover page, but it's blank.

Exercise: Add content to the back cover

Objective: Add text and other items to the back cover.

On the front side of the back cover, you want the copyright, support contacts, and a legal disclaimer. On the back side of the back cover, you want the logo, the item number, and the publication date.

1. Copy the following templates from `DITA-OT/plugins/org.dita.pdf2/xsl/fo/`**front-matter.xsl** to your copy of **front-matter.xsl**:

```
<xsl:template name="createBackCover">
    <xsl:if test="$generate-back-cover">
        <fo:page-sequence master-reference="back-cover" xsl:use-attribute-sets="back-cover">
            <xsl:call-template name="insertBackCoverStaticContents"/>
```

```
            <fo:flow flow-name="xsl-region-body">
                <fo:block-container xsl:use-attribute-sets="__back-cover">
                    <xsl:call-template name="createBackCoverContents"/>
                </fo:block-container>
            </fo:flow>
        </fo:page-sequence>
    </xsl:if>
</xsl:template>
```

```
<xsl:template name="createBackCoverContents">
    <fo:block></fo:block>
</xsl:template>
```

Similar to the front cover templates you just worked with, the first template contains a `<fo:block-container>` and everything you want on the back cover page must go within this `<fo:block-container>`. This block-container in turn calls the second template, which includes an empty (for now) `<fo:block>` element.

Yes, the **back** cover templates go in **front**-matter.xsl. They don't have to, if that mismatch bothers you. You can create a new stylesheet named **back-matter.xsl** and add the templates there.

2. Replace the empty `<fo:block>` in the *createBackCoverContents* template, with the following blocks:

```
<fo:block xsl:use-attribute-sets="__backmatter__copyright">
    <xsl:text>Copyright &#169;</xsl:text>
    <xsl:value-of select="$copyYear"/><xsl:text> </xsl:text>
</fo:block>

<fo:block xsl:use-attribute-sets="__backmatter__text">
    <xsl:text>For support, contact the DITA Users list: </xsl:text>
    <fo:basic-link external-destination="http://tech.groups.yahoo.com/group/dita-users">
    ▶http://tech.groups.yahoo.com/group/dita-users</fo:basic-link>
    <xsl:text>. Also refer to the DITA Open Toolkit 1.8 User Documentation: </xsl:text>
    <fo:basic-link external-destination="http://dita-ot.sourceforge.net/doc/ot-userguide/xhtml/">

    ▶http://dita-ot.sourceforge.net/doc/ot-userguide/xhtml/</fo:basic-link>
    <xsl:text>.</xsl:text>
</fo:block>

<fo:block xsl:use-attribute-sets="__backmatter__text__last">
    <xsl:text>Neither XML Press nor the author is responsible for any damage to
    equipment or loss of data that might occur as a result of downloading and/or installing
    any software mentioned herein. Nor is XML Press or the author responsible
    for any penalties that might be incurred due to illegal download, installation or usage
    of any software mentioned herein.</xsl:text>
</fo:block>

<fo:block xsl:use-attribute-sets="__backmatter__logo">
    <fo:external-graphic src="url(Customization/OpenTopic/common/artwork/logo.png)"/>
</fo:block>

<fo:block xsl:use-attribute-sets="__backmatter__publish">
    <xsl:call-template name="getVariable">
        <xsl:with-param name="id" select="'Published date'"/>
    </xsl:call-template>
    <xsl:text> </xsl:text>
    <xsl:value-of select="$bc.pubDate"/>
</fo:block>
```

```
<fo:block xsl:use-attribute-sets="__backmatter__item">
    <xsl:call-template name="getVariable">
        <xsl:with-param name="id" select="'Item number'"/>
    </xsl:call-template>
    <xsl:text> </xsl:text>
    <xsl:value-of select="$bc.itemNumber"/>
</fo:block>
```

These blocks add the information you need. It works the same way here as on the front cover page. Notice the © in the first <fo:block>. This is an entity. This kind of entity allows you to define a character using that character's numeric code (© is the code for the copyright symbol). Special characters such as copyright symbols, double quotes, foreign currency symbols, and so forth are often referred to using their entity code. Some editors let you enter the character directly with a special key combination. Other editors might require you to enter the entity code in numeric form.

Also notice there is one new variable: *bc.itemNumber*. You need to create that variable.

Finally, two new localization variables are called: *Published date* and *Item number*. You're using localization variables rather than plain text because these terms ("Published on" and "Item #") are likely to need translation. (In fact, all of the text you're adding to the back cover page will probably need to be translated, so you might consider treating all of it as localization variables.) At any rate, you need to add these two new variables to your localization variables file.

3. Add the following to your copy of **root-processing.xsl** just below the definition for your last custom variable:

```
<xsl:variable name="bc.itemNumber" select="//*[contains(@class, ' bookmap/bookid ')]//
►*[contains(@class, ' bookmap/booknumber ')]"/>
```

4. Add the following to your **en.xml** file:

```
<variable id="Published date">Published on</variable>
<variable id="Item number">Item #:</variable>
```

5. Save your changes.

Now you need to add all the attribute sets called by the <fo:block> elements you added.

Exercise: Format content on the back cover

Objective: Format the text and other items on the back cover.

Before you start

Complete the exercise **Add content to the back cover** *(p. 184)*.

Now that you've added content to the back cover, you need to format it using new attribute sets.

1. Add the following attribute sets to your copy of **front-matter-attr.xsl**:

(Or to a new attribute set that you might want to create, **back-matter-attr.xsl**.)

```
<xsl:attribute-set name ="__backmatter__copyright">
   <xsl:attribute name="font-family">Trebuchet MS, Arial Unicode MS, Helvetica
   ►</xsl:attribute>
   <xsl:attribute name="font-size">14pt</xsl:attribute>
   <xsl:attribute name="font-weight">bold</xsl:attribute>
   <xsl:attribute name="space-after">18pt</xsl:attribute>
   <xsl:attribute name="color">#990033</xsl:attribute>
</xsl:attribute-set>


<xsl:attribute-set name ="__backmatter__text">
   <xsl:attribute name="font-family">Book Antiqua, Times New Roman, Times
   ►</xsl:attribute>
   <xsl:attribute name="font-size">10pt</xsl:attribute>
   <xsl:attribute name="font-weight">normal</xsl:attribute>
   <xsl:attribute name="space-after">6pt</xsl:attribute>
   <xsl:attribute name="line-height">14pt</xsl:attribute>
</xsl:attribute-set>


<xsl:attribute-set name ="__backmatter__text__last">
   <xsl:attribute name="font-family">Book Antiqua, Times New Roman, Times
   ►</xsl:attribute>
   <xsl:attribute name="font-size">10pt</xsl:attribute>
   <xsl:attribute name="font-weight">normal</xsl:attribute>
   <xsl:attribute name="space-after">6pt</xsl:attribute>
   <xsl:attribute name="line-height">14pt</xsl:attribute>
</xsl:attribute-set>


<xsl:attribute-set name ="__backmatter__item">
   <xsl:attribute name="font-family">Trebuchet MS, Arial Unicode MS, Helvetica
   ►</xsl:attribute>
   <xsl:attribute name="font-size">9pt</xsl:attribute>
   <xsl:attribute name="font-weight">bold</xsl:attribute>
   <xsl:attribute name="space-before">12pt</xsl:attribute>
   <xsl:attribute name="space-after">3pt</xsl:attribute>
   <xsl:attribute name="text-align">right</xsl:attribute>
   <xsl:attribute name="color">#990033</xsl:attribute>
</xsl:attribute-set>
```

```
<xsl:attribute-set name ="__backmatter__logo">
   <xsl:attribute name="text-align">right</xsl:attribute>
</xsl:attribute-set>
```

```
<xsl:attribute-set name ="__backmatter__publish">
   <xsl:attribute name="font-family">Trebuchet MS, Arial Unicode MS, Helvetica
   ▸</xsl:attribute>
   <xsl:attribute name="font-size">9pt</xsl:attribute>
   <xsl:attribute name="font-weight">normal</xsl:attribute>
   <xsl:attribute name="space-after">3pt</xsl:attribute>
   <xsl:attribute name="text-align">right</xsl:attribute>
</xsl:attribute-set>
```

2. Save your changes, run a build, and check your work.

Everything looks pretty good. However, the logo, item number, and published date are on the same page as the copyright and legal information. You want them at the bottom of the next page. First, let's force the logo, item number and published date to the next page.

3. Add the following attribute to the *__backmatter__text__last* attribute set:

```
<xsl:attribute name="break-after">page</xsl:attribute>
```

This attribute does pretty much what its name suggests; it creates a page break after any text that uses the attribute set.

> **Note** You might be asking yourself, "Wouldn't it work just as well to add a *break-before* attribute to the *__backmatter__logo* attribute set, so that you wouldn't need two attribute sets to handle the two paragraphs on the first back cover page?" Or, "How come I can't add *break-before* to the *__backmatter__logo__container* attribute set?" Good questions. Give it a try if you like. If your results are like mine, neither approach works. Exactly why is complicated; just chalk it up to the mysteries of XSL-FO and the many levels of nesting produced by these XSLT stylesheets. When you're working with pagination, you may need to try several things before you get the results you want.

Now it's time to position the logo, item number, and published date. From working with the front cover, you should remember how to position a block on a specific place on the page using `<fo:block-container>`. If you don't remember, refer to **Place cover page information in a specific location** *(p. 183)* as a refresher.

4. Open your copy of **front-matter.xsl**.

5. Enclose the three `<fo:block>` elements that create the logo, item number and published date within an `<fo:block-container>` element that uses attribute set *__backmatter__logo__container*.

6. In your copy of **front-matter-attr.xsl**, add the following attribute set:

```
<xsl:attribute-set name="__backmatter__logo__container">
    <xsl:attribute name="position">absolute</xsl:attribute>
    <xsl:attribute name="top">8.35in</xsl:attribute>
</xsl:attribute-set>
```

7. Save changes, run a PDF build, and check your work.

Exercise: Adjust pagination of the back cover

Objective: Determine whether the back cover starts on an odd or even page.

You might find that you get a blank odd back cover page, followed by an even back cover page with the copyright, support information and legal disclaimer, followed by another odd back cover page with the logo, item number, and publication date, followed by a blank odd back cover page. If you get extra pages like this (or some other combination), there are two other attribute sets you can use to adjust the result.

1. Copy these two attribute sets from DITA-OT/plugins/org.dita.pdf2/cfg/fo/attrs/**front-matter-attr.xsl** to your copy of **front-matter-attr.xsl** (or **back-matter-attr.xsl**, if you've created that file):

```
<xsl:attribute-set name="back-cover">
    <xsl:attribute name="force-page-count">end-on-even</xsl:attribute>
</xsl:attribute-set>
```

```
<xsl:attribute-set name="__back-cover">
    <xsl:attribute name="break-before">even-page</xsl:attribute>
</xsl:attribute-set>
```

2. In the first attribute set, change the value of *force-page-count* to "no-force".

3. In the second attribute set, change the value of *break-before* to "auto".

4. Save changes, run a PDF build, and check your work.

The result of this combination is an odd back cover page with the copyright, support information, and legal disclaimer followed by an even back cover page with the logo, item number, and publication date. If this is not the result you want either, experiment with *force-page-count* and *break-before* until you get what you want.

Other things you can do

Here are a few additional customizations you might want to make to your plugin. These customizations aren't part of the PDF specifications you're following, but they can be useful (and cool).

Exercise: Insert text based on language

Objective: Automatically insert custom text based on the language of the publication map.

At some point, you might need to insert different text based on the publication language. For example, let's say you've been asked to include the publication language on the back cover of your PDF. The easiest way to do this is to grab the value of the *xml:lang* attribute and use it to determine which language name to output. There are several ways you could go about it, but here is one of the easiest approaches.

> **Note** This method only captures the main language of the document. It doesn't handle a case where some topics have one value of *xml:lang* and some have another.

These steps assume you've added a back cover to your PDF. If you didn't do those exercises, take a few minutes to set that up. Start with the **Create a back cover** *(p. 184)* overview. Alternatively, you can add this information to the front cover page.

1. In your copy of **root-processing.xsl**, add a new variable called *bc.printedLang* as follows:

```
<xsl:variable name="bc.printedLang">
    <xsl:value-of select="//*[contains(@class,' map/map ')]/@xml:lang"/>
</xsl:variable>
```

This variable selects the value of *xml:lang* from your map or bookmap. Notice that by selecting `' map/map '` you will match either a map or a bookmap, whichever is your top level map.

2. In your copy of **front-matter.xsl**, add a new `<fo:block>` as follows:

```
<fo:block xsl:use-attribute-sets="__backmatter__language">
</fo:block>
```

You already added an `<fo:block>` to include the published date. Put the language `<fo:block>` just below that one.

3. In your copy **front-matter-attr.xsl**, copy and paste the *__backmatter__publish* attribute set and change the name to *__backmatter__language*.

4. In **front-matter.xsl**, add the following inside the new `<fo:block>` you just added:

```
<fo:block xsl:use-attribute-sets="__backmatter__language">
<xsl:choose>
    <xsl:when test="$bc.printedLang = 'de'">
        <xsl:text>Deutsch</xsl:text>
    </xsl:when>
    <xsl:when test="$bc.printedLang = 'en'">
        <xsl:text>English</xsl:text>
    </xsl:when>
    <xsl:when test="$bc.printedLang = 'es'">
        <xsl:text>español</xsl:text>
    </xsl:when>
    <xsl:when test="$bc.printedLang = 'fr'">
        <xsl:text>français</xsl:text>
    </xsl:when>
    <xsl:otherwise>
        <xsl:text>[Unspecified]</xsl:text>
    </xsl:otherwise>
</xsl:choose>
</fo:block>
```

> **Important:** Anytime you're dealing with non-western-European languages, it's a good idea to understand literal characters versus numeric character references. See this topic for details: **Literal characters and numeric character references** *(p. 77)*.

For the sake of brevity, there are only a few languages in this example. You'd add an `<xsl:when>` test for every language you publish in. The `<xsl:otherwise>` is a fallback condition when *xml:lang* isn't one of the values you tested for. The "[Unspecified]" text is a visual cue to you to add the new value.

As a matter of XSLT practice, whenever you have a long `<xsl:choose>`, like this one, that reflects a set of potentially open values, it's best to use `<apply-templates>` in a specific mode. This book doesn't go into `<apply-templates>` in general, although there is one example of using it in **Place titles below tables** *(p. 285)*.

5. Save your changes, run a PDF build, and check your work.

Exercise: Eliminate the front cover page

Objective: Omit the front cover from your PDF without eliminating any other frontmatter.

If you don't need a front cover page, you can easily eliminate it.

> **Note** By default, the PDF plugin does not generate a front cover page when you are publishing a single topic.

1. Copy the following variable from
   `DITA-OT/plugins/org.dita.pdf2/cfg/fo/attrs/`**basic-settings.xsl** to your copy of
   basic-settings.xsl:

```
<xsl:variable name="generate-front-cover" select="true()"/>
```

2. Change the value of *select* to "false()".
3. Save changes, run a PDF build, and check your work.

Chapter 10

Titles, body text, and notes

This chapter covers formatting of body text, title, and notes. Start here if you are looking for information on formatting text and don't find a chapter specifically devoted to what you need. For a list of attributes that affect text, refer to **Common attribute sets** *(p. 406)* and **Domain attribute sets** *(p. 421)*.

General text formatting specifications

Let's review the specifications for general text formatting. This includes headings, body text, and notes.

General text

Chapter, appendix titles	28pt Trebuchet normal black, no borders. Autonumber is 16pt Trebuchet regular black.
Level 1 heading font	20pt Trebuchet normal black, no borders
Level 2 heading font	18pt Trebuchet normal dark red
Level 3 heading font	16pt Trebuchet normal dark red, .5in indent
Level 4+ heading font	14pt Trebuchet italic dark red, .5in indent
Section headings	12pt Trebuchet bold dark red
Regular body font	11pt Book Antiqua normal dark gray. Used for all body elements except where noted below.
Small body font	10pt Book Antiqua normal dark gray. Used for *note, info, stepxmp, stepresult, choice.*

| Code samples, system messages | 9pt Consolas normal black, light gray background |
| Line height (leading) | 120% |

Notes

Note label	dark red, bold. Not used on note type="note".
Note borders	left and right, 2pt dark red
Note indents	.5in left and right
Note font	Small body font, justified
Note image	note.png
Custom note label	Best Practice
Label for fastpath notes	Quick access

Correctly formatted headings provide visual cues that tell readers about the relationships between pieces of information. Getting the headings right will take you a long way towards the final look and feel of your PDF. So let's tackle the four heading levels first.

Files you need

If you have completed other exercises, some of these files might already be present in your plugin.

If they are not already part of your plugin, create the following files in `DITA-OT/plugins/com.company.pdf/cfg/fo/attrs/`:

- **basic-settings.xsl**
- **commons-attr.xsl**
- **pr-domain-attr.xsl**
- **static-content-attr.xsl**
- **sw-domain-attr.xsl**
- **task-elements-attr.xsl**

Be sure to add the appropriate `<xsl:import>` statements to **custom.xsl** (attrs).

If they are not already part of your plugin, create the following files in `DITA-OT/plugins/com.company.pdf/cfg/fo/xsl/`:

- **commons.xsl**
- **pr-domain.xsl**
- **tables.xsl**
- **task-elements.xsl**

Be sure to add the appropriate `<xsl:import>` statements to **custom.xsl** (xsl).

> **Important:** Even though you aren't going to use **task-elements.xsl** in this section, you have to create it in your plugin in order to activate the changes you'll make to **task-elements-attr.xsl**. If you don't add it, you won't see those changes in your PDF. And you'll end up using this file later anyway.

For instructions on creating attribute files and XSLT stylesheets in your PDF plugin, refer to **Create an attribute set file in your plugin** *(p. 31)* and **Create an XSLT stylesheet in your plugin** *(p. 32)*.

If they are not already part of your plugin, create the following files in `DITA-OT/plugins/com.company.pdf/cfg/common/vars/`:

- **en.xml**
- **strings-en-us.xml**

For instructions on creating variables and strings files, refer to **Create a localization variables file** *(p. 78)* and **Create a localization strings file** *(p. 79)*.

Be sure to add the appropriate language and locale mapping statements in **strings.xml**. You can find instructions in **Mapping strings files to xml:lang values in the strings.xml file** *(p. 37)*.

Title formatting

How topic title attribute sets work

In most desktop publishing applications, you create a set of heading paragraph formats to indicate the hierarchy of the headings. These paragraph formats are often named something like Heading1, Heading2, Heading3, etc. You then manually assign the paragraph format to create the illusion of hierarchy. It's an illusion because, in reality, everything is at the same level in a flat desktop publishing application file.

As a writer, you have to choose the level for each heading. Logically, Heading2 should always follow Heading1, and Heading3 should always follow Heading2. But because you apply the formatting manually, it's easy to end up with inconsistent hierarchies like this one:

Heading 1

Heading 3

Heading 2

Hierarchies like these won't work in DITA. In a map, you create actual, nested hierarchies by nesting topicrefs within other topicrefs. There's no way to skip a level.

If you've worked through the exercises so far, you can probably guess that headings (or in DITA terminology, *topic titles*) are formatted using attribute sets. There's a separate attribute set for each level, and these attribute sets live in **commons-attr.xsl**. Here they are:

- topic.title
- topic.topic.title
- topic.topic.topic.title
- topic.topic.topic.topic.title
- topic.topic.topic.topic.topic.title
- topic.topic.topic.topic.topic.topic.title

These six attribute sets format the titles for up to six levels of topics in a map:

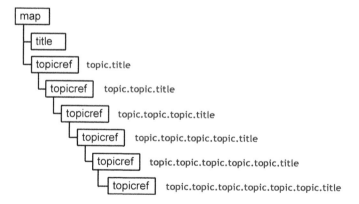

Figure 27: map topic levels and corresponding title attribute sets

> **Note** By default, the title of the map is used on the title page but does not appear elsewhere in the output. The title of the highest-level topicref in the map is considered the top-level title and is styled by the *topic.title* attribute set.

In theory, you can nest topicrefs infinitely, but it rarely makes sense to nest more than a few levels. Six should be more than enough.

Unlike most of the others, these attribute sets are not defined in **commons.xsl**. The PDF plugin defines them in the file **attr-set-reflection.xsl** using a mostly automated process. With very few exceptions, you should never need to change this processing. If you completed the **Change the default alignment for all images** *(p. 309)* exercise, then you're a little bit familiar with this file.

The topic hierarchy is created by your map, and the PDF plugin automatically assigns an attribute set based on this hierarchy. Although these topic title attribute sets are assigned a little differently than other attribute sets, you can edit them just like any of the other attribute sets.

Exercise: Format topic titles

Objective: Determine the appearance of `<title>` elements within topics.

1. Copy the attribute set *topic.topic.title* from `DITA-OT/plugins/org.dita.pdf2/xsl/fo/` **commons-attr.xsl** to your copy of **commons-attr.xsl**.

This attribute set corresponds to the top-level heading in a chapter (not the chapter title itself, which is formatted by *topic.title*). From the specs, you know it needs to be 20pt Trebuchet regular black.

2. Edit the attributes as shown:

```
<xsl:attribute name="space-before">12pt</xsl:attribute>
<xsl:attribute name="space-after">12pt</xsl:attribute>
<xsl:attribute name="font-size">20pt</xsl:attribute>
<xsl:attribute name="font-weight">normal</xsl:attribute>
```

> **Note** There are two *space-before* attributes in this attribute set. Delete one of them so there's no conflict.

You're covered for Trebuchet because this attribute set uses the *common.title* attribute set, which is already set to use the sans-serif font family, which you've set up to be Trebuchet MS (in the exercise **Specify fonts to use** *(p. 83)*. Black is the default color unless you specify otherwise.

Notice this attribute set calls the *common.border__bottom* attribute set. By default, this attribute set defines a border (line) underneath the title, but you don't want a border.

You'd think that removing *common.border__bottom* from the *use-attribute-sets* attribute would do the job, but no. Remember that *common.border__bottom* would still be called in the default copy of *topic.topic.title* in `org.dita.pdf2`. Therefore, you need to override the individual attributes in *common.border__bottom* in your copy of *topic.topic.title*. For details on why this is so, take a look at **How attribute set defaults work** *(p. 56)*.

3. Find the *common.border__bottom* attribute set in `DITA-OT/plugins/org.dita.pdf2/xsl/fo/`**commons-attr.xsl**.

4. Copy the following attributes from it to your *topic.topic.title* attribute set and change the values as shown here:

```
<xsl:attribute name="border-bottom-style">none</xsl:attribute>
<xsl:attribute name="border-bottom-width">0pt</xsl:attribute>
```

These values override the ones in the default *common.border__bottom* attribute set.

5. Follow this process to edit the *topic.topic.topic.title* (for level-2 titles) *topic.topic.topic.topic.title* (for level-3 headings) and *topic.topic.topic.topic.topic.title* (for level-4 titles) attribute sets to match the specifications in **General text formatting specifications** *(p. 193)*.

Notice that *topic.topic.topic.topic.title* and *topic.topic.topic.topic.topic.title* include the *start-indent* attribute, which takes the value of the *side-col-width* variable. Many attribute sets use this variable to create a standard indent. Your template specifications state that level-3 and level-4 headings should be indented .5in from the left margin.

6. Copy the attribute set *topic.topic.title* from `DITA-OT/plugins/org.dita.pdf2/cfg/fo/attrs/`**commons-attr.xsl** to your copy of **commons-attr.xsl**.

This attribute set corresponds to the top-level heading in a chapter (not the chapter title itself, which is formatted by *topic.title*). From the specs, you know it needs to be 20pt Trebuchet regular black.

7. Open `DITA-OT/plugins/org.dita.pdf2/cfg/fo/attrs/`**basic-settings.xsl** and find the *side-col-width* variable.

8. Copy it to your **basic-settings.xsl**.

9. Change its value to .5in.

10. Save changes, run a PDF build, and check your work.

Exercise: Format section titles

Objective: Determine the appearance of `<title>` elements within `<section>` elements.

Now that you've seen how to format topic titles, you can probably guess that you do something similar to format section titles. Your specification says that section titles are 12pt Trebuchet bold dark red.

1. Open `DITA-OT/plugins/org.dita.pdf2/cfg/fo/attrs/` **commons-attr.xsl** and find the *section.title* attribute set.

2. Copy it to your **commons-attr.xsl**.

3. Add or edit the following attributes:

```
<xsl:attribute name="space-before">12pt</xsl:attribute>
<xsl:attribute name="font-size">12pt</xsl:attribute>
<xsl:attribute name="color">#990033</xsl:attribute>
```

Because *section.title* uses the *common.title* attribute set, which is already set up to use the sans-serif font, you don't need to add the *font-family* attribute to this attribute set. Because of the way this attribute set interacts with some other section-related attribute sets, which you will be changing shortly, you need to specify the color here.

> **Note** DITA section titles are not hierarchical. They are meant to be used only as dividers. Therefore, a section title in a level-4 topic looks the same as a section title in a level-2 topic. If your topic titles get smaller and more indented as you move deeper down the hierarchy, you could end up with section titles that are larger than the parent title and have a smaller indent. If this happens, one solution would be to test for the parent topic's level within the *section.title* attribute set and use `<xsl:choose>` inside your attribute definitions to adjust values as needed. The details are beyond the scope of this book, but you can see an example of an attribute that uses `<xsl:choose>` in the *__force__page__count* attribute set in **commons-attr.xsl**.

4. Save changes, run a PDF build, and check your work.

Exercise: Format chapter, appendix, or part titles

Objective: Determine the appearance of chapter, appendix, or part titles.

When you formatted topic titles (in **Format topic titles** *(p. 197)*), remember the *topic.title* attribute set that you didn't do anything with? That's the one that formats chapter titles, and its time has come.

1. Open `DITA-OT/plugins/org.dita.pdf2/xsl/fo/` **commons-attr.xsl** and find the *topic.title* attribute set.

2. Copy it to your **commons-attr.xsl**.

3. Add or edit the following attributes:

```
<xsl:attribute name="border-bottom">0pt solid black</xsl:attribute>
<xsl:attribute name="font-size">28pt</xsl:attribute>
<xsl:attribute name="font-weight">normal</xsl:attribute>
```

Because *topic.title* uses the *common.title* attribute set, which is already set up to use the sans-serif font, you don't need to add the *font-family* attribute to this attribute set. Black is the default color unless you specify otherwise.

Changing *border-bottom* from 3pt to 0pt is the easiest way to eliminate the border.

> **Important:** Why are you leaving "solid" and "black" in the *border-bottom* specification? The truth is, you probably don't need to. But because the specifications for *border-bottom* in the default copy of the *topic.title* attribute set contain values for style and color, you might as well keep them in your own specification. If you delete them, solid and black specifications will still come in from the default copy of *topic.title*, so you'll end up with them anyway. They don't matter because the 0pt width hides the border altogether, but if you keep them in your own *border-bottom* specification, they will be available to you or anyone else who might work with your plugin in the future.

4. Save changes, run a PDF build, and check your work.

Exercise: Format the chapter, appendix, or part autonumber

Objective: Determine the appearance of the number generated for chapter, appendix, or part titles.

By default, the PDF plugin generates a two-part autonumber for chapter, appendix, and part titles. The autonumber consists of a label (the words Chapter, Appendix, Part, or any alternatives you choose) and the number. Each part has its own attribute set.

1. In `DITA-OT/plugins/org.dita.pdf2/cfg/fo/attrs/` **static-content-attr.xsl**, find these two attribute sets:

   - *__chapter__frontmatter__name__container*: formats the label and the number if no separate formatting is specified for the number
   - *__chapter__frontmatter__number__container*: formats the number

     > **Note** The same two attribute sets also apply to appendix and part autonumbering.

   In your PDF, the chapter title is 20pt Trebuchet regular black. You also want the autonumber to be Trebuchet regular black, but you want it to be 16pt.

2. Copy them into your **static-content-attr.xsl**.

3. In _chapter_frontmatter_name_container, add or edit the attributes as follows:

```
<xsl:attribute name="font-size">16pt</xsl:attribute>
<xsl:attribute name="font-weight">normal</xsl:attribute>
<xsl:attribute name="border-before-style">none</xsl:attribute>
<xsl:attribute name="border-after-style">none</xsl:attribute>
<xsl:attribute name="border-before-width">0pt</xsl:attribute>
<xsl:attribute name="border-after-width">0pt</xsl:attribute>
<xsl:attribute name="color">black</xsl:attribute>
<xsl:attribute name="font-family">Book Antiqua, Times New Roman, Times</xsl:attribute>
```

4. In _chapter_frontmatter_number_container edit the attributes as follows:

```
<xsl:attribute name="font-size">16pt</xsl:attribute>
<xsl:attribute name="font-weight">normal</xsl:attribute>
```

5. Save changes, run a PDF build, and check your work.

Exercise: Change the chapter number label and format

Objective: Use a custom label or a custom numbering scheme (such as Roman numerals) for chapter, appendix, or part titles.

In the previous exercise, you changed the appearance of the chapter autonumber. What if you want to change the wording of the label from Chapter or Appendix to something else? What if you want to use Roman numerals?

1. In DITA-OT/plugins/org.dita.pdf2/xsl/fo/**commons.xsl**, find the template that begins
`<xsl:template name="insertChapterFirstpageStaticContent">`.

2. In that template, find these lines:

```
<xsl:call-template name="getVariable">
   <xsl:with-param name="id" select="'Chapter with number'"/>
```

Lines like these are always a dead giveaway that a variable from either **en.xml** or **strings-en-us.xml** is being inserted. In this case, the variable is in **en.xml**. You need to change that variable's value.

3. In your **en.xml**, find the _Chapter with number_ variable:

```
<variable id="Chapter with number">Chapter <param ref-name="number"/></variable>
```

4. Change the value of this variable to whatever you want to use.

5. Return to DITA-OT/plugins/org.dita.pdf2/xsl/fo/**commons.xsl** and find the template immediately after the **_insertChapterFirstpageStaticContent_** template.

You'll have to scroll a bit. The ***insertChapterFirstpageStaticContent*** template is rather long. The template you're looking for begins

```
<xsl:template match="*[contains(@class, ' bookmap/chapter ')] |
           opentopic:map/*[contains(@class, ' map/topicref ')]" mode="topicTitleNumber"
           ►priority="-1">
```

> **Note** If you completed the exercise **Add chapter, appendix, or part numbers to page numbers (Antenna House, XEP)** *(p. 154)*, this template is already in your copy of **commons.xsl**.

6. Copy the template to your **commons.xsl**.

7. In that template, find the highlighted line:

```
<xsl:template match="*[contains(@class, ' bookmap/chapter ')] |
►opentopic:map/*[contains(@class, ' map/topicref ')]"
►mode="topicTitleNumber" priority="-1">
   ...
   <xsl:number format="1" count="*[contains(@class, ' bookmap/chapter ')]"/>
   ...
</xsl:template>
```

The `number format="1"` controls the kind of number being used.

8. Change the format to whatever you want to use.

Options are 1, I, i, A, and a.

9. Repeat this process in the templates for appendix and part titles.

Those templates are immediately below the section you just edited.

> **Note** If you completed the exercise **Add chapter, appendix, or part numbers to page numbers (Antenna House, XEP)** *(p. 154)*, these template are also already in your copy of **commons.xsl**.

10. Save changes, run a PDF build, and check your work.

Exercise: Put the chapter, appendix or part label, number, and title on the same line

Objective: Place all of the parts of a chapter, appendix, or part title on the same line, rather than on separate lines, which is the default.

By default, the chapter label, number, and title are rendered on three separate lines. That's probably not the most common choice of template designers. It's more common to have the label and number on one line and the title on another or all three items on the same line. Here's how to accomplish that.

1. Copy the template named ***insertChapterFirstpageStaticContent*** from `DITA-OT/plugins/org.dita.pdf2/xsl/fo/`**commons.xsl** to your copy of **commons.xsl**.

This template is fairly long so be sure to copy the whole thing.

2. Find the following section inside ***insertChapterFirstpageStaticContent***:

```
<xsl:when test="$type = 'chapter'">
   <fo:block xsl:use-attribute-sets="__chapter__frontmatter__name__container">
      <xsl:call-template name="getVariable">
         <xsl:with-param name="id" select="'Chapter with number'"/>
         <xsl:with-param name="params">
            <number>
               <fo:block xsl:use-attribute-sets="__chapter__frontmatter__number__container">
                  <xsl:apply-templates select="key('map-id', @id)[1]"
                  ▸mode="topicTitleNumber"/>
               </fo:block>
            </number>
         </xsl:with-param>
      </xsl:call-template>
   </fo:block>
</xsl:when>
```

Notice the nested `<fo:block>` elements. The value of the parameter *theParameters* includes an `<fo:block>`, and the call to ***getVariable*** is surrounded by an `<fo:block>`. Even though the `<fo:block>`s are nested, the number still appears on a separate line because that's the way `<fo:block>` works—it always creates a new line. To keep the number on the same line as the "Chapter" text, you need to put it in an `<fo:inline>` element instead.

3. Change the code above as follows:

```
<xsl:when test="$type = 'chapter'">
   <fo:block xsl:use-attribute-sets="__chapter__frontmatter__name__container">
      <xsl:call-template name="getVariable">
         <xsl:with-param name="id" select="'Chapter with number'"/>
         <xsl:with-param name="params">
            <number>
               <fo:inline xsl:use-attribute-sets="__chapter__frontmatter__number__container">
                  <xsl:apply-templates select="key('map-id', @id)[1]"
                  ▸mode="topicTitleNumber"/>
               </fo:inline>
            </number>
         </xsl:with-param>
      </xsl:call-template>
   </fo:block>
</xsl:when>
```

4. Further down in the same template you will find similar sections that handle appendix and part titles. Make the same change to those sections.

5. Save your changes, run a build, and check your work.

The text "Chapter" and the chapter number are on the same line, as are the text "Appendix" and the appendix number and the text "Part" and the part number. If this is what you want, you can stop here. If you also want to have the chapter title on the same line, keep going.

6. Add the following to the code above:

```
<xsl:when test="$type = 'chapter'">
   <fo:block xsl:use-attribute-sets="__chapter__frontmatter__name__container">
      <xsl:call-template name="getVariable">
         <xsl:with-param name="id" select="'Chapter with number'"/>
         <xsl:with-param name="params">
            <number>
               <fo:inline xsl:use-attribute-sets="__chapter__frontmatter__number__container">
                  <xsl:apply-templates select="key('map-id', @id)[1]"
                     ▸mode="topicTitleNumber"/>
               </fo:inline>
               <xsl:text>: </xsl:text>
               <fo:inline xsl:use-attribute-sets="topic.title">
                  <xsl:value-of select="*[contains(@class, ' topic/title ')]"/>
               </fo:inline>
            </number>
         </xsl:with-param>
      </xsl:call-template>
   </fo:block>
</xsl:when>
```

This new `<fo:inline>` element inserts a copy of the chapter title on the same line as "Chapter" and the number using the same attribute set (*topic.title*) used to format the original chapter title. It puts a colon and space between the number and the title.

7. Add the same code to the sections of the ***insertChapterFirstpageStaticContent*** template that process appendix and part titles.

Now you need to hide the original chapter title—the one that's on a separate line. Otherwise the title will be repeated. However, you don't want to remove it altogether because there's some other important processing in that template.

8. In your copy of **commons-attr.xsl**, copy the existing *topic.title* attribute set and paste the copy directly below the original.

9. Rename the copy *topic.title.hide*.

10. Edit *topic.title.hide* as follows:

```
<xsl:attribute-set name="topic.title.hide">
   <xsl:attribute name="border-after-width">0pt</xsl:attribute>
   <xsl:attribute name="space-before">0pt</xsl:attribute>
   <xsl:attribute name="space-after">30pt</xsl:attribute>
   <xsl:attribute name="font-size">2pt</xsl:attribute>
   <xsl:attribute name="font-weight">normal</xsl:attribute>
   <xsl:attribute name="padding-top">0pt</xsl:attribute>
   <xsl:attribute name="keep-with-next.within-column">always</xsl:attribute>
   <xsl:attribute name="border-bottom">0pt solid black</xsl:attribute>
```

```
    <xsl:attribute name="color">#ffffff</xsl:attribute>
    <xsl:attribute name="line-height">4pt</xsl:attribute>
</xsl:attribute-set>
```

What you're doing here is making the original title very small (2pt) and white with no space above it. It's as close to not being there as it can be. The 30pt *space-after* is to create space between the title that's visible and the mini-TOC.

> **Note** There are more elegant ways to eliminate the original title. However, all of those methods change more code to accommodate the other processing associated with titles: bookmarks, index entries, etc. This method may not be the prettiest, but it is the easiest.

11. Copy the three templates named *processTopicChapter*, *processTopicAppendix*, and *processTopicPart* from `DITA-OT/plugins/org.dita.pdf2/xsl/fo/`**commons.xsl** to your copy of **commons.xsl**.

These templates are also fairly long, so be careful to copy them in their entirety.

12. Find the following section, deep inside the *processTopicChapter* template:

```
<xsl:call-template name="insertChapterFirstpageStaticContent">
    <xsl:with-param name="type" select="'chapter'"/>
</xsl:call-template>

<fo:block xsl:use-attribute-sets="topic.title">
    <xsl:call-template name="pullPrologIndexTerms"/>
    <xsl:for-each select="child::*[contains(@class,' topic/title ')]">
        <xsl:apply-templates select="." mode="getTitle"/>
    </xsl:for-each>
</fo:block>
```

13. Change this line to call *topic.title.hide*:

```
<fo:block xsl:use-attribute-sets="topic.title.hide">
```

14. Find the same section in the *processTopicAppendix* template:

```
<xsl:call-template name="insertChapterFirstpageStaticContent">
    <xsl:with-param name="type" select="'appendix'"/>
</xsl:call-template>

<fo:block xsl:use-attribute-sets="topic.title">
    <xsl:call-template name="pullPrologIndexTerms"/>
    <xsl:for-each select="child::*[contains(@class,' topic/title ')]">
        <xsl:apply-templates select="." mode="getTitle"/>
    </xsl:for-each>
</fo:block>
```

15. Change the `<fo:block>` line to call *topic.title.hide*.

16. Find the same section in the ***processTopicPart*** template:

```
<xsl:call-template name="insertChapterFirstpageStaticContent">
    <xsl:with-param name="type" select="'part'"/>
</xsl:call-template>

<fo:block xsl:use-attribute-sets="topic.title">
    <xsl:call-template name="pullPrologIndexTerms"/>
    <xsl:for-each select="child::*[contains(@class,' topic/title ')]">
        <xsl:apply-templates select="." mode="getTitle"/>
    </xsl:for-each>
</fo:block>
```

17. Change the `<fo:block>` line to call *topic.title.hide*.

18. Save changes, run a PDF build, and check your work.

Text formatting

Formatting body text

There's really no such thing as body text when you're formatting DITA topics. Content in the body of a topic might be in a paragraph, a note, a code sample, or some other element, each of which can have its own format.

Some attribute sets let you apply formatting broadly. Examples include *body, body__toplevel, body__secondLevel, conbody*, and *refbody*. You might find these sets are too broad, but give them a chance and see if using one can save effort. A handy thing to do is to assign a noticeable color, such as red, to one of these attribute sets, then test to see exactly what the set affects. You might be able to do a lot of your formatting in one fell swoop and then fine-tune with more specific attribute sets such as *p, note*, or *cmd*.

For example, your template specifications call for most body text to be dark gray. You might add an attribute to set the color to the *body, body__toplevel*, and *body__secondLevel* attribute sets. This will makes almost everything that's not a title dark gray. If there is any body text you want to be some other color, you can use the attribute set specific to that element to format that element's color.

Most of the body attribute sets are in **commons.xsl**, and they are pretty intuitively named. Attribute sets that apply to elements in task topics are in **task-elements-attr.xsl**, and attribute sets that apply to domain elements are in the attribute set files with the same domain names.

If you are converting from an existing, word-processor template, try matching each of your current template styles with the closest DITA element. This will this help you determine which attribute sets you will need to edit.

Here's a simple example of a style-to-element match-up:

Template style	DITA element
Body	p
Bulleted	ul > li
Caution	note type="caution"
CellBody	tbody > entry > p
CellHeading	thead > entry
CodeSample	codeblock
Footnote	fn
Heading1	topic.topic.title
Heading2	topic.topic.topic.title
HeadingRunIn	topic.topic.topic.topic.title
ImageTitle	fig > title
Indented	p > p
Message	msgblock
Note	note type="note"
Numbered1	ol > li
Numbered2	ol > li > ol > li
TableTitle	table > title
Title	title
Warning	note type="warning"

Template style	DITA element
Bold	b
FieldName	uicontrol
FunctionKey	uicontrol
Italics	i
Underlined	u
WindowName	wintitle

Sometimes there is a perfect one-to-one match between template style and DITA element, and sometimes there is not. Overlaps or ambiguities will help you identify instances where the match is not perfect.

For example, you might be using the Numbered paragraph styles for steps in tasks and also for items in ordered lists. Evaluating the situation, you might decide that it's okay to uniformly convert all instances of Numbered1 to either cmd or li. If that won't work, you may need to handle the differences manually, either in the current template before conversion or as a manual step in your conversion. Regardless, an early analysis will help your conversion go more smoothly.

Exercise: Set body font sizes

Objective: Set the font size for particular elements via their corresponding attribute sets.

Your template specification calls for a large body font size of 11pt and a small body font size of 10pt. While you might have to fine-tune individual attribute sets, the plugin already sets up the primary font size across the board. Here's how.

Many of the body attribute sets call the *base-font* attribute set, which uses the *default-font-size* variable. In **basic-settings.xsl**, *default-font-size* is set to 10pt. So by default, most of the body text in your PDF is 10pt. You want it to be 11pt with a few exceptions. You specify that by changing the value of *default-font-size*.

1. Copy the *default-font-size* variable from
DITA-OT/plugins/org.dita.pdf2/cfg/fo/attrs/**basic-settings.xsl** to your copy of
basic-settings.xsl.

2. Change the value of *default-font-size* from 10pt to 11pt.

That takes care of most of the body text. Now you need to deal with the exceptions in your specification. Those exceptions are *note*, *info*, *stepxmp*, *stepresult*, and *choice*.

3. Copy the *note* attribute set from `DITA-OT/plugins/org.dita.pdf2/cfg/fo/attrs/`**commons-attr.xsl** to your copy of **commons-attr.xsl**.

4. Add the *font-size* attribute with a value of 10pt.

5. Copy the *info*, *stepresult*, *stepxmp*, and *choices.choice* attribute sets from `DITA-OT/plugins/org.dita.pdf2/cfg/fo/attrs/`**task-elements-attr.xsl** to your copy of **task-elements-attr.xsl**:

6. Add the *font-size* attribute with a value of 10pt to each of those attribute sets.

7. Save changes, run a PDF build, and check your work.

Exercise: Format code samples and messages

Objective: Determine the appearance of text in the `<codeph>`, `<codeblock>`, `<msgph>`, and `<msgblock>` elements.

You probably use the `<codeph>` and `<codeblock>` elements for code samples. Likewise, you probably use `<msgph>` and `<msgblock>` elements for system messages. You use two different domain attribute set files to format these elements.

1. Copy the *codeph* and *codeblock* attribute sets from `DITA-OT/plugins/org.dita.pdf2/cfg/fo/attrs/`**pr-domain-attr.xsl** to your copy of **pr-domain-attr.xsl**.

2. Add the following attribute to both attribute sets.

```
<xsl:attribute name="font-size">9pt</xsl:attribute>
```

You also want `<codeblock>` to have a light gray background.

3. In the *codeblock* attribute set, change the value of the *background-color* to "#e6e6e6":

4. Add the *background-color* attribute to the *codeph* attribute set, with the same value.

5. Copy the *msgph* and *msgblock* attribute sets from `DITA-OT/plugins/org.dita.pdf2/cfg/fo/attrs/`**sw-domain-attr.xsl** to your copy of **sw-domain-attr.xsl**.

6. Add the following attributes to the *msgph* attribute set:

```
<xsl:attribute name="font-size">9pt</xsl:attribute>
<xsl:attribute name="background-color">#e6e6e6</xsl:attribute>
```

7. Add the following attributes to the *msgblock* attribute set:

```
<xsl:attribute name="font-family">Consolas, Courier New, Courier</xsl:attribute>
<xsl:attribute name="font-size">9pt</xsl:attribute>
<xsl:attribute name="background-color">#e6e6e6</xsl:attribute>
```

8. Save changes, run a PDF build, and check your work.

Exercise: Set the default line height

Objective: Determine the default line height, or leading, that will be used for all text within your PDF, unless overridden by a specific line height setting in a particular attribute set.

Line height is the space between the lines of text in a text block, such as a paragraph. It's also known as *leading* if you're old school. In the PDF plugin, you can set line height as a fixed number, such as 14pt, or as a percentage. The percentage method is handy because it automatically recalculates the appropriate line height based on the font size (11pt text looks nice with a 14pt line height; 18pt text looks a little crowded with 14pt line height). Your template specification calls for a line height of 120%. Fortunately, you don't have to add this element-by-element. You can set the default in one place.

1. Copy the *default-line-height* variable from
DITA-OT/plugins/org.dita.pdf2/xsl/fo/**basic-settings.xsl** to your copy of **basic-settings.xsl**.

2. Change the value of the *default-line-height* variable from 12pt to 120%.

Now you need to apply this attribute to your body text. By default, the plugin uses this variable only in the *section* and *example* attribute sets.

3. Copy the *base-font* attribute set from
DITA-OT/plugins/org.dita.pdf2/xsl/fo/**commons-attr.xsl** to your copy of **commons-attr.xsl**.

4. Add the following attribute and value:

```
<xsl:attribute name="line-height"><xsl:value-of select="$default-line-height"/></xsl:attribute>
```

You'll recall that the three attribute sets that format most of the body text (*body, body__toplevel, body__secondLevel*) all call the *base-font* attribute set, so this is a quick way to apply the default line height almost across the board. As always, you'll have to fine-tune some attribute sets.

5. Save changes, run a PDF build, and check your work.

Exercise: Format trademark, registered trademark, service mark, and copyright symbols

Objective: Determine the appearance of the symbols created using the `<tm>` element as well as the copyright symbol.

Using the `<tm>` element along with the appropriate *tmtype* attribute value, you can create service mark (SM), registered trademark (®), and trademark (™) symbols. By default, those symbols are output superscripted above and smaller than the associated text. If you'd prefer to output them inline, that's an easy change to make, although it's not completely straightforward.

> **Tip:** You can't use `<tm>` to create a copyright symbol (©). For that, you either need to use the literal character or the numeric character reference (© or ©, for instance).

1. Copy the following attribute sets from `DITA-OT/plugins/org.dita.pdf2/cfg/fo/attrs/`**commons-attr.xsl** to your copy of **commons-attr.xsl**:

   tm, tm__content, tm__content__service

   Notice that *tm__content* and *tm__content__service* both specify that the *font-size* should be 75% and 40%, respectively, of the size of the surrounding text. (The reason that the service mark is output at a smaller percentage is that the service mark is actually not a character but the literal text "SM". The characters are smaller to begin with but "SM" starts out at the same size as the surrounding text and so has to be downsized a little more to remain proportional.)

   Notice also that *tm__content* and *tm__content__service* both specify that the *baseline-shift* should be 20% and 50%, respectively, meaning that the trademark and registered trademark symbols are output 20% higher than the surrounding text, while the service mark is output 50% higher. (Again, the difference is due to the fact that trademark and registered trademark are symbols that are offset to begin with, while the service mark is literal text that is not offset to begin with.)

2. Edit *tm__content* and *tm__content__service* as follows:

   ```
   <xsl:attribute name="font-size">100%</xsl:attribute>
   <xsl:attribute name="baseline-shift">0%</xsl:attribute>
   ```

   These values specify that the trademark and registered trademark characters and the service mark text are output at the same size and on the same baseline as the surrounding text.

3. Save your changes and run a build to check your work.

You'll notice that while the service mark is now aligned along the same baseline as the surrounding text, the trademark and registered trademark characters are still superscripted. Remember that these characters are designed to be offset to begin with, so you need to bring them down to the surrounding text's baseline.

4. In *tm__content*, change the value of *baseline-shift* to -20%:

5. Save your changes and run a build to check your work.

The registered trademark is now aligned along the same baseline as the surrounding text but the trademark symbol is still superscripted. This this is due to the way the characters are designed in this font. If you want to align the trademark symbol along the surrounding text baseline, you'll need to create a new attribute set for it, and you'll need to use that new attribute set specifically for the trademark symbol. Fortunately, the code is already set up to make that an easy change.

6. In your **commons-attr.xsl**, add a new attribute set named *tm__content__tm* and add the *font-size* and *baseline-shift* attributes with appropriate values.

You might need to experiment a little bit to find the right percentage for *baseline-shift*.

7. Copy the template that begins with the following from DITA-OT/plugins/org.dita.pdf2/xsl/fo/**commons.xsl** to your copy of **commons.xsl**:

```
<xsl:template match="*[contains(@class, ' topic/tm ')]">
```

This template includes an `<xsl:choose>` that tests for the value of the *tmtype* attribute.

8. Find the `<xsl:when>` test that looks for `tmtype='tm'` and change the resulting text to use the new *tm__content__tm* attribute set:

```
<xsl:when test="@tmtype='tm'">
    <fo:inline xsl:use-attribute-sets="tm__content__tm">&#8482;</fo:inline>
</xsl:when>
```

9. Save your changes and run a build to check your work.

The trademark symbol is now aligned along the same baseline as the surrounding text.

If you're keeping score at home, you've realized that nothing here affects the copyright symbol, because it's not created using `<tm>`. So what do you do if you want to make it the same size and alignment as the surrounding text? For that, there are two approaches.

If you are using font-mapping, continue with step 10 below.

If you are not using font-mapping, your best bet is to wrap the copyright symbol in an inline element such as `<ph>`, assign an outputclass to that element, and conditionalize the corresponding attribute set to format the copyright symbol appropriately.

10. Open your **font-mappings.xml**.

In that file, for each of the three logical fonts (Sans, Serif, and Monospace) there is a section like this:

```
<physical-font char-set="SymbolsSuperscript">
    <font-face>Trebuchet MS, Helvetica, Arial Unicode MS</font-face>
    <baseline-shift>20%</baseline-shift>
    <override-size>smaller</override-size>
</physical-font>
```

This is the section that determines formatting for any special characters rendered by that logical font.

11. Do one of the following:

- If it's okay to affect all characters, not just the copyright symbol, edit `<baseline-shift>` and `<override-size>` as follows for all three logical fonts:

```
<baseline-shift>0%</baseline-shift>
<override-size>inherit</override-size>
```

   Now go to step continue to step **15**.

- If you want to affect only the copyright symbol, remove it from the list of characters affected by the SymbolsSuperscript character set.

   To do so, go to step continue to step **12**.

12. Copy **en.xml** from `DITA-OT/plugins/org.dita.pdf2/cfg/fo/i18n/` to the i18n subfolder of your plugin.

> **Note** This is the only time in these exercises that you'll use one of the i18n files. You don't need to know a lot about them except that they pertain to internationalization and localization, and they define character sets and encodings for different languages.

13. Open **en.xml** and find this section:

```
<alphabet char-set="SymbolsSuperscript">
    <character-set>
        <!-- Copyright -->
        <character>&#169;</character>
        <!-- Registered Trademark -->
        <character>&#174;</character>
        <!-- Trademark -->
        <character>&#8482;</character>
        <!-- Service mark -->
        <character>&#2120;</character>
    </character-set>
</alphabet>
```

This section lists all of the characters affected by the SymbolsSuperscript character set within each logical font. Notice that trademark, registered trademark, and service mark are listed here as well.

Because they have attribute sets that directly affect their formatting, it's best to edit those attribute sets rather than use **en.xml** to change their formatting. You should use **en.xml** only for the copyright symbol.

14. Comment out `<character>&#169;</character>`.

15. Save changes, run a PDF build, and check your work.

The copyright symbol is the same size and aligned along the same baseline as the surrounding text.

Notes formatting

Exercise: Format note labels

Objective: Determine the appearance of the labels that are automatically generated for `<note>` elements, such as "Note" or "Tip".

According to your template specification, note labels should be dark red.

1. Copy the *note__label* attribute set from `DITA-OT/plugins/org.dita.pdf2/cfg/fo/attrs/`**commons-attr.xsl** to your copy of **commons-attr.xsl**.

Notice there are also separate attribute sets for each note type: *note__label__note*, *note__label__notice*, *note__label__tip*, and so on. **commons.xsl** is set up so that the template that places notes contains a top-level `<fo:inline>` that uses *note__label*. Nested within this `<fo:inline>` are separate `<fo:inline>` elements that place each individual note-type label and format it using the more specific attribute set. Unless you need separate formatting for certain note-type labels, you can format them all using the top-level `<fo:inline>`.

2. Add the following attribute to *note__label*:

```
<xsl:attribute name="color">#990033</xsl:attribute>
```

3. Save changes, run a PDF build, and check your work.

Exercise: Change the label for a note type

Objective: Specify the text for labels that are generated for `<note>` elements, such as "Note" or "Tip".

Your template specification defines the label for notes of type **fastpath** to be "Quick access." This is an easy change.

> **Note** There is a separate variable for each note type except **other**. By default, the PDF plugin uses whatever value you assigned to *othertype* as the label for **other** notes. If you don't want a label for **other** notes, there's no way to delete it in the localization variables file, so you have to change the way the element is processed in **commons.xsl**.

> **Important:** The label variable for notes of type **notice** is still found in the **en.xml** localization variables file. To change this variable, copy it to your copy of **en.xml**. You will find all of the other note label variables in the **strings-en-us.xml** file. This seems to be an oversight and may change soon.

1. Copy the *Fastpath* variable from `DITA-OT/xsl/common/`**strings-en-us.xml** to your copy of **strings-en-us.xml**.

2. Enter the new value for the variable.

```
<str name="Fastpath">Quick access</str>
```

3. Save changes, run a PDF build, and check your work.

Repeat these steps for any other localization variables or strings files you've customized.

Exercise: Delete the label for a note type

Objective: Specify that certain note types should not include a text label.

Your specification calls for no label on notes of type "note", because you're going to use an icon instead.

1. Open `DITA-OT/xsl/common/`**strings-en-us.xml** and find the strings similar to the following:

```
<str name="Note">Note</str>
```

There are strings for note, tip, caution, warning, danger, restriction, attention, remember, important, fastpath, and trouble. The string, or variable for notice is found in the localization variables file **en.xml**. You might think you could just delete the value from the string, but that wouldn't work completely.

2. Copy the template that begins with the following from `DITA-OT/plugins/org.dita.pdf2/xsl/fo/` **commons.xsl** to your copy of **commons.xsl**:

```
<xsl:template match="*" mode="placeNoteContent">
```

This code inserts notes, including the label and the divider between the label and the text (by default, a colon). It's rather long, so be sure to copy the whole thing.

3. Find the following section:

```
<xsl:when test="@type='note' or not(@type)">
    <fo:inline xsl:use-attribute-sets="note__label__note">
        <xsl:call-template name="getVariable">
            <xsl:with-param name="id" select="'Note'"/>
        </xsl:call-template>
    </fo:inline>
</xsl:when>
```

This section inserts the label defined in the *Note* variable when the *type* attribute is set to **note** or doesn't exist (if it doesn't exist, the default is the same as setting *type* to **note**). You need to change this section so nothing is inserted.

4. Comment out the entire `<fo:inline>` section:

```
<xsl:when test="@type='note' or not(@type)">
    <!--<fo:inline xsl:use-attribute-sets="note__label__note">
        <xsl:call-template name="getVariable">
            <xsl:with-param name="id" select="'Note'"/>
        </xsl:call-template>
    </fo:inline>-->
</xsl:when>
```

You've eliminated the label, but you're still stuck with the colon. Here's how to get rid of it.

5. Scroll down to the end of the template and find this code:

```
<xsl:call-template name="getVariable">
    <xsl:with-param name="id" select="'#note-separator'"/>
</xsl:call-template>
```

This code inserts the colon after the label. It applies to all notes, regardless of the type. This would normally be convenient, unless you need to handle different note types differently, which you need to do in this case. But it's not that hard to handle this situation.

6. Wrap the `<xsl:call-template>` in an `<xsl:choose>`:

```
<xsl:choose>
    <xsl:when test="@type='note' or not(@type)">
        <!--don't insert separator-->
    </xsl:when>
    <xsl:otherwise>
        <xsl:call-template name="getVariable">
            <xsl:with-param name="id" select="'#note-separator'"/>
        </xsl:call-template>
    </xsl:otherwise>
</xsl:choose>
```

This `<xsl:choose>` statement tests whether the *type* attribute is set to "note" or doesn't exist. If either of these conditions is true, you don't want the separator, so you do nothing. In every other case, you want the separator, so the `<xsl:call-template>` is executed.

7. Save changes, run a PDF build, and check your work.

Exercise: Change or delete note icons

Objective: Use an icon other than the default **warning.gif** for warning notes, or use no icon at all.

The Open Toolkit treats `<note>` elements differently depending on whether there is an associated icon or not. Having an associated icon means that there is a path to an icon in the localization variable—for example, *warning Note Image Path*—that corresponds to the note type. If there is an associated icon, the Open Toolkit creates a two-column table with the icon on the left and the note text on the right. The style for the note text comes primarily from the *note* attribute set. However, if there is an icon, the note text is also styled by the *note__text__entry* attribute set in addition to the *note* attribute set. The table that surrounds everything is styled using the *note__table* attribute set.

By default, the plugin uses some basic images as note icons. You may have other icons you'd rather use, or you may not want icons at all.

`DITA-OT/plugins/com.company.pdf/cfg/common/artwork/cfg/common/artwork` contains note icons and other images. By keeping your icons here, you can easily move the entire plugin and images as a single, compact package.

1. Copy all of the variables similar to the following from `DITA-OT/plugins/org.dita.pdf2/cfg/common/vars/`**en.xml** to your copy of **en.xml**:

```
<!-- Image path to use for a note of "note" type. -->
<variable id="note Note Image Path">Configuration/OpenTopic/cfg/common/artwork/hand.gif</variable>
```

There are variables for all the note types: note, tip, caution, danger, restriction, attention, remember, important, fastpath, notice, trouble, and other.

2. Do one of the following:

- To change the icon, enter the new path and image file name for the appropriate variable(s).

  For example, to use an icon named **mywarning.png** for warning notes, the new path would be:

  ```
  <variable id="warning Note Image
  Path">Customization/OpenTopic/common/artwork/mywarning.png</variable>ble>
  ```

  > **Important:** Notice that the new path includes "Customization", rather than "Configuration", and that the cfg path level is no longer there.

- To remove the icon, delete the value for the appropriate variable(s).

  For example:

  ```
  <variable id="warning Note Image Path"/>
  ```

3. Save changes, run a PDF build, and check your work.

Repeat these steps for any other note icons you want to replace or remove. Then repeat the whole thing for each localization variables file you've customized.

Exercise: Format note text

Objective: Determine the appearance of text in the <note> element.

Before you start

Complete the exercise **Change or delete note icons** *(p. 217)*, which pertains to notes without icons.

Your template specification says that note text should be 9pt, justified. And it should be indented from the surrounding body text .5in on the left and right.

1. Copy the *note* attribute set from DITA-OT/plugins/org.dita.pdf2/cfg/fo/attrs/**commons-attr.xsl**, to your copy of **commons-attr.xsl**.

2. Add the following attributes and values:

```
<xsl:attribute name="font-size">9pt</xsl:attribute>
<xsl:attribute name="margin-right">.5in</xsl:attribute>
<xsl:attribute name="margin-left">.5in</xsl:attribute>
<xsl:attribute name="text-align">justify</xsl:attribute>
```

3. Save changes, run a PDF build, and check your work.

Exercise: Add left and right borders to notes

Objective: Add a vertical line to the left and right of notes.

Your template specifications call for a 2pt dark red double-line border to the left and right of notes.

1. If it is not already present, copy the *note* attribute set from `DITA-OT/plugins/org.dita.pdf2/cfg/fo/attrs/`**commons-attr.xsl** to your copy of **commons-attr.xsl**.

2. Add the following attributes and values to the *note* attribute set:

```
<xsl:attribute name="border-left">2pt double #990033</xsl:attribute>
<xsl:attribute name="border-right">2pt double #990033</xsl:attribute>
<xsl:attribute name="padding-left">5pt</xsl:attribute>
<xsl:attribute name="padding-right">5pt</xsl:attribute>
```

The padding adds a little space between the border and the text, which looks a lot better.

3. Save changes, run a PDF build, and check your work.

Exercise: Use a custom label for other-type notes

Objective: Specify a custom label for use with `<note>` elements whose *type* attribute is set to "other".

The *type* attribute on the note element has a large set of pre-defined values you can choose from, each which has a default label. If you need a note type that's not in the list, set the value of *type* to "other" and the value of *type* to the label you want. That works fine if you want the label to be the same as the value of *type*. But what if you want the label to be different?

Your template specification calls for a custom note label: "Best Practice." Here's how to set that up.

1. If it is not already present copy the template that begins with the following from `DITA-OT/plugins/org.dita.pdf2/xsl/fo/`**commons.xsl** to your copy of **commons.xsl**:

```
<xsl:template match="*" mode="placeNoteContent">
```

This template contains an `<xsl:choose>` statement with an `<xsl:when>` test for each note type. Within each test, the template generates an `<fo:inline>` statement with the attribute set for that note type, then calls a template that inserts a variable specific to that note type. The last `<xsl:when>` test is for `type="other"`. This is the section you need to edit.

2. Find the section that begins `<xsl:when test="@type='other'">`.

By default, the stylesheet is set up to insert the value of the *othertype* attribute as the note prefix and if there is no value, to insert the value of the *type* attribute instead. You want to change this to insert your custom text instead of the value of the *othertype* attribute when that value is "bestpractice".

3. Add the highlighted code as shown below:

```
<xsl:when test="@type='other'">
    <fo:inline xsl:use-attribute-sets="note__label__other">
        <xsl:choose>
            <xsl:when test="@othertype='bestpractice'">
                <xsl:text>Best Practice</xsl:text>
            </xsl:when>
            <xsl:when test="@othertype">
                <xsl:value-of select="@othertype"/>
            </xsl:when>
```

4. Save your changes, run a build, and check your work.

This works like a charm, but there's a potential problem with this approach. Hardcoding the label text like this makes translation difficult. A more flexible approach is to use a localization variable.

5. Open your **strings-en-us.xml** and add the following variable:

```
<str name="Best Practice">Best practice</str>
```

6. Return to the *placeNoteContent* template in **commons.xsl** and replace the code you added in Step 3 with the highlighted code shown below:

```
<xsl:when test="@type='other'">
    <fo:inline xsl:use-attribute-sets="note__label__other">
        <xsl:choose>
            <xsl:when test="@othertype='bestpractice'">
                <xsl:call-template name="getVariable">
                    <xsl:with-param name="id" select="'Best Practice'"/>
                </xsl:call-template>
            </xsl:when>
            <xsl:when test="@othertype">
                <xsl:value-of select="@othertype"/>
            </xsl:when>
            <xsl:otherwise>
                <xsl:text>[</xsl:text>
                    <xsl:value-of select="@type"/>
                <xsl:text>]</xsl:text>
            </xsl:otherwise>
        </xsl:choose>
    </fo:inline>
</xsl:when>
```

When `othertype="bestpractice"`, this code inserts the *Best Practice* variable value instead of the value of *othertype*.

Notice this test comes before the more general test for the presence of *othertype*. The `<xsl:when>` tests are performed in order. Once the first match is found, all subsequent tests are skipped. So a note with `othertype="bestpractice"` satisfies the first `<xsl:when>` test, and the label will be used as the variable value. If the first two tests were reversed, a note with `othertype="bestpractice"` would fit the first condition, and the note label would be set to the value of *othertype* instead of the variable value.

7. Save changes, run a PDF build, and check your work.

Be sure to add the *Best Practice* string to any other strings files in your plugin.

Other things you can do

Here are a few additional customizations you might want to make to your plugin. These customizations aren't part of the PDF specifications you're following, but they can be useful (and cool).

Exercise: Mark non-standard items for resolution

Objective: Apply an attribute to an element in a specific context to make it easy to spot when that element has been used in a non-standard way.

Your template specifications call for four heading levels. But there's nothing to keep a writer from creating more levels than that. Since the DITA standard allows more than four heading levels, you might want to flag this situation in the output. That way, writers can easily see when they've nested too deeply.

By applying special formatting to elements that are used in a non-standard way, you can create visual cues in this situation and other situations where the DITA standard allows something that your style standards don't. This approach, of course, is only one of many ways to handle content model violations. A much more sophisticated approach would be to use an application such as **Acrolynx** to catch these violations during the authoring process.

Here's one way to mark topic titles that exceed the four-level limit.

1. Open `DITA-OT/plugins/org.dita.pdf2/cfg/fo/attrs/`**commons-attr.xsl** and copy the *topic.topic.topic.topic.topic.topic.title* attribute set to your **commons-attr.xsl**.

You remember that the other five topic-title-related attribute sets format the chapter heading and the four allowed topic-level headings. This attribute set formats a fifth, disallowed, topic-level heading. If an author creates a map that includes this level, you want to flag that heading.

2. Add the following attribute to that attribute set:

```
<xsl:attribute name="color">#ff00ff</xsl:attribute>
```

This turns the disallowed level-five heading magenta, making it easy to spot in a PDF. (If you actually use magenta in your PDF [!], pick a different color.)

3. Save changes, run a PDF build, and check your work.

Instead of marking with a color, you could also set the *text-decoration* attribute to "strikethrough", "underline", or "overline". You just want to make the text instantly noticeable.

> **Important:** There's a caveat with this method. If you're going to use it, you need to be sure you have robust quality checks in place to catch any of this flagged text before publication. You don't want your readers to get PDFs with struck-through, magenta topic titles!

Exercise: Eliminate the chapter, appendix, or part label and number

Objective: Specify that your plugin should not automatically generate a label or number for chapter, appendix, or part titles.

If you don't want labels and autonumbering on chapter titles, it's easy to remove them.

1. To eliminate the label, copy the following variables from `DITA-OT/plugins/org.dita.pdf2/cfg/common/vars/`**en.xml** to your copy of **en.xml**:

- *Chapter with number*
- *Appendix with number*
- *Part with number*

2. Delete the text (Chapter, Appendix, Part) from the variables, leaving just the *number* parameter:

```
<variable id="Chapter with number"><param ref-name="number"/></variable>
```

If you haven't made any changes to the *number* parameters in the ***insertChapterFirstpageStaticContent*** template, you can just delete the *number* parameter if you don't want the chapter/appendix/part number either.

However, if you completed any of the previous exercises that involved the *number* parameters in the *insertChapterFirstpageStaticContent* template, then the *number* parameter now contains the topic title as well, and if you just delete it, you'll be left with no title. In that case, you need to do something else to eliminate the number.

3. If it is not already present, copy the *insertChapterFirstpageStaticContent* template from `DITA-OT/plugins/org.dita.pdf2/xsl/fo/`**commons.xsl** to your copy of **commons.xsl**.

4. In that template, find the section that begins with the following:

```
<xsl:when test="$type = 'chapter'">
```

5. In that section, find the `<number>` element.

This element corresponds to the number param in the *Chapter with number* variable.

If you completed the **Put the chapter, appendix or part label, number, and title on the same line** *(p. 202)* or **Use a specific title level in the running header** *(p. 172)* exercises, then the `<number>` now has two `<fo:inline>` sections. The first creates a marker that keeps track of the current chapter number and also outputs the current chapter number; and the second outputs the chapter title. You want to keep the second `<fo:inline>` but not the first.

6. Comment out the first `<fo:inline>` section in `<number>` and the `<xsl:text>` line as shown:

```
<number>
    <!--<fo:inline xsl:use-attribute-sets="__chapter__frontmatter__number__container">
        <xsl:apply-templates select="key('map-id', @id)[1]" mode="topicTitleNumber"/>
    </fo:inline>
    <xsl:text>: </xsl:text>-->
    <fo:inline xsl:use-attribute-sets="topic.title">
        <xsl:for-each select="*[contains(@class,' topic/title ')]">
            <xsl:apply-templates select="." mode="getTitle"/>
        </xsl:for-each>
    </fo:inline>
</number>
```

(Your code might look slightly different depending on which exercises you've completed.)

7. Repeat this process to comment out the number for appendices and parts as well.

You will find the `<xsl:when>` tests for appendix and part immediately below the one for chapter.

8. Save changes, run a PDF build, and check your work.

Exercise: Format note labels as sideheads

Objective: Place note labels to the left of and aligned with the first line of a note rather than inline with the note text.

By default, the `org.dita.pdf2` plugin puts notes with icons in a two-column table, with the icon in the left column and the text in the right column. It puts notes with labels in a single text block, with the label output as an inline element.

Suppose you want to set up notes with labels so they look like the following:

> **Important:**   If you add any of the domain attr
> **sw-domain-attr.xsl**, **ui-domain-a**
> domain elements, you must add

In that case, you need to follow the same model as notes with icons.

This exercise uses the code found in the original **commons.xsl**. If you completed the **Delete the label for a note type** *(p. 215)* exercise, you'll need to make a few changes to some of the following steps to accommodate the code changes made in that exercise.

1. Copy the template that begins with the following from DITA-OT/plugins/org.dita.pdf2/xsl/fo/**commons.xsl** to your copy of **commons.xsl**:

```
<xsl:template match="*[contains(@class,' topic/note ')]" mode="setNoteImagePath">
```

Call this template the "get path" template (in your mind).

2. Copy the template that begins with the following from DITA-OT/plugins/org.dita.pdf2/xsl/fo/**commons.xsl** to your copy of **commons.xsl**:

```
<xsl:template match="*[contains(@class,' topic/note ')]">
```

Call this template the "table note" template (yes, in your mind).

3. Scroll through the "table note" template until you find the following code:

```
<xsl:template match="*[contains(@class,' topic/note ')]">
   <xsl:variable name="noteImagePath">
      <xsl:apply-templates select="." mode="setNoteImagePath"/>
   </xsl:variable>
   <xsl:choose>
   . . .
```

This code creates a variable named *noteImagePath*. To define a value for that variable, the code points to a template that has the *mode* attribute set to the value "setNoteImagePath". The value of *mode* (highlighted in the code above) matches the *mode* attribute in the "get path " template. The mode mechanism allows an element to be processed by different templates, depending on the value of the *mode* attribute on a `<xsl:apply-templates>` or `<xsl:call-template>` statement.

In the "get path" template, the first `<xsl:when>` tests to see if the note has a type value. If so, the template selects that value. If not, it uses the default value for that type of note. It then concatenates the value with the text "Note Image Path" to create a localization variable. For example, if the type value is "trouble" the concatenated result is *trouble Note Image Path*, which you might recognize as the localization variable of the same name in **en.xml**. For now, the template just holds this information; it doesn't actually use it anywhere.

The "table note" template creates a variable named *noteImagePath* and assigns it a value that is the path in the *[note type] Note Image Path* localization variable that the inline note template just came up with. Moving on to the `<xsl:choose>`, the first `<xsl:when>` is invoked if that localization path variable actually contains a value. If so, the template creates a two-column table. It puts the icon specified by the value of *noteImagePath* into the first column and the note text into the second column. If not, the plugin invokes a different template that begins with the following:

```
<xsl:template match="*" mode="placeNoteContent">
```

You worked with that template in the exercise **Use a custom label for other-type notes** *(p. 219)*.

Now, though, you want to create the two-column table in both cases—whether there is an image path or not. When the value of *noteImagePath* is blank or empty, you want to put the label in the left column. When *noteImagePath* has a value, you want to put the icon in the left column.

4. In your copy of **commons.xsl**, go to the "table note" template.

```
...
<xsl:choose>
   <xsl:when test="not($noteImagePath = '')">
      <fo:table xsl:use-attribute-sets="note__table">
         <fo:table-column xsl:use-attribute-sets="note__image__column"/>
         <fo:table-column xsl:use-attribute-sets="note__text__column"/>
         <fo:table-body>
            <fo:table-row>
               <fo:table-cell xsl:use-attribute-sets="note__image__entry">
                  <fo:block>
                     <fo:external-graphic src="url('{concat($artworkPrefix,
                     ▶ $noteImagePath)}')" xsl:use-attribute-sets="image"/>
                  </fo:block>
               </fo:table-cell>
               <fo:table-cell xsl:use-attribute-sets="note__text__entry">
                  <xsl:apply-templates select="." mode="placeNoteContent"/>
               </fo:table-cell>
            </fo:table-row>
         </fo:table-body>
      </fo:table>
   </xsl:when>
   <xsl:otherwise>
      <xsl:apply-templates select="." mode="placeNoteContent"/>
   </xsl:otherwise>
</xsl:choose>
```

5. Cut the entire `<fo:table>` section (highlighted above) and paste it directly below `</xsl:variable>`:

```
...
</xsl:variable>
<fo:table xsl:use-attribute-sets="note__table">
    <fo:table-column xsl:use-attribute-sets="note__image__column"/>
    <fo:table-column xsl:use-attribute-sets="note__text__column"/>
    <fo:table-body>
...
```

You now have a two-column table that will be created in all situations, which is what you want. But you want the content of the first column to be the icon if there is a path in **en.xml** or the label if there is no path. So you're going to need an `<xsl:choose>` in the first column. The good news is, most of the code you need is already in the stylesheet, you just need to move it around.

6. Add an `<xsl:choose>` statement in the first `<fo:table-cell>` as shown below:

```
<fo:table-cell xsl:use-attribute-sets="note__image__entry">
    <xsl:choose>
        <xsl:when>
        </xsl:when>
        <xsl:otherwise>
        </xsl:otherwise>
    </xsl:choose>
```

7. Edit the `<xsl:when>` statement to be the following:

```
<xsl:when test="not($noteImagePath = '')">
```

8. Cut the `<fo:block>` section and paste it inside the `<xsl:when>`:

```
<xsl:when test="not($noteImagePath = '')">
    <fo:block>
        <fo:external-graphic src="url({concat($artworkPrefix, $noteImagePath)})"
        ▸ xsl:use-attribute-sets="image"/>
    </fo:block>
</xsl:when>
```

Now you need to deal with the `<xsl:otherwise>`, which will output the appropriate label. Again, the code is already here, you just need to move it. This time it's a big chunk.

9. Scroll up to the template that begins with the following line:

```
<xsl:template match="*" mode="placeNoteContent">
```

(If this template is not already in your copy of **commons.xsl**, copy it from `DITA-OT/plugins/org.dita.pdf2/xsl/fo/`**commons.xsl**.)

Within that template there is an `<xsl:choose>` block that contains 13 `<xsl:when>` statements, one for each note type.

10. Copy the entire `<xsl:choose>` block.

11. Paste it inside the `<xsl:otherwise>` that you added in step 6.

12. Wrap the entire `<xsl:choose>` in an `<fo:block>` that uses the *note__label* attribute set:

```
<xsl:otherwise>
    <fo:block xsl:use-attribute-sets="note__label">
        <xsl:choose>
            <xsl:when test="@type='note' or not(@type)">
            . . .
        </xsl:choose>
    </fo:block>
</xsl:otherwise>
```

13. In the `<xsl:template match="*" mode="placeNoteContent">` template, comment out the entire `<xsl:choose>` statement that you originally copied:

```
<xsl:template match="*" mode="placeNoteContent">
    <fo:block xsl:use-attribute-sets="note">
        <xsl:call-template name="commonattributes"/>
            <fo:inline xsl:use-attribute-sets="note__label">
            <!--<xsl:choose>
                <xsl:when test="@type='note' or not(@type)">
                    <fo:inline xsl:use-attribute-sets="note__label__note">
                        <xsl:call-template name="getVariable">
                            <xsl:with-param name="id" select="'Note'"/>
                        </xsl:call-template>
                    </fo:inline>
                </xsl:when>
                . . .
            </xsl:choose>-->
            . . .
```

If you leave this code in place, you'll get two labels.

Because you're putting the label in a separate table cell, you probably don't want the separator (:).

14. In the `<xsl:template match="*" mode="placeNoteContent">` template, comment out this section of code:

```
<!--<xsl:call-template name="getVariable">
    <xsl:with-param name="id" select="'#note-separator'"/>
</xsl:call-template>-->
```

15. Also comment out the text line that inserts a blank, which is just below the previous section:

```
</fo:inline>
<!--<xsl:text> </xsl:text>-->
<xsl:apply-templates>
```

16. In the "get path" template, comment out the `<xsl:choose>` block right after the table:

```
        ...
        </fo:table-body>
    </fo:table>

    <!--<xsl:choose>
        <xsl:when test="not($noteImagePath = '')">
        </xsl:when>
        <xsl:otherwise>
            <xsl:apply-templates select="." mode="placeNoteContent"/>
        </xsl:otherwise>
    </xsl:choose>-->
</xsl:template>
```

17. Save changes, run a PDF build, and check your work.

All notes now appear in a two-column table. If there is an icon specified for a note type in **en.xml**, that icon appears in the left column. If no icon is specified, the corresponding label appears in the left column. Most likely, the note text overruns the outside page margin. Play around with the appropriate attribute sets to set the margin correctly. By now, you should be able to figure out which attribute sets to adjust.

> **Tip:** You can also use basic XSL-FO list formatting, and treat the note label like the list bullet. After you work through the chapter on **Lists** *(p. 239)*, see if you can give it a try!

Exercise: Rotate text

Objective: Rotate text a specified number of degrees.

Instead of the usual left-to-right or right-to-left orientation of text, you might want to spice things up a bit by rotating text to appear vertically. Vertical text is especially useful in table headings where the heading is long but the cell contents are brief. A vertical heading saves space. Consider **Figure 28**.

The column heading in the second column is much wider than the actual column contents. Look how much space you'd save if you rotated that column header, like in the third column.

As of the publication date of this book, the DITA Open Toolkit does not support the DITA *rotate* attribute. When that support becomes available, it might or might not resemble the code here, but it would be a good idea to update your plugin to reflect the official processing at that point.

	Average results June-December	Average results June-December
Group A	43.80%	43.80%
Group B	57.10%	57.10%
Group C	48.90%	48.90%

Figure 28: Table with rotated header cells

1. In your DITA content, ensure that each entry you want to rotate is marked `@rotate="1"`.

   The *rotate* attribute is new in DITA 1.3. It is not available in earlier versions of DITA. If you are working with an earlier version of DITA, you can use *outputclass* instead.

2. Copy the following template from `DITA-OT/plugins/org.dita.pdf2/xsl/fo/`**tables.xsl** to your copy of **tables.xsl**:

```
<xsl:template match="*[contains(@class, ' topic/thead ')]/*[contains(@class, ' topic/row ')]
▶/*[contains(@class, ' topic/entry ')]">
   <fo:table-cell xsl:use-attribute-sets="thead.row.entry">
      <xsl:call-template name="commonattributes"/>
      <xsl:call-template name="applySpansAttrs"/>
      <xsl:call-template name="applyAlignAttrs"/>
      <xsl:call-template name="generateTableEntryBorder"/>
         <fo:block xsl:use-attribute-sets="thead.row.entry__content">
            <xsl:apply-templates select="." mode="ancestor-start-flag"/>
               <xsl:call-template name="processEntryContent"/>
            <xsl:apply-templates select="." mode="ancestor-end-flag"/>
         </fo:block>
   </fo:table-cell>
</xsl:template>
```

Be careful to get the right template; there are several with similar *match* attributes.

This template calls the *thead.row.entry__content* attribute set, which, as you know from working with tables, is used to format the content of cells in the table header row.

3. Edit the template as shown:

```
<xsl:template match="*[contains(@class, ' topic/thead ')]/*[contains(@class, ' topic/row ')]
▶/*[contains(@class, ' topic/entry ')]">
   <fo:table-cell xsl:use-attribute-sets="thead.row.entry">
      <xsl:call-template name="commonattributes"/>
      <xsl:call-template name="applySpansAttrs"/>
      <xsl:call-template name="applyAlignAttrs"/>
      <xsl:call-template name="generateTableEntryBorder"/>
      <xsl:choose>
         <xsl:when test="@rotate='1'">
            <fo:block-container reference-orientation="90" inline-progression-dimension
            ▶="2in">
               <fo:block xsl:use-attribute-sets="thead.row.entry__content">
                  <xsl:apply-templates select="." mode="ancestor-start-flag"/>
                     <xsl:call-template name="processEntryContent"/>
                  <xsl:apply-templates select="." mode="ancestor-end-flag"/>
               </fo:block>
            </fo:block-container>
         </xsl:when>
         <xsl:otherwise>
            <fo:block xsl:use-attribute-sets="thead.row.entry__content">
               <xsl:apply-templates select="." mode="ancestor-start-flag"/>
                  <xsl:call-template name="processEntryContent"/>
               <xsl:apply-templates select="." mode="ancestor-end-flag"/>
            </fo:block>
         </xsl:otherwise>
      </xsl:choose>
   </fo:table-cell>
</xsl:template>
```

In other words, use the rotated `<fo:block-container>` element only when the *rotate* attribute of an entry is set to "1". Otherwise, use the original `<fo:block>` element.

`<fo:block-container>` is really handy. If you worked through the **Place cover page information in a specific location** *(p. 183)* exercise, you learned that you can use `<fo:block-container>` to position text in a precise place on the page. And you can also use it to rotate text.

As you can see in **Figure 29**, rotation is counter-clockwise from the beginning of the text.

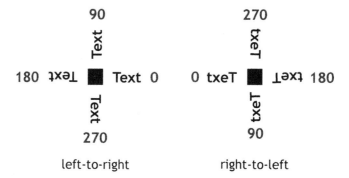

Figure 29: Degrees of text rotation

The *reference-orientation* attribute defines the degrees of rotation you want. Options are 0 (default left-to-right-text), 90, 180, 270, -90, -180, and -270. Other degrees of rotation, such as 33, are not currently possible. This is a limitation of the XSL-FO specification, which allows only these values.

DITA 1.3 defines only two values for *rotate*: 0 (no rotation) and 1 (90 degrees counter-clockwise), which limits you to a *reference-orientation* value of 90. If you use *outputclass*, you can use any of the values listed above for *reference-orientation*, but your content will not be using the DITA 1.3 feature.

The *inline-progression-dimension* attribute defines the amount of vertical space the rotated text can use. In this example, the header row can be a maximum of 2 inches high. A column heading longer than 2 inches will wrap. If you don't specify an *inline-progression-dimension* amount, the text takes up all the available space—probably way more than you want!

4. Save changes, run a PDF build, and check your work.

If you decide to use rotated text like this, you'll probably need to fine-tune the vertical alignment of the table header row cells.

Text rotation isn't limited to table cells. Although *rotate* is available only on the `<entry>` element, you can rotate almost anything using *outputclass*. You could get spiffy and place chapter titles along the outside edge of the page or use rotated text to create the illusion of tabs. The sky's the limit! As long as you rotate in multiples of 90.

Exercise: Include a topic's ID in the PDF

Objective: Output a topic's original ID just below the title.

You might find it helpful to include topic IDs in your PDF. For example, if you send PDFs out for review, it would be nice to have an ID everyone can use to locate a topic quickly. If you're using a content management system (CMS), where topics might be easier to find using IDs instead of titles, having the ID handy is especially helpful. In this exercise, you will place the topic ID right below the title.

1. Copy the following template from `DITA-OT/plugins/org.dita.pdf2/xsl/fo/`**commons.xsl** to your copy of **commons.xsl**:

```
<xsl:template match="*" mode="commonTopicProcessing">
    <xsl:apply-templates select="*[contains(@class, ' topic/title ')]"/>
    <xsl:apply-templates select="*[contains(@class, ' topic/prolog ')]"/>
    <xsl:apply-templates select="*[not(contains(@class, ' topic/title ')) and
                                    not(contains(@class, ' topic/prolog ')) and
                                    not(contains(@class, ' topic/topic '))]"/>
    <xsl:apply-templates select="." mode="buildRelationships"/>
    <xsl:apply-templates select="*[contains(@class,' topic/topic ')]"/>
    <xsl:apply-templates select="." mode="topicEpilog"/>
</xsl:template>
```

2. Between the `<apply-template>` elements that process the title and the prolog, add the following line:

```
<xsl:template match="*" mode="commonTopicProcessing">
  <xsl:apply-templates select="*[contains(@class, ' topic/title ')]"/>
  <xsl:text> (Topic ID: </xsl:text><xsl:value-of select="@oid"/><xsl:text>)</xsl:text>
  <xsl:apply-templates select="*[contains(@class, ' topic/prolog ')]"/>
  <xsl:apply-templates select="*[not(contains(@class, ' topic/title ')) and
                                  not(contains(@class, ' topic/prolog ')) and
                                  not(contains(@class, ' topic/topic '))]"/>
  <xsl:apply-templates select="." mode="buildRelationships"/>
  <xsl:apply-templates select="*[contains(@class,' topic/topic ')]"/>
  <xsl:apply-templates select="." mode="topicEpilog"/>
</xsl:template>
```

Notice that you're using the *oid* attribute for the topic ID. You might wonder why not use the *id* attribute, since that's the topic's ID, right? That's true, but at the beginning of the pdf creation process, the DITA Open Toolkit creates a merged file named **[map name]_MERGED.xml**. This is the file that gets processed to create your PDF. The Toolkit assigns new IDs to each topic to ensure that all IDs in the merged file are unique. This step is necessary because two independently developed topics in the input map could have the same ID, which would play havoc with related links and cross-references.

Therefore, if you used *id*, you'd end up with system-generated, unrecognizable IDs in your PDF, which would not be helpful. Fortunately, the Open Toolkit preserves the original topic ID in the *oid* attribute for situations like this.

3. Save changes, run a PDF build, and check your work.

Exercise: Add a prefix to a topic title

Objective: Add a text, variable, or image prefix to a topic title so that it is part of the title itself and appears in the Table of Contents, bookmarks, related links, and other locations.

You might want to add a text prefix to a topic title, such as the "Exercise:" prefix found in front of task titles in this book. As you might imagine, topic titles are processed in many different places in the PDF plugin; they're processed as part of the topic itself, as part of the Table of Contents, as part of the PDF bookmarks, and so on. You could append this prefix in any of several places, but it wouldn't actually become part of the title—you wouldn't see it in the Table of Contents or PDF bookmarks, for example. There is only one place where you can add this prefix so that it actually becomes part of the title.

But before you add that prefix, you should create it as a localization variable. It's probably going to change from language to language, so you don't want to hardcode it

1. Add the following variable to your copy of **en.xml**:

```
<variable id="Task prefix">Exercise:</variable>
```

2. In `DITA-OT/plugins/org.dita.pdf2/xsl/fo/`**commons.xsl**, do a search for "topic/title".

You'll find a lot of places where titles are processed, and most of them include some code like this:

```
<xsl:apply-templates select="." mode="getTitle"/>
```

Without going into a lot of detail, the *mode* attribute means that the template is applied according to rules or processing set up elsewhere, probably in a template that uses the mode "getTitle".

3. Do a search for "getTitle" until you find the following:

```
<xsl:template match="*" mode="getTitle">
    <xsl:choose>
        <!--add keycol here once implemented-->
        <xsl:when test="@spectitle">
            <xsl:value-of select="@spectitle"/>
        </xsl:when>
        <xsl:otherwise>
            <xsl:apply-templates/>
        </xsl:otherwise>
    </xsl:choose>
</xsl:template>
```

This is the template that retrieves the topic title. Most of the other places in the PDF plugin that process titles point back to this template. If you want to append anything to the title and make it part of the title in all contexts, this is the place to do it. And `<xsl:apply-templates/>` is the line you need to work with.

4. Copy this template into your **commons.xsl**.

5. Add the prefix to the title as shown:

```
<xsl:template match="*" mode="getTitle">
    <xsl:choose>
    <!-- add keycol here once implemented-->
        <xsl:when test="@spectitle">
            <xsl:value-of select="@spectitle"/>
        </xsl:when>
        <xsl:otherwise>
            <xsl:call-template name="getVariable">
                <xsl:with-param name="id" select="'Task prefix'"/>
            </xsl:call-template>
            <xsl:text> </xsl:text>
            <xsl:apply-templates/>
        </xsl:otherwise>
    </xsl:choose>
</xsl:template>
```

6. Save your changes, run a build, and check your work.

All of your titles now have the "Exercise:" prefix on them, including chapter titles. That's probably not what you wanted, so you need to narrow down the cases where the prefix is included. In this exercise, you want to have the prefix only on task titles, so you need to add that condition to the code.

7. Within the `<xsl:otherwise>` statement, add another `<xsl:choose>` statement as shown:

```
<xsl:template match="*" mode="getTitle">
    <xsl:choose>
    <!-- add keycol here once implemented-->
        <xsl:when test="@spectitle">
            <xsl:value-of select="@spectitle"/>
        </xsl:when>
        <xsl:otherwise>
            <xsl:choose>
                <xsl:when test="parent::*[contains(@class,' task/task ')]">
                    <xsl:call-template name="getVariable">
                        <xsl:with-param name="id" select="'Task prefix'"/>
                    </xsl:call-template>
                    <xsl:text> </xsl:text>
                </xsl:when>
                <xsl:otherwise>
                    <xsl:apply-templates/>
                </xsl:otherwise>
            </xsl:choose>
        </xsl:otherwise>
    </xsl:choose>
</xsl:template>
```

The `<xsl:when>` statement checks to see if the parent of the current node (that is, the title) is a task. If so, it appends the prefix. Otherwise, it simply outputs the title as-is.

8. Save changes, run a PDF build, and check your work.

Only your task titles have the "Exercise:" prefix on them. This prefix is actually part of the title, so it appears in the Table of Contents, PDF bookmarks, cross-references, related links, and anywhere else the task title is used.

If you want to add different prefixes for different topic types or select a prefix based on the value of an attribute such as *outputclass*, you can add more `<xsl:when>` statements to check for those conditions.

You can also append an image:

```
<xsl:when test="parent::*[contains(@class,' task/task ')]">
    <fo:external-graphic src="url(Customization/OpenTopic/common/artwork/task.png)"/>
    <xsl:apply-templates/>
</xsl:when>
```

or an attribute value:

```
<xsl:when test="parent::*[contains(@class,' task/task ')]">
    <xsl:value-of select="@outputclass"/><xsl:apply-templates/>
</xsl:when>
```

as the prefix using this approach.

Adding page breaks manually

Objective: Use a processing instruction and a small plugin to insert page breaks.

When we move to a more automated publishing process using the DITA Open Toolkit, we need to let go of our attachment to the intricate, hand-crafted page formatting that we all perfected while working in GUI desktop publishing applications. That's not to say that PDFs generated from the Open Toolkit can't be well-formatted and attractive—they certainly can, and that's what this book is all about. However, trying to do manual, ad hoc adjustments to font sizes, line spacing, table column widths, and the like is almost impossible when transforming DITA content to PDFs using the Open Toolkit.

One tweak that many writers are fond of in desktop publishing is manually adjusting page breaks, either by inserting extra carriage returns to force text to the next page, or inserting actual page breaks where desired. This is all well and good, but it means that the page flow is always going to need a lot of babysitting. Even if you adjust page breaks as the very last step before publishing, what happens during the next round of updates when content is added or removed? That's right—the page flow changes and has to be manually readjusted before publishing. And on and on.

A much better alternative to manual page breaks is careful use of the *keep-with* attribute to ensure that elements flow logically with preceding or following elements. There is an example of how to do this in **Select specific elements by context** *(p. 70)*. While the result might not always be 100% perfect, it's a good bet no user is going to throw a manual aside because a page occasionally breaks in an odd place. They, will, however, have a less than optimal user experience if the manual is missing instructions that were never written because the writer was spending her time adjusting page breaks instead. Honestly, wouldn't you rather have writers spend their time and expertise on creating solid, accurate content than on adding and removing page breaks?

That said, on the rare occasion when you must insert a manual page break, there is a way. Radu Coravu at Syncro Soft (makers of the oXygen XML editor) has posted instructions for creating a small Open Toolkit plugin that enables you to add a page break processing instruction (PI) in locations where you want to insert a manual page break and then have the Open Toolkit process that PI as a page break.

You can find instructions for creating the plugin on the **oXygen website**.

Exercise: Generate line numbers in codeblock elements

Objective: Automatically generate line numbers for `<codeblock>` elements.

It's often helpful to include line numbers in `<codeblock>` elements so that you can refer to specific lines when explaining or describing the code. The PDF plugin now offers this capability. Line numbers would be useful in `<msgblock>` as well, but they are not currently available.

1. Copy the following templates from `DITA-OT/plugins/org.dita.pdf2/xsl/fo/`**pr-domain.xsl** to your copy of **pr-domain.xsl**:

```
<xsl:variable name="codeblock.wrap" select="false()"/>
<xsl:template match="node()" mode="codeblock.generate-line-number" as="xs:boolean">
    <xsl:sequence select="false()"/>
</xsl:template>


<xsl:template match="*[contains(@class,' pr-d/codeblock ')]">
    <xsl:call-template name="generateAttrLabel"/>
    <fo:block xsl:use-attribute-sets="codeblock">
        <xsl:call-template name="commonattributes"/>
        <xsl:call-template name="setFrame"/>
        <xsl:call-template name="setScale"/>
        <xsl:call-template name="setExpanse"/>
        <xsl:variable name="codeblock.line-number" as="xs:boolean">
            <xsl:apply-templates select="." mode="codeblock.generate-line-number"/>
        </xsl:variable>
        <xsl:choose>
            <xsl:when test="$codeblock.wrap or $codeblock.line-number">
                <xsl:variable name="content" as="node()*">
                    <xsl:apply-templates/>
                </xsl:variable>
                <xsl:choose>
                    <xsl:when test="$codeblock.line-number">
                        <xsl:variable name="buf" as="document-node()">
                            <xsl:document>
                                <xsl:processing-instruction name="line-number"/>
                                <xsl:apply-templates select="$content"
                                ▸ mode="codeblock.line-number"/>
                            </xsl:document>
                        </xsl:variable>
                        <xsl:variable name="line-count" select="count($buf/descendant::
                        ▸processing-instruction('line-number'))"/>
                        <xsl:apply-templates select="$buf" mode="codeblock">
                            <xsl:with-param name="line-count" select="$line-count"
                            ▸ tunnel="yes"/>
                        </xsl:apply-templates>
                    </xsl:when>
                    <xsl:otherwise>
                        <xsl:apply-templates select="$content" mode="codeblock"/>
                    </xsl:otherwise>
                </xsl:choose>
            </xsl:when>
            <xsl:otherwise>
                <xsl:apply-templates/>
            </xsl:otherwise>
        </xsl:choose>
    </fo:block>
</xsl:template>


<xsl:template match="text()" mode="codeblock.line-number"
            name="codeblock.line-number" priority="10">
    <xsl:param name="text" select="." as="xs:string"/>
    <xsl:variable name="head" select="substring($text, 1, 1)"/>
    <xsl:variable name="tail" select="substring($text, 2)"/>
    <xsl:value-of select="$head"/>
    <xsl:if test="$head = '&#xA;'">
        <xsl:processing-instruction name="line-number"/>
    </xsl:if>
    <xsl:if test="$tail">
        <xsl:call-template name="codeblock.line-number">
```

```
            <xsl:with-param name="text" select="$tail"/>
        </xsl:call-template>
    </xsl:if>
</xsl:template>
```

```
<xsl:template match="processing-instruction('line-number')"
              mode="codeblock" priority="10">
    <xsl:param name="line-count" as="xs:integer" tunnel="yes"/>
    <xsl:variable name="line-number" select="count(preceding::processing-instruction
    ▸('line-number')) + 1" as="xs:integer"/>
    <fo:inline xsl:use-attribute-sets="codeblock.line-number">
        <xsl:for-each select="string-length(string($line-number)) to string-length
        ▸(string($line-count)) - 1">
            <xsl:text> </xsl:text>
        </xsl:for-each>
        <xsl:value-of select="$line-number"/>
    </fo:inline>
</xsl:template>
```

You won't make changes to all of these templates, but they must all be present in your copy of **pr-domain.xsl**.

2. In the first template, change both values to "true()":

```
<xsl:variable name="codeblock.wrap" select="true()"/>
<xsl:template match="node()" mode="codeblock.generate-line-number" as="xs:boolean">
    <xsl:sequence select="true()"/>
</xsl:template>
```

3. Save changes, run a PDF build, and check your work.

To format the line numbers, use the *codeblock.line-number* attribute set in **pr-domain.attr.xsl**.

Chapter 11

Lists

Lists are a clear and concise way to display multiple items. In many cases, lists are easier to read and follow than a paragraph. The PDF plugin supports detailed formatting for both unordered and ordered lists.

This chapter explains how the list-related attribute sets work together and how to use them to format your lists. It also explains how to edit or create new bullets, customize numbering formats, create a checklist, and format a definition list as a list rather than as a table.

For a list of attributes that affect lists, refer to **List attribute sets** *(p. 433)*.

Hints on working with list-related attribute sets

Whenever an element has any kind of autonumbering—a number or bullet, you'll see a general pattern of associated attribute sets. Here are simplified examples:

```
ol
ol.li
ol.li__body
ol.li__content
ol.li__label
ol.li__label__content
```

or

```
steps
steps.step
steps.step__body
steps.step__content
steps.step__label
steps.step__label__content
```

The items in the patterned list affect the element in a hierarchical manner, from most general to most specific (see **Figure 30** for the attribute sets associated with unordered lists. There is a parallel set for ordered lists).

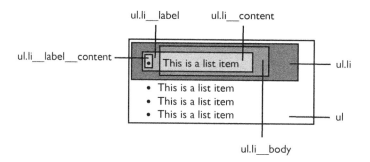

Figure 30: List-related attribute sets hierarchy

In **Figure 30**, the *ul* attribute set affects everything in an unordered list. The *ul.li* attribute set affects each li element. The *ul.li__body* attribute set affects only certain aspects of the li elements. This might seem to be hair-splitting, and it is—it's hard to tease the effects of these two attribute sets apart.

The attribute sets you'll probably deal with the most are *ul.li__content*, *ul.li__label*, and *ul.li__label__content* and their equivalents for ol lists.

Generally, list items have a number or bullet, a space, and the text. The *ul.li__label__content* attribute set formats the number or bullet. The *ul.li__label* attribute set formats the space between the bullet or number and the text. The *ul.li__content* attribute set formats the list item text.

So, for example, any attributes you add to the *ul* attribute set apply to the entire unordered list: the label, the list item text, and the spacing above and below the list and between list items. Any attributes you add to the *ul.li* attribute set apply only to list items within unordered lists. Attributes you add to the *ul.li__content* attribute set apply only to list item text, not to the label. Attributes you add to the *ul.li__label__content* attribute set apply only to list item labels, not to the list item text.

These layers let you be as general or as specific as you want when formatting lists.

> **Tip:** When choosing the characters to use for bullets in your lists, be sure to test your output with the fonts you have chosen. Many fonts do not support all of the possible bullet characters available. For example, the typeface used for this book, Minion Pro, does not support the checkbox character described in the specifications below, and the typeface described in the specifications, Book Antiqua, does not support the ballot box character. As long as you try out different possibilities, you should be fine.

List specifications

Here are your specifications for lists.

Bulleted and numbered lists

Top-level ordered list numbering	1, 2, 3...
Second-level ordered list numbering	i, ii, iii...
Top-level bulleted list bullet	▪ black square (■ or ■)
Second-level bulleted list bullet	— em-dash (Ᾱ or —)
Checklist (custom list type)	☐ empty checkbox (□ or □)

Files you need

If you have completed other exercises, some of these files might already be present in your plugin.

If not already present, create this file in `DITA-OT/plugins/com.company.pdf/cfg/fo/attrs/`:

- **lists-attr.xsl**

Be sure to add the appropriate `xsl:import` statement to **custom.xsl** (attrs).

If not already present, create these files in `DITA-OT/plugins/com.company.pdf/cfg/fo/xsl/`:

- **lists.xsl**
- **tables.xsl**

Be sure to add the appropriate `xsl:import` statements to **custom.xsl** (xsl).

For instructions on creating attribute files and XSLT stylesheets in your PDF plugin, refer to **Create an attribute set file in your plugin** *(p. 31)* and **Create an XSLT stylesheet in your plugin** *(p. 32)*.

If not already present, create this file in `DITA-OT/plugins/com.company.pdf/cfg/common/vars/`:

- **en.xml**

For instructions on creating localization variables files, refer to **Create a localization variables file** *(p. 78)*.

Exercise: Create multiple bullet formats for unordered lists

Objective: Create additional bullet schemes for additional levels of unordered list, such as for second- or third-level list items.

According to your spec, you need to use the Unicode "black square" ((■)) for top-level bullet list items and the Unicode "em-dash" (—) for nested bullet list items.

1. Open `DITA-OT/plugins/org.dita.pdf2/cfg/common/vars/`**en.xml** and copy the *Unordered List bullet* variable:

   ```
   <variable id="Unordered List bullet">•</variable>
   ```

2. Paste it into your **en.xml**.
3. Delete the bullet character and enter `&#9632;` as the variable value.

   You could just type (■) directly into the variable value, but to be on the safe side and make sure you always get the character you intend, you might want to use the Unicode.

   > **Note** To find the Unicode value for a character, you can use the Character Map (Windows, Mac and UNIX all have this functionality in some form). Most XML editors also include some kind of character map as well. You can also refer to unicode.org for the complete Unicode character set. Or a web search for "Unicode *character name*" will usually find the value. One helpful site is **http://www.fileformat.info/info/unicode/**.

4. Below the *Unordered List bullet* variable, add a second variable:

   ```
   <variable id="Unordered List bullet nested">&#8121;</variable>
   ```

5. Copy the following template from `DITA-OT/plugins/org.dita.pdf2/xsl/fo/`**lists.xsl** to your copy of **lists.xsl**:

   ```
   <xsl:template match="*[contains(@class, ' topic/ul ')]/*[contains(@class, ' topic/li ')]">
       <fo:list-item xsl:use-attribute-sets="ul.li">
           <xsl:apply-templates select="*[contains(@class,' ditaot-d/ditaval-startprop ')]"
           ▶ mode="flag-attributes"/>
           <fo:list-item-label xsl:use-attribute-sets="ul.li__label">
               <fo:block xsl:use-attribute-sets="ul.li__label__content">
                   <fo:inline>
                       <xsl:call-template name="commonattributes"/>
                   </fo:inline>
                   <xsl:call-template name="getVariable">
                       <xsl:with-param name="id" select="'Unordered List bullet'"/>
   ```

```
            </xsl:call-template>
          </fo:block>
      </fo:list-item-label>
      <fo:list-item-body xsl:use-attribute-sets="ul.li__body">
          <fo:block xsl:use-attribute-sets="ul.li__content">
              <xsl:apply-templates/>
          </fo:block>
      </fo:list-item-body>
    </fo:list-item>
</xsl:template>
```

6. Paste it into your **lists.xsl**.

You can see where it calls the *Unordered List bullet* variable. Sometimes you want to call this variable, and sometimes you want to call your new *Unordered List bullet nested* variable. Sounds like a case for either `<xsl:if>` or `<xsl:choose>`. Because you only have two conditions, you could use `<xsl:if>`. But down the road, you might want a third bullet character, and then you'd need to switch over to `<xsl:choose>`, so just go with that from the start.

7. Add an `<xsl:choose>` just above the `<xsl:call-template>` line that inserts the *Unordered List bullet* variable, as shown here:

```
... </fo:inline>
<xsl:choose>
    <xsl:when>
    </xsl:when>
    <xsl:otherwise>
    </xsl:otherwise>
</xsl:choose>
<xsl:call-template name="getVariable">
```

8. Cut the entire `<xsl:call-template>` section and paste it inside `<xsl:when>`.

9. Paste it again inside `<xsl:otherwise>`.

10. Edit the `<xsl:call-template>` within `<xsl:when>` to use the *Unordered List bullet nested* variable:

```
<fo:block xsl:use-attribute-sets="ul.li__label__content">
    <fo:inline>
        <xsl:call-template name="commonattributes"/>
    </fo:inline>
    <xsl:choose>
        <xsl:when>
            <xsl:call-template name="getVariable">
                <xsl:with-param name="id" select="'Unordered List bullet nested'"/>
            </xsl:call-template>
        </xsl:when>
        <xsl:otherwise>
            <xsl:call-template name="getVariable">
                <xsl:with-param name="id" select="'Unordered List bullet'"/>
            </xsl:call-template>
        </xsl:otherwise>
    </xsl:choose>
</fo:block>
```

Now you've got the `<xsl:when>` and `<xsl:otherwise>` statements set up, but the `<xsl:when>` doesn't include a test yet. **Figure 31** shows the structure of a nested unordered list.

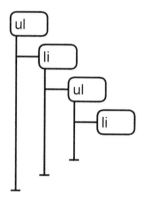

Figure 31: Nested unordered list structure

So what you're looking for is an `<li>` element with another `<li>` element as its ancestor, making it a nested `<li>`. If the `<li>` element has no `<li>` ancestor, that means it's a top-level `<li>`.

11. Change `<xsl:when>` to

```
<xsl:when test="ancestor::*[contains(@class, ' topic/li ')]">
```

> **Note** This tests for any number of `<li>` elements as ancestor, meaning that all levels of `<li>` below the top one are going to use the nested bullet. You'd have to be more specific with your test if you wanted to use one bullet for the first level of nesting, another for the second level of nesting, and so forth.

12. Save changes, run a PDF build, and check your work.

Reality check

This approach doesn't make any distinction between whether the ancestor `<li>` is within a `<ul>` or an `<ol>`. So, if you nest a `<ul>` inside an `<ol>`, the top level of the `<ul>` uses *Unordered List bullet nested* instead of *Unordered List bullet*. This might be what you want. However, if you want the top level of the `<ul>` to use *Unordered List bullet* when it's nested within an `<ol>`, and *Unordered List bullet nested* when it's nested within another `<ul>`, you need to use a more specific `<xsl:when>` test. Take a look at the code in **Create multiple numbering formats for ordered lists** *(p. 245)* as an example of how you might do this.

Exercise: Create multiple numbering formats for ordered lists

Objective: Create numbering schemes for additional levels of ordered list.

According to your specification, top-level numbered list items should use 1, 2, 3, etc. and nested numbered list items should use i, ii, iii, etc. To figure out what you need to do to make this happen, let's look at how ordered list numbering is set up.

1. Open **en.xml** and find the *Ordered List Number* variable:

```
<variable id="Ordered List Number"><param ref-name="number"/>. </variable>
```

Notice how this variable doesn't actually contain a character, like the variable for the unordered list bullet? Instead, it contains a parameter followed by a period. Keep that in mind.

2. Copy the following template from `DITA-OT/plugins/org.dita.pdf2/xsl/fo/`**lists.xsl** to your copy of **lists.xsl**:

```
<xsl:template match="*[contains(@class, ' topic/ol ')]/*[contains(@class, ' topic/li ')]">
    <fo:list-item xsl:use-attribute-sets="ol.li">
      <xsl:apply-templates select="*[contains(@class,' ditaot-d/ditaval-startprop ')]"
    ▸ mode="flag-attributes"/>
      <fo:list-item-label xsl:use-attribute-sets="ol.li__label">
          <fo:block xsl:use-attribute-sets="ol.li__label__content"> .
              <fo:inline>
                  <xsl:call-template name="commonattributes"/>
              </fo:inline>
              <xsl:call-template name="getVariable">
                  <xsl:with-param name="id" select="'Ordered List Number'"/>
                  <xsl:with-param name="params">
                      <number>
                          <xsl:choose>
                              <xsl:when test="parent::*[contains(@class, ' topic/ol ')]
                              ▸/parent::*[contains(@class, ' topic/li ')]/parent::
                              ▸*[contains(@class, ' topic/ol ')]">
                                  <xsl:number format="a"/>
                              </xsl:when>
                              <xsl:otherwise>
                                  <xsl:number/>
                              </xsl:otherwise>
                          </xsl:choose>
                      </number>
                  </xsl:with-param>
              </xsl:call-template>
          </fo:block>
      </fo:list-item-label>
      <fo:list-item-body xsl:use-attribute-sets="ol.li__body">
          <fo:block xsl:use-attribute-sets="ol.li__content">
              <xsl:apply-templates/>
          </fo:block>
      </fo:list-item-body>
```

```
      </fo:list-item>
   </xsl:template>
```

3. Just after the call to the *Ordered List Number* variable, note the `<number>` section:

```
<number>
   <xsl:choose>
      <xsl:when test="parent::*[contains(@class, ' topic/ol ')]/
      ▸parent::*[contains(@class, ' topic/li ')]/
      ▸parent::*[contains(@class, ' topic/ol ')]">
         <xsl:number format="a"/>
      </xsl:when>
      <xsl:otherwise>
         <xsl:number/>
      </xsl:otherwise>
   </xsl:choose>
</number>
```

4. Within that section, look at the `<xsl:number>` element.

`<xsl:number>` is a complex XSL feature, but in its simplest form, it generates a sequence number for the element being processed (`<li>` in this case). When called for the first `<li>` in a list, the value will be 1, for the second it will be 2, and so forth. By default, it returns Arabic numerals (1, 2, 3, etc.), but you can specify other formats. Therefore, you don't need to figure out the numbering, you just need to define the format. By default, top-level ordered list items already use Arabic numerals. You only need to deal with the nested ordered list items.

5. Look at the `<xsl:when>` statement inside the `<number>` element:

```
<xsl:when test="parent::*[contains(@class, ' topic/ol ')]/
       ▸parent::*[contains(@class, ' topic/li ')]/parent::*[contains(@class, ' topic/ol ')]">
   <xsl:number format="a"/>
</xsl:when>
```

This condition tests to see if the `<li>` is within an `<ol>` that's within an `<li>` within an `<ol>`—in other words, an `<ol>` within an `<ol>`. See the line `<xsl:number format="a"/>`? That's what you need to change.

6. Change the value of *format* to "i" (valid formats are: 1, 01, A, a, I, i):

```
<xsl:number format="i"/>
```

7. Save changes, run a PDF build, and check your work.

If you want to change the format for the top-level list items, just add the format attribute to `<xsl:number>` in the `<xsl:otherwise>` block.

Exercise: Create a checklist

Objective: Create an additional kind of list type—a checklist—based on the standard unordered list.

Checklists are common in documentation. That is, a list where there's a little box to check beside each list item rather than a bullet or a number. Your PDF specification calls for one, but out of the box, the PDF plugin doesn't create checklists.

Your first clue comes from the sentence "little box ... rather than a bullet." You can create a special kind of unordered list that uses a box as the bullet. Remember the exercise where you created a new bullet variable (the em-dash) and assigned it to nested unordered list items? Well, you're going to do something very similar here.

> **Note** You're not specializing any elements in this book. You could, of course, specialize `<ul>` to `<cl>` or some such thing (the element name is your choice), but this approach uses the basic `<ul>` element without any specialization.

1. Add the *Checklist bullet* variable to your copy of **en.xml**:

```
<variable id="Checklist bullet">&#9633;</variable>
```

`&#9633;` is the Unicode "white square" character (□), expressed as a numeric character reference. An even better choice might be the Unicode "ballot box" character (☐) but it's not available in the font you're using, Book Antiqua.

2. If it is not already present, copy the following template from `DITA-OT/plugins/org.dita.pdf2/xsl/fo/`**lists.xsl** to your copy of **lists.xsl**.

If you completed the **Create multiple bullet formats for unordered lists** *(p. 242)* exercise, you already copied this template into your plugin and modified it.

```
<xsl:template match="*[contains(@class, ' topic/ul ')]/*[contains(@class, ' topic/li ')]">
    <fo:list-item xsl:use-attribute-sets="ul.li">
        <xsl:apply-templates select="*[contains(@class,' ditaot-d/ditaval-startprop ')]"
         ▶ mode="flag-attributes"/>
        <fo:list-item-label xsl:use-attribute-sets="ul.li__label">
            <fo:block xsl:use-attribute-sets="ul.li__label__content">
                <fo:inline>
                    <xsl:call-template name="commonattributes"/>
                </fo:inline>
                <xsl:call-template name="getVariable">
                    <xsl:with-param name="id" select="'Unordered List bullet'"/>
                </xsl:call-template>
            </fo:block>
        </fo:list-item-label>
        <fo:list-item-body xsl:use-attribute-sets="ul.li__body">
            <fo:block xsl:use-attribute-sets="ul.li__content">
```

```
                <xsl:apply-templates/>
            </fo:block>
        </fo:list-item-body>
    </fo:list-item>
</xsl:template>
```

3. Paste it into your **lists.xsl**.

4. Add the section highlighted below.

```
<fo:block xsl:use-attribute-sets="ul.li__label__content">
    <fo:inline>
        <xsl:call-template name="commonattributes"/>
    </fo:inline>
    <xsl:choose>
        <xsl:when test="../@outputclass='checklist'">
            <fo:inline font-size="18pt">
                <xsl:call-template name="getVariable">
                    <xsl:with-param name="id" select="'Checklist bullet'"/>
                </xsl:call-template>
            </fo:inline>
        </xsl:when>
        <xsl:otherwise>
            <xsl:call-template name="getVariable">
                <xsl:with-param name="id" select="'Unordered List bullet'"/>
            </xsl:call-template>
        </xsl:otherwise>
    </xsl:choose>
</fo:block>
```

This `<xsl:when>` statement tests to see if the *outputclass* attribute of this `<li>` element's parent (`<ul>`) has the value "checklist". If so, it uses the checklist bullet.

This example assumes you did not complete **Create multiple bullet formats for unordered lists** *(p. 242)*. If you did, add the highlighted code above as the first `<xsl:when>` test in the existing `<xsl:choose>`.

5. Add `outputclass="checklist"` to a `<ul>` element in your document.

6. Save changes, run a PDF build, and check your work.

Other things you can do

Here are a few additional customizations you might want to make to your plugin. These customizations aren't part of the PDF specifications you're following, but they can be useful (and cool).

Exercise: Change the amount of space between a number/bullet and text

Objective: Increase or decrease the amount of space between a list item's number or bullet and its text.

Although each kind of list (bullet, numbered, etc.) has its own attribute set, there is a single attribute set (*provisional-distance-between-starts*) that determines the amount of space between the bullet or number and the text.

Here is where you will find the attribute sets you need to change spacing in the following contexts:

Footnotes	*fn__body* in **commons-attr.xsl**
Lists	*ul*, *ol*, *sl* in **lists-attr.xsl**
Mini-TOC	*__toc__mini__list* in **toc-attr.xsl**
Related-links	*related-links.ol* and *related-links.ul* in **links-attr.xsl**
Steps	*steps-unordered*, *steps*, *substeps*, and *choices* in **task-elements-attr.xsl**

All of the attribute sets that format `<step>` and `<related-links>` elements use the *ol* attribute sets. Unless you need step or related-link elements to have different formatting from ordered lists, you only need to edit the *ol* attribute sets in **lists-attr.xsl**.

1. When you locate the right attribute set for what you want to change, copy and paste it into your copy of the corresponding attribute set file.
2. Change the value of *provisional-distance-between-starts*.

    By default, this value is "5mm"; you can use another unit of measurement, such as points (pt) or inches (in), if you prefer.

3. Save changes, run a PDF build, and check your work.

Exercise: Use an image as a bullet

Objective: Use an image rather than a text symbol as the bullet for unordered lists.

1. Create the image you want to use as a bullet, name it **bullet.png**, and place it in `DITA-OT/plugins/com.company.pdf/cfg/common/artwork/`.

2. Add the *Bullet Image Path* variable to your copy of **en.xml**:

```
<variable id="Bullet Image Path">Customization/OpenTopic/common/artwork/bullet.png</variable>
```

3. Copy the template that starts with the following from `DITA-OT/plugins/org.dita.pdf2/xsl/fo/`**lists.xsl** to your copy of **lists.xsl**:

```
<xsl:template match="*[contains(@class, ' topic/ul ')]/*[contains(@class, ' topic/li ')]">
```

This is the code you customized earlier to add a different bullet for nested unordered list items in the **Create multiple bullet formats for unordered lists** *(p. 242)* exercise. You can expand this section to accommodate an image bullet.

4. Just after the line above, add the following:

```
<xsl:template match="*[contains(@class, ' topic/ul ')]/*[contains(@class, ' topic/li ')]">
    <xsl:variable name="bulletImagePath">
        <xsl:call-template name="getVariable">
            <xsl:with-param name="id" select="'Bullet Image Path'"/>
        </xsl:call-template>
    </xsl:variable>
```

5. Just above the line `<xsl:when test="ancestor::*[contains(@class, ' topic/li ')]">`, add the highlighted code:

```
<xsl:choose>
    <xsl:when test="not($bulletImagePath = '')">
        <fo:external-graphic src="url({$bulletImagePath})" xsl:use-attribute-sets="image"/>
    </xsl:when>
    <xsl:when test="ancestor::*[contains(@class, ' topic/li ')]">
        <xsl:call-template name="getVariable">
            <xsl:with-param name="id" select="'Unordered List bullet nested'"/>
        </xsl:call-template>
    </xsl:when>
...
```

This condition tests to see if there is a value for *Bullet Image Path* in **en.xml**. If there is, it inserts that graphic as the bullet. If not, the conditions that follow are tested in order, and the code inserts either ▪ or — or creates a checklist.

6. Save changes, run a PDF build, and check your work.

Exercise: Format definition lists as a list instead of a table

Objective: Output definition lists (`<dl>`) in list format rather than in the default table format.

Although the element tag <dl> stands for "definition **list**," the Open Toolkit formats <dl> using a table. Using a table makes it easier to render the term and its definition on the same line, and a table makes it easier to align the definitions.

However, if you really want definition lists to be formatted as lists, you can make that happen. You can follow the steps below to make the changes to the base templates, or you can integrate Eliot Kimber's dl-as-dl plugin, found on **GitHub**.

1. In DITA-OT/plugins/org.dita.pdf2/xsl/fo/**tables.xsl**, find the section that begins with the comment <!--Definition list-->.

There are seven templates in this section. For clarity, here is a list of the first lines of each template and what they will be called throughout this exercise.

The *dl* template:

```
<xsl:template match="*[contains(@class, ' topic/dl ')]">
```

The *dlhead* template:

```
<xsl:template match="*[contains(@class, ' topic/dl ')]/*[contains(@class, ' topic/dlhead ')]">
```

The *dthd* template:

```
<xsl:template match="*[contains(@class, ' topic/dlhead ')]/*[contains(@class, ' topic/dthd ')]">
```

The *ddhd* template:

```
<xsl:template match="*[contains(@class, ' topic/dlhead ')]/*[contains(@class, ' topic/ddhd ')]">
```

The *dlentry* template:

```
<xsl:template match="*[contains(@class, ' topic/dlentry ')]">
```

The *dt* template:

```
<xsl:template match="*[contains(@class, ' topic/dt ')]">
```

The *dd* template:

```
<xsl:template match="*[contains(@class, ' topic/dd ')]">
```

2. Copy the first five templates into your copy of **tables.xsl**.

The *dt* and *dd* templates don't include any table-related code, so you do not need to change them for this exercise.

3. Edit the *dl* template to comment out the table-related code, like this:

```
<xsl:template match="*[contains(@class, ' topic/dl ')]">
    <!--<fo:table xsl:use-attribute-sets="dl">-->
        <xsl:call-template name="commonattributes"/>
        <xsl:apply-templates select="*[contains(@class, ' topic/dlhead ')]"/>
        <!--<fo:table-body xsl:use-attribute-sets="dl__body">-->
            <xsl:choose>
                <xsl:when test="contains(@otherprops,'sortable')">
                    <xsl:apply-templates select="*[contains(@class, ' topic/dlentry ')]">
                        <xsl:sort select="opentopic-func:getSortString(normalize-space
                                ( opentopic-func:fetchValueableText(*
                                ▶[contains(@class, ' topic/dt ')]) ))"
                                lang="{$locale}"/>
                    </xsl:apply-templates>
                </xsl:when>
                        <xsl:otherwise>
                            <xsl:apply-templates select="*[contains(@class,
                            ▶' topic/dlentry ')]"/>
                        </xsl:otherwise>
            </xsl:choose>
        <!--</fo:table-body>-->
    <!--</fo:table>-->
</xsl:template>
```

4. Edit the *dlhead* template to comment out the table-related code, like this:

```
<xsl:template match="*[contains(@class, ' topic/dl ')]/*[contains(@class, ' topic/dlhead ')]">

    <!--<fo:table-header xsl:use-attribute-sets="dl.dlhead">-->
        <xsl:call-template name="commonattributes"/>
        <!--<fo:table-row xsl:use-attribute-sets="dl.dlhead__row">-->
            <xsl:apply-templates/>
        <!--</fo:table-row>-->
    <!--</fo:table-header>-->
</xsl:template>
```

5. Edit the *dthd* template to comment out the table-related code, like this:

```
<xsl:template match="*[contains(@class, ' topic/dlhead ')]/*[contains(@class, ' topic/dthd ')]">

    <!--<fo:table-cell xsl:use-attribute-sets="dlhead.dthd__cell">-->
        <xsl:call-template name="commonattributes"/>
        <fo:block xsl:use-attribute-sets="dlhead.dthd__content">
            <xsl:apply-templates/>
        </fo:block>
    <!--</fo:table-cell>-->
</xsl:template>
```

6. Edit the *ddhd* template to comment out table-related code, like this:

```
<xsl:template match="*[contains(@class, ' topic/dlhead ')]/*[contains(@class, ' topic/ddhd ')]">

    <!--<fo:table-cell xsl:use-attribute-sets="dlhead.ddhd__cell">-->
        <xsl:call-template name="commonattributes"/>
        <fo:block xsl:use-attribute-sets="dlhead.ddhd__content">
            <xsl:apply-templates/>
        </fo:block>
    <!--</fo:table-cell>-->
</xsl:template>
```

7. Edit the *dlentry* template to comment out the table-related code like this:

```
<xsl:template match="*[contains(@class, ' topic/dlentry ')]">
   <!--<fo:table-row xsl:use-attribute-sets="dlentry">-->
   <xsl:call-template name="commonattributes"/>
      <fo:table-cell xsl:use-attribute-sets="dlentry.dt">
         <xsl:apply-templates select="*[contains(@class, ' topic/dt ')]"/>
      </fo:table-cell>
      <fo:table-cell xsl:use-attribute-sets="dlentry.dd">
         <xsl:apply-templates select="*[contains(@class, ' topic/dd ')]"/>
      </fo:table-cell>
   <!--</fo:table-row>-->
</xsl:template>
```

8. In the same template, change the `<fo:table-cell>` elements to `<fo:block>`, like this:

```
<xsl:template match="*[contains(@class, ' topic/dlentry ')]">
   <!--<fo:table-row xsl:use-attribute-sets="dlentry">-->
   <xsl:call-template name="commonattributes"/>
      <fo:block xsl:use-attribute-sets="dlentry.dt">
         <xsl:apply-templates select="*[contains(@class, ' topic/dt ')]"/>
      </fo:block>
      <fo:block xsl:use-attribute-sets="dlentry.dd">
         <xsl:apply-templates select="*[contains(@class, ' topic/dd ')]"/>
      </fo:block>
   <!--</fo:table-row>-->
</xsl:template>
```

9. Save changes, run a PDF build, and check your work.

The attribute sets you need to format definition lists are in **tables-attr.xsl**. You should be able to find them, copy them to your plugin, and modify them by now, right?

Chapter 12

Task topics

Formatting tasks is a lot like formatting text—find the attribute set that corresponds to the element you need to format and change it. If you worked through **Titles, body text, and notes** *(p. 193)*, you should be able to make 99% of the changes needed to handle tasks. But there are a few things that are unique to tasks, and this chapter covers them.

This chapter covers formatting task and step labels, formatting step numbers, and creating and using a count of steps in a task.

For a list of attributes that affect tasks, refer to **Task element attribute sets** *(p. 444)*.

Task topic specifications

Here are your specifications for tasks.

Tasks

Optional step label	(Optional)
Step section labels	12pt Trebuchet bold dark red prereq: "Before you start" context: "About this task" steps: "Procedure" steps-unordered: "Procedure" result: ""Reality check"" example: "Show me" tasktroubleshooting: "Help me fix it" postreq: "What to do next"

Prereq, context, result, example, postreq, tasktroubleshooting text format	11pt Book Antiqua normal dark gray
Info, stepxmp, stepresult, choices, substeps text format	10 pt Book Antiqua normal dark gray
Step numbers	11pt Trebuchet bold dark red. Arabic number with a parenthesis after: 1)
Substep numbers	10pt Trebuchet bold dark red. Roman numeral with a period after: i.

Files you need

If you have completed other exercises, some of these files might already be present in your plugin.

If not already present, create the following files in `DITA-OT/plugins/com.company.pdf/cfg/fo/attrs/`:

- **task-elements-attr.xsl**
- **commons-attr.xsl**

Be sure to add the appropriate `<xsl:import>` statements to **custom.xsl** (attrs).

If not already present, create the following file in `DITA-OT/plugins/com.company.pdf/cfg/fo/xsl/`:

- **task-elements.xsl**

Be sure to add the appropriate `<xsl:import>` statement to **custom.xsl** (xsl).

For instructions on creating attribute files and XSLT stylesheets in your PDF plugin, refer to **Create an attribute set file in your plugin** *(p. 31)* and **Create an XSLT stylesheet in your plugin** *(p. 32)*.

If not already present, create the following files in `DITA-OT/plugins/com.company.pdf/cfg/common/vars/`:

- **en.xml**
- **strings-en-us.xml**

For instructions on creating variables and strings files, refer to **Create a localization variables file** *(p. 78)* and **Create a localization strings file** *(p. 79)*.

Be sure to add the appropriate language and locale mapping statements in **strings.xml**. You can find instructions in **Mapping strings files to xml:lang values in the strings.xml file** *(p. 37)*.

Task step formatting tips

If you completed the **Set body font sizes** *(p. 208)* exercise, the changes you made affected the size of text in task steps. After reading **Formatting body text** *(p. 206)*, you may also have changed the color.

The attribute sets you need to format task elements are all in **task-elements-attr.xsl**. The name of the attribute set generally matches the name of the element it formats.

Some elements, such as `<cmd>`, `<info>`, `<stepxmp>`, `<stepresult>`, and `<tutorialinfo>` are straightforward. Task sections, like `<prereq>`, `<context>`, `<result>`, `<example>`, and `<postreq>` are a little more complicated because their processing interacts with the processing for sections in general, outside of tasks. The **Format task section labels and text** *(p. 261)* exercise can help you work through that little minefield.

The trickier thing to understand is that `<steps>`, `<substeps>`, `<steps-unordered>`, and `<choices>` are treated like ordered or unordered lists. In fact, if you look at **task-elements-attr.xsl**, you'll see that many of the attribute sets call list-related attribute sets such as *ol.li* and *ol.li__label*. The attribute sets that format these elements work in layers, just like the list-related attribute sets. So, the particular part of the `<step>` or `<substep>` or `<choice>` that you want to format determines which attribute set you edit.

Take a look at the **Hints on working with list-related attribute sets** *(p. 239)* overview for an explanation of how these layers work together.

Exercise: Change the label for optional steps

Objective: Specify the automatic label for steps marked `importance="optional"`.

1. Copy the following variable from `DITA-OT/plugins/org.dita.pdf2/cfg/common/vars/`**en.xml** to your copy of **en.xml**:

```
<!--Label for a step with importance="optional"-->
<variable id="Optional Step">Optional:</variable>
```

> **Note** There is also an *Optional* string in the localization strings files. That string is not used by the `org.dita.pdf2` plugin.

2. Change the value of the variable as shown:

```
<variable id="Optional Step">(Optional) </variable>
```

3. Repeat this process for any other language-specific variables files you've customized.

Using strings vs localization variables for task labels

As of this book's publication, the DITA Open Toolkit offers you the option to use either localization variables or strings to create labels for sections within steps.

The localization variables file **en.xml** in the `org.dita.pdf2` plugin includes this section:

```
<variable id="Task Prereq"></variable>
<variable id="Task Context"></variable>
<variable id="Task Steps"></variable>
<variable id="#steps-unordered-label"></variable>
<variable id="Task Result"></variable>
<variable id="Task Example"></variable>
<variable id="Task Postreq"></variable>
```

None of these variables has a value. The `org.dita.pdf2` plugin uses these variables if they have a value; if not, it looks for the following strings, which are found in files such as **strings-en-us.xml** (notice that the naming is similar):

```
<str name="task_context">About this task</str>
<str name="task_example">Example</str>
<str name="task_postreq">What to do next</str>
<str name="task_prereq">Before you begin</str>
<str name="task_procedure">Procedure</str>
<str name="task_procedure_unordered">Procedure</str>
<str name="task_results">Results</str>
```

Now, open **task-elements.xsl** and find this template:

```
<xsl:template match="*[contains(@class, ' task/prereq ')]">
    <fo:block xsl:use-attribute-sets="prereq">
        <xsl:call-template name="commonattributes"/>
        <xsl:apply-templates select="." mode="dita2xslfo:task-heading">
            <xsl:with-param name="use-label">
                <xsl:apply-templates select="." mode="dita2xslfo:retrieve-task-heading">
                    <xsl:with-param name="pdf2-string">Task Prereq</xsl:with-param>
                    <xsl:with-param name="common-string">task_prereq</xsl:with-param>
                </xsl:apply-templates>
            </xsl:with-param>
        </xsl:apply-templates>
        <fo:block xsl:use-attribute-sets="prereq__content">
            <xsl:apply-templates/>
        </fo:block>
    </fo:block>
</xsl:template>
```

This template (and the similar ones that process `<context>`, `<postreq>`, etc.) calls both the *Task Prereq* variable from **en.xml** and the *task_prereq* string from **strings-en-us.xml**. How does the DITA-OT know which one to choose? Well, at the end of **task-elements.xsl**, there is another template:

```
<!-- Set up to allow string retrieval based on the original PDF2 string;
     if not found, fall back to the common string -->
<xsl:template match="*" mode="dita2xslfo:retrieve-task-heading">
   <xsl:param name="pdf2-string"/>
   <xsl:param name="common-string"/>
   <xsl:variable name="retrieved-pdf2-string">
      <!-- By default, will return the lookup value -->
      <xsl:call-template name="getVariable">
         <xsl:with-param name="id" select="$pdf2-string"/>
      </xsl:call-template>
   </xsl:variable>
   <xsl:choose>
      <xsl:when test="$retrieved-pdf2-string!=$pdf2-string and $retrieved-pdf2-string!=''">
         <xsl:value-of select="$retrieved-pdf2-string"/>
      </xsl:when>
      <xsl:otherwise>
         <xsl:call-template name="getVariable">
            <xsl:with-param name="id" select="$common-string"/>
         </xsl:call-template>
      </xsl:otherwise>
   </xsl:choose>
</xsl:template>
```

As the comment suggests, this template tests to see if the variable in **en.xml** both exists in your plugin *and* has a value. If so, the DITA OT uses it. If not, the DITA OT uses the value in the equivalent string in your copy of **strings-en-us-xml**.

So in short, if you want to use labels from a localization variables file in your plugin (such as **en.xml**), make sure those variables exist in the variables file and that they have a value; and make sure the corresponding string exists in the strings file but does not have a value. And vice-versa.

Exercise: Change variables for task sections

Objective: Specify labels for sections within steps such as `<prereq>`, `<context>`, and `<result>`.

Before you start

By default, task section labels don't appear in PDFs. To make them appear, add the following parameter to your ANT build file (and yes, the value has to be in all caps):

```
<property name="args.gen.task.lbl" value="YES"/>
```

These steps assume you have reviewed **Using strings vs localization variables for task labels** *(p. 258)* and decided which to use: localization variables or strings. Now you need to change the default labels.

1. Copy the task labels from either `DITA-OT/plugins/org.dita.pdf2/cfg/common/vars/`**en.xml** or `DITA-OT/xsl/common/`**strings-en-us.xml**, depending on which you want to use.:

```
<!--Labels for task sections, now reused from common-->
<variable id="Task Prereq"></variable>
<variable id="Task Context"></variable>
<variable id="Task Steps"></variable>
<variable id="#steps-unordered-label"></variable>
<variable id="Task Result"></variable>
<variable id="Task Example"></variable>
<variable id="Task Postreq"></variable>
```

Figure 32: en.xml

-OR-

```
<str name="task_context">About this task</str>
<str name="task_example">Example</str>
<str name="task_postreq">What to do next</str>
<str name="task_prereq">Before you begin</str>
<str name="task_procedure">Procedure</str>
<str name="task_procedure_unordered">Procedure</str>
<str name="task_results">Results</str>
```

Figure 33: strings-en-us.xml

2. Paste your new strings into either your **en.xml** or your **strings-en-us.xml** file.
3. Enter the new values:

```
<!--Labels for task sections, now reused from common-->
<variable id="Task Prereq">Before you start</variable>
<variable id="Task Context">About this task</variable>
<variable id="Task Steps">Procedure</variable>
<variable id="#steps-unordered-label">Procedure</variable>
<variable id="Task Result">Reality check</variable>
<variable id="Task Example">Show me</variable>
<variable id="Task Postreq">What to do next</variable>
```

Figure 34: en.xml

-OR-

```
<str name="task_context">About this task</str>
<str name="task_example">Show me</str>
<str name="task_postreq">What to do next</str>
<str name="task_prereq">Before you start</str>
<str name="task_procedure">Procedure</str>
<str name="task_procedure_unordered">Procedure</str>
<str name="task_results">Reality check</str>
```

Figure 35: strings-en-us.xml

4. Repeat this process for any other language-specific strings files you've customized.

5. Save changes, run a PDF build, and check your work.

Exercise: Format task section labels and text

Objective: Change the appearance of the automatic labels and text within task sections.

1. Open `DITA-OT/plugins/org.dita.pdf2/xsl/fo/`**task-elements.xsl** and find this template:

```
<xsl:template match="*[contains(@class, ' task/prereq ')]">
   <fo:block xsl:use-attribute-sets="prereq">
      <xsl:call-template name="commonattributes"/>
      <xsl:apply-templates select="." mode="dita2xslfo:task-heading">
         <xsl:with-param name="use-label">
            <xsl:apply-templates select="." mode="dita2xslfo:retrieve-task-heading">
               <xsl:with-param name="pdf2-string">Task Prereq</xsl:with-param>
               <xsl:with-param name="common-string">task_prereq</xsl:with-param>
            </xsl:apply-templates>
         </xsl:with-param>
      </xsl:apply-templates>
      <fo:block xsl:use-attribute-sets="prereq__content">
         <xsl:apply-templates/>
      </fo:block>
   </fo:block>
</xsl:template>
```

This is the template that processes the `<prereq>` element.

> **Note** As you probably guessed, the easiest way to find this section is to search for "Task Prereq," which is the name of the variable that inserts the label. And as you probably also guessed, there are similar sections in **task-elements.xsl** for the other task section labels.

The label occurs inside an `<fo:block>` that uses the *prereq* attribute set. So this attribute set applies to the `<prereq>` element and the label.

Within that `<fo:block>` is another `<fo:block>`, which formats the content of `<prereq>` using the *prereq__content* attribute set. You'll want to add appropriate attributes to this attribute set as well, so the content of `<prereq>` doesn't end up looking like the label.

2. Open `DITA-OT/plugins/org.dita.pdf2/cfg/fo/attrs/`**task-elements-attr.xsl** and find the *prereq* attribute set.

Notice that this attribute set is empty but uses the *section* attribute set, as do the *context*, *result*, and *postreq* attribute sets. Since you want all step section labels to look the same, you might think you could just use the *section* attribute set and get it all done at once. But that's a slippery slope.

The problem is that the Open Toolkit uses the *section* attribute set to format text in sections outside of tasks. If you made the *section* attribute set 12pt Trebuchet bold dark red, text within sections outside of tasks would also be 12pt Trebuchet bold dark red, and you don't want that! So, unfortunately, you need to add the desired attributes to each applicable attribute set.

The Open Toolkit formats task elements using a strange combination of attributes, some of which also apply to sections and examples outside of tasks. It can be hard to tease everything apart, so you may need to use some trial and error to get the right results.

3. Copy the *prereq*, *context*, *result*, *task.example*, *task.example__content*, and *postreq* attribute sets from **task-elements-attr.xsl** to your copy of **task-elements-attr.xsl**.

4. Add the following to those attribute sets:

```
<xsl:attribute name="color">#990033</xsl:attribute>
<xsl:attribute name="font-family">sans-serif</xsl:attribute>
<xsl:attribute name="font-size">12pt</xsl:attribute>
<xsl:attribute name="font-weight">bold</xsl:attribute>
```

Don't add this to the *steps* or *steps-unordered* attribute sets. They work differently.

5. In `DITA-OT/plugins/org.dita.pdf2/cfg/fo/attrs/`**task-elements-attr.xsl**, find the *prereq__content* attribute set.

Notice it's empty and uses the *section__content* attribute set. So do *context__content*, *result__content*, and *postreq__content*. If you want all three of these sections to be formatted the same, you can just modify *section__content*. This time it's okay to use the one-fell-swoop approach because the *section__content* attribute set only affects the content of sections within tasks. If you need each section to be formatted differently, you'll need to add attributes to each attribute set: *context__content*, *result__content*, and *postreq__content*. Let's keep things simple for now.

6. If it is not already present, copy the *section__content* attribute set from `DITA-OT/plugins/org.dita.pdf2/cfg/fo/attrs/`**commons-attr.xsl** to your copy of **commons-attr.xsl**.

7. Add the following attributes to *section__content*:

```
<xsl:attribute name="font-size"><xsl:value-of select="$default-font-size"/></xsl:attribute>
<xsl:attribute name="line-height"><xsl:value-of select="$default-line-height"/>
▶</xsl:attribute>
<xsl:attribute name="color">#8a8a8a</xsl:attribute>
<xsl:attribute name="font-family">serif</xsl:attribute>
<xsl:attribute name="font-weight">normal</xsl:attribute>
```

8. Save your changes, run a PDF build, and check your work.

The labels are all 12pt Trebuchet bold dark red, as expected. The text in `<prereq>`, `<context>`, `<result>` and `<postreq>` is 11pt Book Antiqua normal dark gray, also as expected. But the text in `<example>` is 12pt Trebuchet bold dark red, which is not right. That's because text in `<example>` is formatted by the *task.example__content* attribute set, which uses the *example__content* attribute set, not the *section__content* attribute set. Unlike the *section__content* attribute set, you don't want to add attributes to *example__content*, because they will affect text in examples outside of tasks.

9. Add the following to the *task.example__content*:

```
<xsl:attribute name="font-size"><xsl:value-of select="$default-font-size"/></xsl:attribute>
<xsl:attribute name="line-height"><xsl:value-of select="$default-line-height"/>
►</xsl:attribute>
<xsl:attribute name="color">#8a8a8a</xsl:attribute>
<xsl:attribute name="font-family">serif</xsl:attribute>
<xsl:attribute name="font-weight">normal</xsl:attribute>
```

10. Save your changes, run a PDF build, and check your work.

Now all the labels and text are correctly formatted. But there is a border around <example> that you need to get rid of. An <example> element within a task uses the *task.example* attribute set, which calls the *example* attribute set, which calls the *common.border* attribute set. You have to override some of those border attributes in *task.example*.

11. Edit the *task.example* attribute set as shown:

```
<xsl:attribute-set name="task.example" use-attribute-sets="example">
    <xsl:attribute name="color">#000000</xsl:attribute>
    <xsl:attribute name="font-family">sans-serif</xsl:attribute>
    <xsl:attribute name="font-size">12pt</xsl:attribute>
    <xsl:attribute name="font-weight">bold</xsl:attribute>
    <xsl:attribute name="border-top-style">none</xsl:attribute>
    <xsl:attribute name="border-bottom-style">none</xsl:attribute>
    <xsl:attribute name="border-left-style">none</xsl:attribute>
    <xsl:attribute name="border-right-style">none</xsl:attribute>
</xsl:attribute-set>
```

12. Save changes, run a PDF build, and check your work.

Exercise: Format labels for the step section

Objective: Specify the appearance of the step number or other label information.

In **Format labels for the step section** *(p. 263)*, I mentioned that the templates for <steps-unordered> and <steps> process their section labels differently from the other task sections, such as <context> and <prereq>.

For example, compare the template that processes `<prereq>`:

```
<xsl:template match="*[contains(@class, ' task/prereq ')]">
    <fo:block xsl:use-attribute-sets="prereq">
       <xsl:call-template name="commonattributes"/>
       <xsl:apply-templates select="." mode="dita2xslfo:task-heading">
          <xsl:with-param name="use-label">
             <xsl:apply-templates select="." mode="dita2xslfo:retrieve-task-heading">
                <xsl:with-param name="pdf2-string">Task Prereq</xsl:with-param>
                <xsl:with-param name="common-string">task_prereq</xsl:with-param>
             </xsl:apply-templates>
          </xsl:with-param>
       </xsl:apply-templates>
       <fo:block xsl:use-attribute-sets="prereq__content">
          <xsl:apply-templates/>
       </fo:block>
    </fo:block>
</xsl:template>
```

with the one that processes `<steps>`:

```
<xsl:template match="*[contains(@class, ' task/steps ')]">
    <xsl:choose>
        <xsl:when test="$GENERATE-TASK-LABELS='YES'">
          <fo:block>
             <xsl:apply-templates select="." mode="dita2xslfo:task-heading">
                <xsl:with-param name="use-label">
                   <xsl:apply-templates select="." mode="dita2xslfo:retrieve-task-heading">
                      <xsl:with-param name="pdf2-string">Task Steps</xsl:with-param>
                      <xsl:with-param name="common-string">task_procedure</xsl:with-param>
                   </xsl:apply-templates>
                </xsl:with-param>
             </xsl:apply-templates>
             <fo:list-block xsl:use-attribute-sets="steps">
                <xsl:call-template name="commonattributes"/>
                <xsl:apply-templates/>
             </fo:list-block>
          </fo:block>
        </xsl:when>
        <xsl:otherwise>
           <fo:list-block xsl:use-attribute-sets="steps">
              <xsl:call-template name="commonattributes"/>
              <xsl:apply-templates/>
           </fo:list-block>
        </xsl:otherwise>
    </xsl:choose>
</xsl:template>
```

In the first template, there is an outer `<fo:block>` that contains the label and is formatted by the *prereq* attribute set and within that `<fo:block>`, an inner `<fo:block>` that contains the prereq content and is formatted by the *prereq__content* attribute set. In the second template, there is an outer `<fo:block>` that contains the label and does not have an associated attribute set, and within that `<fo:block>` an `<fo:list-block>` that contains the individual steps. **Figure 36** shows how these attribute sets affect the steps section.

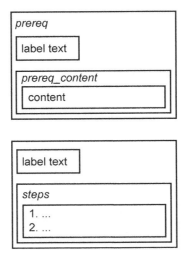

Figure 36: Attribute sets that affect the steps section

As you can see, any attributes that you add to the *steps* attribute set are not applied to the label for the steps section. You need to do a little extra work for this section.

1. Copy the template that processes `<steps>`, shown above, from `DITA-OT/plugins/org.dita.pdf2/xsl/fo/`**task-elements.xsl** to your copy of **task-elements.xsl**.

2. To the first `<fo:block>`, add the following

```
<xsl:when test="$GENERATE-TASK-LABELS='YES'">
    <fo:block xsl:use-attribute-sets="steps__label">
```

3. Open your copy of **task-elements-attr.xsl** and create a new attribute set named *steps__label*:

```
<xsl:attribute-set name="steps__label" use-attribute-sets="section">
    <xsl:attribute name="color">#990033</xsl:attribute>
    <xsl:attribute name="font-family">sans-serif</xsl:attribute>
    <xsl:attribute name="font-size">12pt</xsl:attribute>
    <xsl:attribute name="font-weight">bold</xsl:attribute>
</xsl:attribute-set>
```

You're setting this attribute set up exactly like the ones you changed in **Format labels for the step section** *(p. 263)*.

4. Save changes, run a PDF build, and check your work.

If necessary, repeat these steps for the template that processes `<steps-unordered>`.

Exercise: Format step and substep numbers

Objective: Specify the numbering format and following character for step and substep numbers, as well as the appearance of step and substep numbers.

1. If it is not already present, copy the template that processes steps from `DITA-OT/plugins/org.dita.pdf2/xsl/fo/`**task-elements.xsl** to your copy of **task-elements.xsl**.

The template begins with this line:

```
<xsl:template match="*[contains(@class, ' task/steps ')]/*[contains(@class, ' task/step ')]">
```

This code processes steps like a list, which is exactly what they are. If you worked through the chapter on lists, you remember that you use a localization variable to create the list number (**Create multiple numbering formats for ordered lists** *(p. 245)*). You do the same thing with step numbering. Out of the box, the plugin uses the same variable for step numbering as for list numbering: *Ordered List Number*. But you want step numbers to be different from list numbers, which means you need a new variable.

2. Within the pasted template, find this line:

```
<xsl:with-param name="id" select="'Ordered List Number'"/>
```

3. Change the variable name to *Step Number*:

```
<xsl:with-param name="id" select="'Step Number'"/>
```

4. In your **en.xml**, add a new variable named *Step Number*:

```
<variable id="Step Number"><param ref-name="number"/>)</variable>
```

5. Copy the *steps.step__label__content* attribute set from `DITA-OT/plugins/org.dita.pdf2/cfg/fo/attrs/`**task-elements-attr.xsl** to your copy of **task-elements-attr.xsl**.

This attribute set formats the step label.

6. Add the following attributes and values:

```
<xsl:attribute name="font-family">sans-serif</xsl:attribute>
<xsl:attribute name="font-size"><xsl:value-of select="$default-font-size"/></xsl:attribute>
<xsl:attribute name="font-weight">bold</xsl:attribute>
<xsl:attribute name="color">#990033</xsl:attribute>
```

7. Copy the template that processes substeps from
`DITA-OT/plugins/org.dita.pdf2/xsl/fo/`**task-elements.xsl** to your copy of **task-elements.xsl**.

The template begins with this line:

```
<xsl:template match="*[contains(@class, ' task/substeps ')]/*[contains(@class, ' task/substep ')]">
```

Unlike the template that processes steps, this template doesn't use a variable for the number. It just uses `<xsl:number>`. The `<xsl:number>` element in this template includes the numbering format and the punctuation: `<xsl:number format="a) "/>`.

8. Edit `<xsl:number>` as follows:

```
<xsl:number format="i. "/>
```

9. Copy the *substeps.substep__label__content* attribute set from
`DITA-OT/plugins/org.dita.pdf2/cfg/fo/attrs/`**task-elements-attr.xsl** to your copy of
task-elements-attr.xsl.

This attribute set formats the substep label.

10. Add the following attributes and values:

```
<xsl:attribute name="font-family">sans-serif</xsl:attribute>
<xsl:attribute name="font-size">10pt</xsl:attribute>
<xsl:attribute name="font-weight">bold</xsl:attribute>
<xsl:attribute name="color">#990033</xsl:attribute>
```

11. Save changes, run a PDF build, and check your work.

Other things you can do

Here are a few additional customizations you might want to make to your plugin. These customizations aren't part of the PDF specifications you're following, but they can be useful (and cool).

Exercise: Count the number of steps in a task

Objective: Add a line at the beginning of a task to tell readers how many steps are in the task.

Some usability experts claim that readers often think they've reached the end of a task when they reach the end of the page. If the task spans more than one page, readers may miss the final steps, causing their process to fail. Whether or not you agree, you might be asked to add some hints to make sure readers complete all the steps in each task. One such hint is a count of the steps in the task.

1. Copy the template that begins with the following line
   `DITA-OT/plugins/org.dita.pdf2/xsl/fo/`**task-elements.xsl** to your copy of **task-elements.xsl**:

```
<xsl:template match="*[contains(@class, ' task/steps ')]">
```

2. Add the following code:

```
<xsl:template match="*[contains(@class, ' task/steps ')]">
   <xsl:variable name="actual-step-count" select="count(*[contains(@class,' task/step ')])
   ▶"/>
   <fo:block>
      <xsl:if test="$actual-step-count &gt; 4">
         <xsl:text>This task has </xsl:text><xsl:value-of select="count(*
         ▶[contains(@class,' task/step ')])"/>
         <xsl:text> steps.</xsl:text>
      </xsl:if>
   </fo:block>
```

This code creates a variable named *actual-step-count*. Its value is the number of step elements within the task, calculated using the count function. The `<xsl:if>` statement tests to see if there are more than four steps; if not, the "This task has X steps" text is not output. You can change this number as you like. To output the statement regardless of the number of steps, change *actual-step-count* to 1.

3. Save changes, run a PDF build, and check your work.

As an alternative to hardcoding text such as "This task has…", consider creating a localization variable and using it instead. In this example, the English value of *Number of Steps* might be "Steps in this task: ":

```
<xsl:if test="$actual-step-count>4">
   <xsl:call-template name="getVariable">
      <xsl:with-param name="id" select="'Number of Steps'"/>
   </xsl:call-template>
   <xsl:value-of select="count(*[contains(@class,' task/step ')])"/>
</xsl:if>
```

Exercise: Create links to steps within the same task

Objective: Create links from one step to another within the same task that include a dynamic step number.

While it's a best practice to write tasks so they flow without interruption from the first step to the last, sometimes you need to create a looping task. When you need to do this, it's best to create a cross-reference so that if the steps are re-numbered, the step reference changes automatically.

1. Within a task, add an *id* attribute and value to the step you want to create the reference to.

   The *id* should be on the `<step>` element, and it can be anything you want. It only needs to be unique within the current topic, so you can keep it short and simple.

```
<task id="id1234567890">
   ...
   <step id="selfpay1">
      <cmd>Complete the Self-Pay Information screen as follows.</cmd>
   </step>
   ...
```

2. Within the same task, create the cross-reference like this example:

```
<step>
   <cmd>Select the patient's insurance carrier.</cmd>
   <info>
      <p>If the patient does not have insurance, click Self-Pay and go to
      ▸ step <xref href="#./selfpay1" type="li"/>.</p>
   </info>
</step>
```

   Notice that the first part of the *href* is simply the "self" axis (a period) and the second part of the *href* is the id of the step being referenced. (The use of the period is a new DITA 1.3 feature. If you're using an earlier version of DITA, you'll need to use the syntax `<xref href="#id1234567890/selfpay1" type="step"/>`.)

   Also notice the *type* attribute. It's very important, because you're going to use it as a trigger to process these cross-references differently from others. (You don't *have* to process them differently, but more than likely you don't want the standard "on page xx" verbiage with these links.)

3. Copy the following template from `DITA-OT/plugins/org.dita.pdf2/xsl/fo/`**links.xsl** to your copy of **links.xsl**:

```
<xsl:template name="insertPageNumberCitation">
   <xsl:param name="isTitleEmpty"/>
   <xsl:param name="destination"/>
   <xsl:param name="element"/>

   <xsl:choose>
      <xsl:when test="not($element) or ($destination = '')">
         <fo:inline xsl:use-attribute-sets="link__page">
            <xsl:call-template name="getVariable">
               ...
```

   This section contains the `<xsl:choose>` that includes the content of either the *Page* or the *On the page* localization variables after the link title, depending on various conditions. And those variables, in turn, are what create the "on page xx" verbiage you want to exclude.

4. Before the first `<xsl:when>` test, add a new test as follows:

```
<xsl:template name="insertPageNumberCitation">
   <xsl:param name="isTitleEmpty"/>
   <xsl:param name="destination"/>
   <xsl:param name="element"/>

   <xsl:choose>
      <xsl:when test="@type = 'li'">
         <!--do not output page number-->
      </xsl:when>
      <xsl:when test="not($element) or ($destination = '')">
         <fo:inline xsl:use-attribute-sets="link__page">
            <xsl:call-template name="getVariable">
               . . .
```

5. Save your changes and run a build to check your work.

The step-looping links no longer have any page numbers or verbiage.

If you're always going to say something like "Go to step xx," then you should use a localization variable. Why? Well, remember that when translating, every word costs money. If you use this phrase 100 times in content that you translate, that's 300 words you're paying to translate...297 of them unnecessarily. By turning the phrase into a localization variable, you only pay to translate those three words once per language.

When making this text into a localization variable, you'll need to consider case, position of the text and other aspects that might be context- and language-specific. For example, if the variable text begins with a lower-case letter, you'll always need to use it in a context where lower-case is appropriate.

6. Copy the template that begins with the following line from `DITA-OT/plugins/org.dita.pdf2/xsl/fo/`**links.xml** to your copy of **links.xml**:

```
<xsl:template match="*[contains(@class,' topic/xref ')]" name="topic.xref">
```

This is the template that processes basic `<xref>` links (as opposed to related links or family links).

7. Within that template, find this section:

```
<fo:basic-link xsl:use-attribute-sets="xref">
   <xsl:call-template name="buildBasicLinkDestination">
      <xsl:with-param name="scope" select="@scope"/>
      <xsl:with-param name="format" select="@format"/>
      <xsl:with-param name="href" select="@href"/>
   </xsl:call-template>

   <xsl:choose>
      <xsl:when test="not(@scope = 'external' or @format = 'html')
      ▶and not($referenceTitle = '')">
      . . .
```

The `<xsl:choose>` code processes the links differently depending on the properties of the `<xref>` element. You want to add a new test to process the link when it's a li-type link.

8. Add the following as the first `<xsl:when>` test within `<xsl:choose>`:

```
<xsl:choose>
   <xsl:when test="@type = 'li'">
      <fo:inline xsl:use-attribute-sets="xref_step">
         <xsl:call-template name="getVariable">
            <xsl:with-param name="id" select="'Go to step'"/>
         </xsl:call-template>
      </fo:inline>
      <xsl:copy-of select="$referenceTitle"/>
   </xsl:when>
```

This tests right away to see if the `<xref>` element's *type* attribute is set to *step*. If so, it outputs the value of the *Go to step* localization variable followed by the link content, which is the step number. Notice that you put the template call inside an `<fo:inline>` element that uses the *xref_step* attribute set. This lets you format the variable text differently from the step number. If you want them to have the same formatting, you don't need the `<fo:inline>` element.

> **Note** This approach assumes that having `type="li"` on an `<xref>` supersedes any other attributes you might have on the `<xref>`. If you use other attributes on the xref to control the link's appearance, those attributes will be ignored. If you anticipate using `type="li"` along with other attributes, you'll need to take a more complex approach, such as making the test for `type="li"` be a subtest of some of the other conditions.

You don't yet have a *Go to step* localization variable, so you need to add it. You also need to add the *xref_step* attribute set.

9. Add the following variable to your copy of **en.xml**:

```
<variable id="Go to step">Go to step </variable>
```

10. In your copy of **links-attr.xsl**, add the *xref_step* attribute set.

```
<xsl:attribute-set name="xref_step">
   <xsl:attribute name="color">black</xsl:attribute>
   <xsl:attribute name="font-style">normal</xsl:attribute>
</xsl:attribute-set>
```

You can, of course, add any other attributes you want. You probably want at least these two because the block that contains the text you're formatting with *xref_step* is formatted by the *xref* attribute set, which in turn uses the *common.link* attribute set. *Common.link* includes the *color* and *font-style* attributes, so you likely want to override those attributes in *xref_step*.

11. Review your content. If you previously had hardcoded text such as "Go to step" preceding the step links, you'll need to remove that text.

12. Save changes, run a PDF build, and check your work.

The step-type links are output with the text contained in *Go to step* followed by the step number of the referenced step. The *Go to step* text is plain black, while the step number is formatted using your regular link formatting.

Exercise: Add task troubleshooting templates

Objective: Create new templates to process `<steptroubleshooting>` and `<tasktroubleshooting>`.

DITA 1.3 offers two new elements for use in tasks to include troubleshooing information. You can use `<steptroubleshooting>` to include step-specific troublsehooting information. (Compare it to `<stepresult>`). You can use `<tasktroubleshooting>` to include troubleshooting information that applies to the entire task. (Compare it to `<result>`.)

There is no specific processing in the DITA Open Toolkit for these elements. That doesn't mean they don't show up in your PDF, only that they use the same fall-through processing as several other task elements. If you want to be specific about how these elements are processed and formatted, you need to add templates just for those elements. It's easy.

1. In `DITA-OT/plugins/org.dita.pdf2/xsl/fo/`**task-elements.xsl**, find the template that processes `<result>` elements.

   It begins with:

```
<xsl:template match="*[contains(@class, ' task/result ')]">
```

2. Copy the entire template and paste it into your **task-elements.xsl**.
3. Edit it as follows:

```
<xsl:template match="*[contains(@class, ' task/tasktroubleshooting ')]">
    <fo:block xsl:use-attribute-sets="tasktroubleshooting">
        <xsl:call-template name="commonattributes"/>
        <xsl:apply-templates select="." mode="dita2xslfo:task-heading">
            <xsl:with-param name="use-label">
                <xsl:apply-templates select="." mode="dita2xslfo:retrieve-task-heading">
                    <xsl:with-param name="pdf2-string">Task Troubleshooting</xsl:with-param>
                    <xsl:with-param name="common-string">task_troubleshooting</xsl:with-param>
                </xsl:apply-templates>
            </xsl:with-param>
        </xsl:apply-templates>
        <fo:block xsl:use-attribute-sets="tasktroubleshooting__content">
            <xsl:apply-templates/>
        </fo:block>
    </fo:block>
</xsl:template>
```

If you worked through **Format task section labels and text** *(p. 261)*, you'll remember how these templates work.

Notice that the template uses attribute sets named *tasktroubleshooting* and *tasktroubleshooting__content*. It also uses a localization variable named *Task Troubleshooting* and a localization string named *task_troubleshooting*. Of course, none of these exist yet. You'll add them in the next exercise.

4. Next, in `DITA-OT/plugins/org.dita.pdf2/xsl/fo/`**task-elements.xsl**, find the template that processes `<stepxmp>` elements.

It begins with:

```
<xsl:template match="*[contains(@class, ' task/stepxmp ')]">
```

5. Copy the entire template and paste it into your **task-elements.xsl**.

6. Edit it as follows:

```
<xsl:template match="*[contains(@class, ' task/steptroubleshooting ')]">
    <fo:block xsl:use-attribute-sets="steptroubleshooting">
        <xsl:call-template name="commonattributes"/>
        <xsl:apply-templates/>
    </fo:block>
</xsl:template>
```

This new template uses an attribute set named *steptroubleshooting*. It doesn't exist either. You'll add it in the next exercise.

Complete the next exercise, **Add task troubleshooting attribute sets and variables** *(p. 273)*.

Exercise: Add task troubleshooting attribute sets and variables

Objective: Create new attribute sets and localization variables for `<steptroubleshooting>` and `<tasktroubleshooting>`.

Before you start

Complete the previous exercise, **Add task troubleshooting templates** *(p. 272)*.

Now that you've added templates specifically for `<steptroubleshooting>` and `<tasktroubleshooting>`, you need to add the attribute sets and variables used by those templates so that you can format these elements as you want.

1. In `DITA-OT/plugins/org.dita.pdf2/cfg/fo/attrs/`**task-elements-attr.xsl**, copy the *result* attribute set.

If you completed **Format task section labels and text** *(p. 261)*, that attribute set probably looks like this:

```
<xsl:attribute-set name="result" use-attribute-sets="section">
   <xsl:attribute name="color">#990033</xsl:attribute>
   <xsl:attribute name="font-family">sans-serif</xsl:attribute>
   <xsl:attribute name="font-size">12pt</xsl:attribute>
   <xsl:attribute name="font-weight">bold</xsl:attribute>
</xsl:attribute-set>
```

2. Paste it into your **task-elements-attr.xsl**.

3. Rename the pasted attribute set *tasktroubleshooting*:

```
<xsl:attribute-set name="tasktroubleshooting" use-attribute-sets="section">
```

4. In `DITA-OT/plugins/org.dita.pdf2/cfg/fo/attrs/`**task-elements-attr.xsl**, copy the *result__content* attribute set.

This attribute set is empty by default.

5. Paste it into your **task-elements-attr.xsl**.

6. Rename the pasted attribute set *tasktroubleshooting__content*.

7. In `DITA-OT/plugins/org.dita.pdf2/cfg/fo/attrs/`**task-elements-attr.xsl**, copy the *stepxmp* attribute set.

This attribute set is empty by default.

8. Paste it into your **task-elements-attr.xsl**.

9. Rename the pasted attribute set *steptroubleshooting*.

If you have opted to use localization variables for task tables, go to step 10. If you have opted to use strings, go to step 11.

10. Add the following variable to your **en.xml**:

```
<variable id="Task Troubleshooting">Help me fix it</variable>
```

11. Add the following string to your **strings-en-us.xml**:

```
<str name="task_troubleshooting">Help me fix it</str>
```

12. Save changes, run a PDF build, and check your work.

Chapter 13

Tables

The DITA `<table>` element uses the CALS table model, which is a standard for representing tables in XML. DITA also includes a table type specific to DITA, `<simpletable>`, which is best used for tables that have a regular structure, don't need a title, and don't need much in the way of formatting.

On the other hand, a `<table>` element can take advantage of formatting attributes such as *rowsep*, *colsep*, *colwidth*, *align*, and others to give the table a precise appearance.

This chapter explains how to use some of these attributes. It explains how to format text in table header and body cells and how to format and position the table caption. And there's even an exercise to automatically number `<simpletable>` cells if you ever need to do that.

For a list of attributes that affect tables, refer to **Table attribute sets** *(p. 451)*.

Table specifications

Here are your specifications for tables.

Tables

Table text font	10pt Book Antiqua normal dark gray
Table cell text padding	6pt all around
Table header row background color	dark red
Table header row font	11pt Trebuchet bold white
Table title	12pt Trebuchet bold dark red, below table

Table frame	2pt dark red
Table rules	2pt dark red under the heading row; 1pt dark gray all other rows and columns
Table title numbering	Includes "Table" chapter/appendix number, table number (Table: 1-1). Table numbering restarts with each chapter/appendix.
Table column widths	respect when specified

Files you need

If you have completed other exercises, some of these files might already be present in your plugin.

If not already present, create the following files in `DITA-OT/plugins/com.company.pdf/cfg/fo/attrs/`:

- **commons-attr.xsl**
- **tables-attr.xsl**
- **tables-attr_fop.xsl** (needed only if you're using FOP and need attributes in this file)

Be sure to add the appropriate `xsl:import` statements to **custom.xsl** (attrs).

If not already present, create the following files in `DITA-OT/plugins/com.company.pdf/cfg/fo/xsl/`:

- **tables.xsl**
- **tables_fop.xsl** (needed only if you're using FOP and need templates in this file)

Be sure to add the appropriate `xsl:import` statements to **custom.xsl** (xsl).

For instructions on creating attribute files and XSLT stylesheets in your PDF plugin, refer to **Create an attribute set file in your plugin** *(p. 31)* and **Create an XSLT stylesheet in your plugin** *(p. 32)*.

If not already present, create the following file in `DITA-OT/plugins/com.company.pdf/cfg/common/vars/`:

- **en.xml**

For instructions on creating localization variables fles, refer to **Create a localization variables file** *(p. 78)*.

Fun with table rules

Formatting table rules is one of those things that is just a little harder than it needs to be. There are 13 attribute sets just for table frames and rules.

Before talking about table rule attribute sets, it's important to understand the distinction between table frames and table rules. The table frame is the outside border of the table. It is controlled by the *frame* attribute on the table element, which has these possible values: all, bottom, sides, top, topbot, none.

Table rules are the inner vertical and horizontal lines that separate rows and columns. These are controlled by the *rowsep* and *colsep* attributes, which you can use on the following elements: `<table>`, `<tgroup>`, `<row>` (*rowsep* only), and `<entry>`. *Rowsep* and *colsep* have the values "1" (show the rule) or "0" (don't show the rule). You can use *rowsep* and *colsep* on the `<table>` element or on individual `<row>` or `<entry>` elements. Refer to the DITA specification or other DITA resources for a more information on *rowsep* and *colsep*.

So, back to the attribute sets. One reason there are so many is that there is a separate attribute set for each value of the *frame* attribute on the table element except "none". To make things a little more complicated, the attribute set that corresponds to the value "sides" (*table__tableframe__sides*) calls two other attribute sets (*table__tableframe__right*, *table__tableframe__left*), and those two attribute sets in turn call *common.border__right* and *common.border__left*, respectively.

You may have seen this diagram earlier, but here it is again to show how the table attribute sets are related:

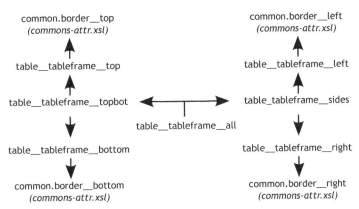

There are other table-rule-related attribute sets. You don't need all the details about them. Just be aware that by default, all of them ultimately point back to *common.border__top*, *common.border__bottom*, *common.border__left*, or *common.border__right*, which are all found in **commons-attr.xsl**. These attribute sets also affect any borders you might place under `<title>`, `<titlealt>`, `<examples>`, `<lq>`, or

`<draft-comment>`. Unless you're lucky enough to want exactly the same kind of border in all these places, odds are you're going to have to do some overriding.

Let's look at a few examples of doing just that.

Table border overrides

The combination of the *frame* attribute on a `<table>` element, and the *colsep*, and *rowsep* attributes on a `<table>`, `<row>` or `<entry>` elements gives you a certain amount of flexibility with table rules.

This flexibility can create confusion when you need to create exceptions to the overall ruling for individual rows or entries in a table, particularly if you need to override rules created by the *frame* attribute.

Say you set `frame="all"` for the table. By default, the PDF plugin creates a table with a 1pt black solid lines on all four sides. Now, suppose you want the header row to have no side rules, like this

State	Capitol	Population
Alabama	Montgomery	374,536
Alaska	Juneau	33,275
Arizona	Phoenix	4,192,887

After a little investigation (from looking at **The topic.fo file** *(p. 49)*, for example), you find out that when you set `frame="all"`, the resulting code that's created and sent to your PDF renderer is this:

```
<fo:table table-layout="fixed" width="100%" space-before="5pt" space-after="5pt"
border-before-style="solid" border-before-width="1pt" border-before-color="black"
border-after-style="solid" border-after-width="1pt" border-after-color="black"
border-end-style="solid" border-end-width="1pt" border-end-color="black"
border-start-style="solid" border-start-width="1pt" border-start-color="black">...
```

You think to yourself, "For the header row, all I need to do is override those *border-width* attributes with a value of 0pt, and I should be set." Unfortunately, it's not that easy. The table header row uses the *thead.row* attribute set. If you edit that attribute set like this:

```
<xsl:attribute-set name="thead.row">
   <xsl:attribute name="border-start-width">0pt</xsl:attribute>
   <xsl:attribute name="border-end-width">0pt</xsl:attribute>
</xsl:attribute-set>
```

you'll find that your side borders are still there. Why is that?

It's because of the way the PDF plugin draws tables. It first draws the outline of the table and uses the *frame* attribute to determine which of the four sides should have a visible border. Next, it draws the header and its row and individual entries. Finally, it draws the body and each of its rows and individual entries. As it draws each row, the plugin looks for *rowsep* values, either on the `<table>` element or on the current `<row>` element. Based on what it finds, the plugin either creates or does not create a visible line. Similarly, within each row, the plugin looks for *colsep* values either on `<table>` or on the current `<entry>`, and based on what it finds, it either creates or does not create column separators for that entry.

So in this example, there is already a 1pt black solid border on the sides of the header row by the time the plugin starts drawing the header row itself (because `frame="all"`). The plugin cannot erase this border with a simple override like the one shown in the *thead.row* attribute set above. You actually have to draw additional side borders on the header row that overlay the 1pt solid black ones and hide them, which means you need the full complement of attributes in the *thead.row* attribute set to match the ones created on the `<fo:table>` element by `frame="all"`:

```
<xsl:attribute-set name="thead.row">
   <xsl:attribute name="border-end-style">solid</xsl:attribute>
   <xsl:attribute name="border-end-width">1pt</xsl:attribute>
   <xsl:attribute name="border-end-color">white</xsl:attribute>
   <xsl:attribute name="border-start-style">solid</xsl:attribute>
   <xsl:attribute name="border-start-width">1pt</xsl:attribute>
   <xsl:attribute name="border-start-color">white</xsl:attribute>
</xsl:attribute-set>
```

If you fail to override every single one of the side border-related attributes on the `<fo:table>` element, you'll get a side border somewhere on the table header row.

The same is not true for overriding individual row and column rules, because the rules for those items are drawn when the row or entry itself is processed, instead of at the very beginning of table processing.

Exercise: Force table column widths to be respected

Objective: Force columns in a table to be rendered precisely at the width indicated within the topic.

If you specify column widths in a table using `<colspec>`, you might be surprised by how those tables are rendered in a PDF. Two of the common PDF renderers respect your column widths, while the third forces the table to span from margin to margin, possibly stretching your columns (I won't name names).

The *pgwide* attribute on tables allows you to manually override the default margin-to-margin span and specify that the table's total width should be the sum of the widths of all its columns, instead of spanning from margin to margin. When `pgwide="1"`, the table's width equals the combined specified width of its columns. When `pgwide="0"`, the table stretches from margin to margin, usually making the columns wider than their specified amount (though the widths remain proportional).

So the most out-of-the-box solution for that one PDF renderer is simply to make sure that all of your tables specify pgwide="1". But what if you always want your table columns to retain the widths specified by authors, and you don't want to require them to specify pgwide="1" on every single table? In that case, you'll need to change the code in **tables.xsl**.

1. Copy this template from DITA-OT/plugins/org.dita.pdf2/xsl/fo/**tables.xsl** to your copy of **tables.xsl**:

```
<xsl:template match="*[contains(@class, ' topic/tgroup ')]" name="tgroup">
    <xsl:if test="not(@cols)">
        <xsl:call-template name="output-message">
            <xsl:with-param name="id" select="'PDFX006E'"/>
        </xsl:call-template>
    </xsl:if>
    <xsl:variable name="scale" as="xs:string?">
        <xsl:call-template name="getTableScale"/>
    </xsl:variable>
    <xsl:variable name="table" as="element()">
        <fo:table xsl:use-attribute-sets="table.tgroup">
            <xsl:call-template name="commonattributes"/>
            <xsl:call-template name="displayAtts">
                <xsl:with-param name="element" select=".."/>
            </xsl:call-template>
            <xsl:if test="(parent::*/@pgwide) = '1'">
                <xsl:attribute name="start-indent">0</xsl:attribute>
                <xsl:attribute name="end-indent">0</xsl:attribute>
                <xsl:attribute name="width">auto</xsl:attribute>
            </xsl:if>
            <xsl:apply-templates/>
        </fo:table>
    </xsl:variable>
    <xsl:choose>
        <xsl:when test="exists($scale)">
            <xsl:apply-templates select="$table" mode="setTableEntriesScale"/>
        </xsl:when>
        <xsl:otherwise>
            <xsl:copy-of select="$table"/>
        </xsl:otherwise>
    </xsl:choose>
</xsl:template>
```

The bold code sets the indent and width attributes for the table based on the value of *pgwide*. If you want these values to be in effect regardless of the value of *pgwide*, or even if *pgwide* is not present, you need to edit the <xsl:if> test so that they are applied in all cases.

2. Change the <xsl:if> test as shown:

```
<xsl:if test="(parent::*/@pgwide)='1' or (parent::*/@pgwide)='0'
    or not((parent::*/@pgwide))">
    <xsl:attribute name="start-indent">0</xsl:attribute>
    <xsl:attribute name="end-indent">0</xsl:attribute>
    <xsl:attribute name="width">auto</xsl:attribute>
</xsl:if>
```

3. Save changes, run a PDF build, and check your work.

Exercise: Change or delete the label for a table title

Objective: Change the label text (such as **Table:**) for a table title, or delete the label altogether.

1. Copy the following variables from `DITA-OT/plugins/org.dita.pdf2/cfg/common/vars/`**en.xml** to your copy of **en.xml**:

```
<variable id="Table.title">Table <param ref-name="number"/>: <param ref-name="title"/>
►</variable>
<variable id="Table Number">Table <param ref-name="number"/></variable>
```

These two variables are similar, but not quite the same. To understand how they work, take a look at **Including titles and autonumbers separately in table and figure links** *(p. 316)*.

Be aware that changes to these variables also affect the format of links to tables.

2. Do one of the following:

- To change the label, enter the new value in the *Table.title* variable:

```
<variable id="Table.title">Grid <param ref-name="number"/>:
► <param ref-name="title"/></variable>
```

- To delete the label, remove the current value:

```
<variable id="Table.title"><param ref-name="number"/>: <param ref-name="title"/></variable>
```

3. Repeat this process for any other language-specific variables files you've customized.

4. Save changes, run a PDF build, and check your work.

Exercise: Format the table heading row

Objective: Determine the appearance of the heading row in a table.

According to the template specifications, the table header row background color should be dark red, and the font should be 11pt Trebuchet bold white. Here's how to set that up.

1. Copy the *thead.row.entry* and *thead.row.entry__content* attribute sets from `DITA-OT/plugins/org.dita.pdf2/cfg/fo/attrs/`**tables-attr.xsl** to your copy of **tables-attr.xsl**

2. In *thead.row.entry*, add the *background-color* attribute with a value of "#990033":

```
<xsl:attribute-set name="thead.row.entry">
  <xsl:attribute name="background-color">#990033</xsl:attribute>
</xsl:attribute-set>
```

3. Add the following attributes and values to *thead.row.entry__content*:

```
<xsl:attribute name="font-family">Trebuchet MS, Arial Unicode MS, Helvetica</xsl:attribute>
<xsl:attribute name="font-size">11pt</xsl:attribute>
<xsl:attribute name="color">#ffffff</xsl:attribute>
```

This attribute set uses two other attribute sets: *common.table.body.entry* sets 3pt margins around the text, and *common.table.head.entry* sets the heading text to bold. Because the *common.table.head.entry* attribute set specifies bold text, you don't need to specify it here.

4. Save changes, run a PDF build, and check your work.

Exercise: Format table cell text

Objective: Determine the appearance of text within table cells.

Based on the PDF specifications, table cell text is 10pt Book Antiqua regular dark gray. You also want 6pt of padding all around the text.

1. Copy the attribute set *tbody.row.entry__content* from DITA-OT/plugins/org.dita.pdf2/cfg/fo/attrs/**tables-attr.xsl** to your copy of **tables-attr.xsl**.

2. Add the following attributes and values to *tbody.row.entry__content*:

```
<xsl:attribute name="font-family">Book Antiqua, Times New Roman, Times</xsl:attribute>
<xsl:attribute name="font-size">10pt</xsl:attribute>
<xsl:attribute name="color">#8A8A8A</xsl:attribute>
```

This attribute set uses another attribute set, *common.table.body.entry*, which sets 3pt margins around the text. You want 6pt margins, so you need to override the attributes in *common.table.body.entry* with the same attributes in *tbody.row.entry__content*.

3. Add the following attributes and values to *tbody.row.entry__content*:

```
<xsl:attribute name="space-before">6pt</xsl:attribute>
<xsl:attribute name="space-after">6pt</xsl:attribute>
<xsl:attribute name="start-indent">6pt</xsl:attribute>
<xsl:attribute name="end-indent">6pt</xsl:attribute>
```

4. Save changes, run a PDF build, and check your work.

Exercise: Format table rules

Objective: Determine the appearance of the rules (lines) in a table.

The PDF specification says that table frames and the rule under the table heading row should be 2pt dark red, but all other table rules should be 1pt dark gray.

> **Note** These steps describe one way to override the specifications in the *common.border__top*, *common.border__bottom*, *common.border__left*, and *common.border__right* attribute sets, which all table-rule-related attribute sets ultimately use. There are other methods as well, so feel free to explore what works best for you.

1. Add the following new attribute sets to your copy of **commons-attr.xsl**:

```
<xsl:attribute-set name="table.frame__top">
    <xsl:attribute name="border-top-style">solid</xsl:attribute>
    <xsl:attribute name="border-top-width">2pt</xsl:attribute>
    <xsl:attribute name="border-top-color">#990033</xsl:attribute>
</xsl:attribute-set>

<xsl:attribute-set name="table.frame__bottom">
    <xsl:attribute name="border-bottom-style">solid</xsl:attribute>
    <xsl:attribute name="border-bottom-width">2pt</xsl:attribute>
    <xsl:attribute name="border-bottom-color">#990033</xsl:attribute>
</xsl:attribute-set>

<xsl:attribute-set name="table.frame__right">
    <xsl:attribute name="border-right-style">solid</xsl:attribute>
    <xsl:attribute name="border-right-width">2pt</xsl:attribute>
    <xsl:attribute name="border-right-color">#990033</xsl:attribute>
</xsl:attribute-set>

<xsl:attribute-set name="table.frame__left">
    <xsl:attribute name="border-left-style">solid</xsl:attribute>
    <xsl:attribute name="border-left-width">2pt</xsl:attribute>
    <xsl:attribute name="border-left-color">#990033</xsl:attribute>
</xsl:attribute-set>
```

You'll use these new attribute sets for table frames. This way, your changes won't interfere with border settings for titles, examples, or any other elements.

2. Just after those new attribute sets, add the following new attribute sets:

```
<xsl:attribute-set name="table.rule__top">
    <xsl:attribute name="border-top-style">solid</xsl:attribute>
    <xsl:attribute name="border-top-width">1pt</xsl:attribute>
    <xsl:attribute name="border-top-color">#8A8A8A</xsl:attribute>
</xsl:attribute-set>

<xsl:attribute-set name="table.rule__bottom">
    <xsl:attribute name="border-bottom-style">solid</xsl:attribute>
    <xsl:attribute name="border-bottom-width">1pt</xsl:attribute>
    <xsl:attribute name="border-bottom-color">#8A8A8A</xsl:attribute>
```

```
</xsl:attribute-set>

<xsl:attribute-set name="table.rule__right">
    <xsl:attribute name="border-right-style">solid</xsl:attribute>
    <xsl:attribute name="border-right-width">1pt</xsl:attribute>
    <xsl:attribute name="border-right-color">#8A8A8A</xsl:attribute>
</xsl:attribute-set>

<xsl:attribute-set name="table.rule__left">
    <xsl:attribute name="border-left-style">solid</xsl:attribute>
    <xsl:attribute name="border-left-width">1pt</xsl:attribute>
    <xsl:attribute name="border-left-color">#8A8A8A</xsl:attribute>
</xsl:attribute-set>
```

You'll use these new attribute sets for table rules. This way your table rule settings won't interfere with border settings for titles, examples, or any other elements.

3. Copy the attribute sets in the left column of the following table from `DITA-OT/plugins/org.dita.pdf2/cfg/fo/attrs/`**tables-attr.xsl** to your copy of **tables-attr.xsl**:

attribute set	uses new attribute set
__tableframe__top	table.rule__top
__tableframe__bottom	table.rule__bottom
thead__tableframe__bottom	table.frame__bottom
__tableframe__left	table.rule__left
__tableframe__right	table.rule__right
table__tableframe__top	table.frame__top
table__tableframe__bottom	table.frame__bottom
table__tableframe__right	table.frame__right
table__tableframe__left	table.frame__left

4. For each attribute, change the value of the *use-attribute-sets* attribute from the current value to the value shown in the right column of the table above.

For example, by default the *__tableframe__top* attribute set uses the *common.border__top* attribute set. But so do some other attribute sets. If you make changes to *common.border__top*, you'll end up changing things you don't want to. So by editing *__tableframe__top* so it uses a different attribute set, *table.rule__top*, any changes you make to *table.rule__top* will affect only the things you want it to:

If you don't understand why you can't just delete *use-attribute-sets* from these attribute sets and add the attributes you need, take a look at **How attribute set defaults work** *(p. 56)*.

5. Save changes, run a PDF build, and check your work.

Exercise: Format the table title

Objective: Determine the appearance of a table's caption.

Based on the PDF specifications, you want the table caption to be 12pt Trebuchet bold dark red.

1. Copy the *table.title* attribute set from `DITA-OT/plugins/org.dita.pdf2/cfg/fo/attrs/`**tables-attr.xsl** to your copy of **tables-attr.xsl**.

2. Add or change the following attributes:

```
<xsl:attribute name="font-size">12pt</xsl:attribute>
<xsl:attribute name="color">#990033</xsl:attribute>
```

This attribute set also calls the *base-font* attribute set, which sets the default font size you're overriding here, and the *common.title* attribute set. In the **Specify fonts to use** *(p. 83)* exercise, you set the font in *common.title* to Trebuchet MS, so you don't need to specify the font again in *table.title* unless you want to override the setting in *common.title*.

3. Save changes, run a PDF build, and check your work.

Exercise: Place titles below tables

Objective: Place table captions below tables rather than above, which is the default.

By default, the PDF plugin places table titles above the table. It would be great if it were as easy to move them under the table as it is to move figure titles (see **Place titles above images** *(p. 305)*), but it isn't. Here's what you need to do.

1. Add the following to **custom.xsl** (xsl) after all of the `<xsl:import>` statements:

```
<xsl:template match="*[contains(@class,' topic/table ')]/*[contains(@class,' topic/title ')]"
/>
```

By default, the PDF plugin processes the title before the rest of the table because `<title>` precedes `<tgroup>` within the `<table>` element. Therefore, the default is to place the title above the table. This line blocks that order-based processing. At this point, if you were to generate a PDF, there wold be no titles for any of your tables.

2. Copy the following template from `DITA-OT/plugins/org.dita.pdf2/xsl/fo/`**tables.xsl** to your copy of **tables.xsl**

```
<xsl:template match="*[contains(@class, ' topic/table ')]/*[contains(@class, ' topic/title ')]
►">
   <fo:block xsl:use-attribute-sets="table.title">
      <xsl:call-template name="commonattributes"/>
      <xsl:call-template name="getVariable">
         <xsl:with-param name="id" select="'Table.title'"/>
         <xsl:with-param name="params">
            <number>
               <xsl:apply-templates select="." mode="table.title-number"/>
            </number>
            <title>
               <xsl:apply-templates/>
            </title>
         </xsl:with-param>
      </xsl:call-template>
   </fo:block>
</xsl:template>
```

(Notice this is the same template that you blocked with the line in **custom.xsl**.)

3. Change the first line to:

```
<xsl:template match="*[contains(@class, ' topic/table ')]/*[contains
►(@class, ' topic/title ')]" mode="titleBelow">
```

Using *mode* allows the node (in this case, the `<title>` element in a table) to be processed more than once. In this case, you're setting up a processing instance that can be called by setting the *mode* attribute to **titleBelow**.

4. If it is not present, copy this template from `DITA-OT/plugins/org.dita.pdf2/xsl/fo/`**tables.xsl** to your copy of **tables.xsl**:

```
<xsl:template match="*[contains(@class, ' topic/tgroup ')]" name="tgroup">
   <xsl:if test="not(@cols)">
      <xsl:call-template name="output-message">
         <xsl:with-param name="id" select="'PDFX006E'"/>
      </xsl:call-template>
   </xsl:if>
   <xsl:variable name="scale" as="xs:string?">
      <xsl:call-template name="getTableScale"/>
   </xsl:variable>
   <xsl:variable name="table" as="element()">
      <fo:table xsl:use-attribute-sets="table.tgroup">
         <xsl:call-template name="commonattributes"/>
         <xsl:call-template name="displayAtts">
            <xsl:with-param name="element" select=".."/>
         </xsl:call-template>
```

```
        <xsl:if test="(parent::*/@pgwide) = '1'">
            <xsl:attribute name="start-indent">0</xsl:attribute>
            <xsl:attribute name="end-indent">0</xsl:attribute>
            <xsl:attribute name="width">auto</xsl:attribute>
        </xsl:if>
        <xsl:apply-templates/>
    </fo:table>
    <xsl:apply-templates select="preceding-sibling::*[contains(@class, ' topic/title ')]"
    ▶ mode="titleBelow"/>
    </xsl:variable>
    <xsl:choose>
        <xsl:when test="exists($scale)">
            <xsl:apply-templates select="$table" mode="setTableEntriesScale"/>
        </xsl:when>
        <xsl:otherwise>
            <xsl:copy-of select="$table"/>
        </xsl:otherwise>
    </xsl:choose>
</xsl:template>
```

5. Add the line highlighted above.

The `<apply-templates>` line selects the preceding sibling of `<tgroup>` that has the class `topic/title`, which just happens to be the `<title>`. And it sets the mode to **titleBelow**, which causes the XSL processor to trigger the template you modified in Step 3.

At a high level, here's what happened. You set up the title processing to happen twice, using the mode. You suppressed the default processing (before table) with the line in **custom.xsl**. Then you added a second processing instance with the new `<apply-templates>` line, which selects the `<title>` and outputs it after the `<tgroup>`.

6. In the above template, find the following line:

```
<xsl:variable name="table" as="element()">
```

7. Edit this line to read as follows:

```
<xsl:variable name="table" as="element()*">
```

8. Open your copy of **tables-attr.xsl** and find the *table.title* attribute set.

9. Change the value of *keep-with-next.within-column* to "auto".

10. Add the *keep-with-previous.within.column* attribute, with a value of "always".

```
<xsl:attribute name="keep-with-previous.within-column">always</xsl:attribute>
```

You're adding this attribute because you moved the caption below the table, and you always want it to stay with the table. The *keep-with-next* attribute no longer applies, though it would be applicable if you were leaving the caption above the table. But since you can't delete the *keep-with-next* attribute, you can inactivate it by changing the value to "auto". Then you add the *keep-with-previous.within.column* attribute, which does keep the title with the table above it.

If you don't remember why you can't just delete the *keep-with-next* attribute, take a look at **How attribute set defaults work** *(p. 56)*.

11. Save changes, run a PDF build, and check your work.

Exercise: Add chapter, appendix or part numbers to table titles

Objective: Use the *getChapterPrefix* template method to number table captions.

Out of the box, the PDF plugin doesn't add chapter, part, or appendix numbers to table titles. Fortunately, it's easy to add them yourself. You will use the *getChapterPrefix* template to add chapter numbers to table titles (If you've already done the **Add chapter, appendix, or part numbers to TOC page numbers** *(p. 336)* exercise, then you already have this template in your plugin, and you can skip Step 1).

1. If it is not already present, copy the entire *getChapterPrefix* template into **custom.xsl** (xsl).

You can find this template in the **GetChapterPrefix template** *(p. 481)* appendix.

> **Note** You could copy this template into any of the .xsl files you've added to your plugin, including **tables.xsl**, but it's probably best to copy it into **custom.xsl**. That way you can see right away it's a custom template that didn't come with the PDF plugin.

2. Copy the following template from DITA-OT/plugins/org.dita.pdf2/xsl/fo/**tables.xsl** to your copy of **tables.xsl**.

```
<xsl:template match="*[contains(@class, ' topic/table ')]/*[contains(@class, ' topic/title ')
►]">
    <fo:block xsl:use-attribute-sets="table.title">
    <xsl:call-template name="commonattributes"/>
    <xsl:call-template name="getVariable">
        <xsl:with-param name="id" select="'Table.title'"/>
        <xsl:with-param name="params">
            <number>
                <xsl:apply-templates select="." mode="table.title-number"/>
            </number>
            <title>
                <xsl:apply-templates/>
            </title>
        </xsl:with-param>
    </xsl:call-template>
    </fo:block>
</xsl:template>
```

This template processes table titles. (If you moved titles below tables, you will also see the attribute mode="titleBelow" on the first line.)

3. Add the highlighted code to the `<number>` section, as shown below:

```
<number>
   <xsl:call-template name="getChapterPrefix"/>
   <xsl:text>-</xsl:text>
   <xsl:apply-templates select="." mode="table.title-number"/>
</number>
```

4. Save **tables.xsl**, run a build, and check your work.

At this point, tables are numbered consecutively from the beginning of the book to the end. If that's what you want, you can stop here. If you need numbering to restart with each chapter, keep going.

5. Edit the `<number>` element in the code above as shown:

```
<number>
   <xsl:call-template name="getChapterPrefix" />
   <xsl:text>-</xsl:text>
   <xsl:choose>
      <xsl:when test="count(ancestor-or-self::*[contains(@class, ' topic/topic')]
      ▸[position()=last()][count(preceding-sibling::*
      ▸[contains(@class, ' topic/topic')]) &gt; 0])">
         <xsl:value-of select="count(./preceding::*
         ▸[contains(@class, ' topic/fig ')]
         ▸[child::*[contains(@class, ' topic/title ')]][ancestor-or-self::*
         ▸[contains(@class, ' topic/topic')][position()=last()]]) -
         ▸count(ancestor-or-self::*[contains(@class, ' topic/topic')]
         ▸[position()=last()]/preceding-sibling::*[contains(@class, ' topic/topic')]
         ▸//*[contains(@class, ' topic/fig ')][child::*
         ▸[contains(@class, ' topic/title ')]])+1"/>
      </xsl:when>
      <xsl:otherwise>
         <xsl:value-of select="count(./preceding::*[contains(@class, ' topic/fig ')]
         ▸[child::*[contains(@class, ' topic/title ')]][ancestor-or-self::
         ▸*[contains(@class, ' topic/topic')][position()=last()]])+1"/>
      </xsl:otherwise>
   </xsl:choose>
</number>
```

(Delete the line `<xsl:apply-templates select="." mode="table.title-number"/>`.)

> **Note** When you copy the code from this example and see a line-break marker (▸), make sure you don't leave any space between the end of the previous line and the character that follows the marker. Yes, those really are very long test and select attribute values.

6. Save changes, run a PDF build, and check your work.

Other things you can do

Here are a few additional customizations you might want to make to your plugin. These customizations aren't part of the PDF specifications you're following, but they can be useful (and cool).

Exercise: Add a bottom table rule when a table breaks across a page

Objective: Specify that the last row of a table that breaks across a page should have a bottom rule.

By default, the PDF plugin omits row rules when a table breaks across a page. Some writers use this to cue readers that the table continues on the next page. Other writers prefer to keep the row rules. Fortunately, this is easy to change.

1. Copy the attribute sets *table__tableframe__all* and *table__tableframe__topbot* from DITA-OT/plugins/org.dita.pdf2/cfg/fo/attrs/**tables-attr.xsl** to your copy of **tables-attr.xsl**.

   When a table breaks across a page, it essentially becomes two tables, at least for the purposes of creating bottom and top rules. If your table does not have top and bottom rules (`frame="sides"` or `frame="none"`), then the issue is moot. However, if the table has rules on all four sides (`frame="all"`) or on the top and bottom (`frame="topbot"`), you need to edit these two attribute sets.

2. Add the following attributes to both attribute sets.

   ```
   <xsl:attribute name="border-before-width.conditionality">retain</xsl:attribute>
   <xsl:attribute name="border-after-width.conditionality">retain</xsl:attribute>
   ```

3. Save changes, run a PDF build, and check your work.

Exercise: Number table cells

Objective: Apply autonumbering to table cells.

When the parts of a diagram need to be labeled, many writers use numbers for the labels so the diagram doesn't contain any text to be translated. The labels can then be used as keys to the text, which is placed in an ordered list below the diagram. If the list is long, it's a real space-saver to present it in a two-column table such as **Figure 37**:

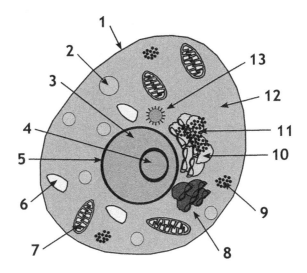

1. Cell membrane	2. Lysosome
3. Nucleus	4. Nucleolus
5. Nuclear membrane	6. Vacuole
7. Mitochondria	8. Golgi body
9. Ribosomes	10. Smooth ER
11. Rough ER	12. Cytoplasm
13. Centrosome	

Figure 37: Example of numbered table cells for diagram caption

> **Note** For obvious reasons, it's going to be a lot easier to maintain a table where the entries are numbered across each row, rather than down one column, then down the other. If you number down one column, then down the other, you'll have to rearrange the entire table each time you need to add an item. For this reason, stick with the row-across approach.

1. Copy the template that begins with the code below from DITA-OT/plugins/org.dita.pdf2/xsl/fo/**tables.xsl** to your copy of **tables.xsl**:

```
<xsl:template match="*[contains(@class, ' topic/strow ')]/*[contains(@class,
▶ ' topic/stentry ')]">
    <fo:table-cell xsl:use-attribute-sets="strow.stentry">
        <xsl:call-template name="commonattributes"/>
        <xsl:variable name="entryCol" select="count(preceding-sibling::*[contains(@class,
        ▶ ' topic/stentry ')]) + 1"/>
        <xsl:variable name="frame" as="xs:string">
            <xsl:variable name="f" select="ancestor::*[contains(@class, ' topic/simpletable ')]
            ▶[1]/@frame"/>
                <xsl:choose>
                    <xsl:when test="$f">
                        <xsl:value-of select="$f"/>
                    </xsl:when>
                    <xsl:otherwise>
                        <xsl:value-of select="$table.frame-default"/>
                    </xsl:otherwise>
                </xsl:choose>
        </xsl:variable>
        . . .
```

In this exercise, you'll work with `<simpletable>`. You can tell because the elements in question are `<strow>` and `<stentry>` rather than `<row>` and `<entry>`. Why `<simpletable>`? Because the example above is a `<fig>`, and you can't put `<table>` inside `<fig>`, so `<simpletable>` it is.

2. Add the following variable definitions to the code:

```
<xsl:template match="*[contains(@class, ' topic/strow ')]/*[contains(@class, ' topic/stentry
')]">
    <xsl:variable name="current-row-count" select="number(count(../preceding-sibling::
    ▶*[contains(@class, ' topic/strow ')])+1)"/>
    <xsl:variable name="preceding-row-count" select="number(count(../preceding-sibling::
    ▶*[contains(@class, ' topic/strow ')]))"/>
    <xsl:variable name="preceding-entry" select="count(preceding-sibling::
    ▶*[contains(@class,' topic/stentry ')])"/>
    </xsl:variable>
    <xsl:variable name="left-cell-count" select="$current-row-count + $preceding-row-count"/>
    <xsl:variable name="right-cell-count" select="$left-cell-count + 1"/>
    <fo:table-cell xsl:use-attribute-sets="strow.stentry">
        <xsl:call-template name="commonattributes"/>
        <xsl:variable name="entryCol" select="count(preceding-sibling::
        ▶*[contains(@class, ' topic/stentry ')]) + 1"/>
        <xsl:variable name="frame">
```

What are these variables for? You want the left column to be numbered 1, 3, 5, etc. You want the right column to be numbered 2, 4, 6, etc. This translates to, "In the left column, number the first cell to match the current row number. Number the second cell to match the current row number plus one.

Number the third cell to match the current row number plus two. And so on." That means the left column should be numbered as [(current row number)+(number of preceding rows)].

The *preceding-row-count* variable counts the number of preceding rows. The *current-row-count* variable counts the number of preceding rows and adds one. Added together you get *left-cell-count*, which is the number for the left column. The right-column numbering is a lot simpler. It's just the number in the corresponding left column plus one. That's how the *right-cell-count* variable is set.

What is the *preceding-entry* variable? You'll get to that.

3. Now find this section of code in the same template:

```
<xsl:choose>
   <xsl:when test="number(ancestor::*[contains(@class, ' topic/simpletable ')]
   ►[1]/@keycol) = $entryCol">
      <fo:block xsl:use-attribute-sets="strow.stentry__keycol-content">
         <xsl:apply-templates/>
      </fo:block>
   </xsl:when>
   <xsl:otherwise>
      <fo:block xsl:use-attribute-sets="strow.stentry__content">
         <xsl:apply-templates/>
      </fo:block>
   </xsl:otherwise>
</xsl:choose>
```

This is the code that actually outputs the cell contents. You need to add your new variables to it. But you need to make a distinction between a cell in the left column and one in the right column. That sounds like a job for `<xsl:choose>`. For each `<stentry>`, you want to test to see if there is a preceding `<stentry>` in the row. If so, you need to use the *right-cell-count* variable. If not, you need to use the *left-cell-count* variable.

This is where the *preceding-entry* variable comes in. It counts the number of preceding `<stentry>` elements in the row.

4. Edit the `<xsl:otherwise>` section like this:

```
<xsl:otherwise>
   <xsl:choose>
      <xsl:when test="$preceding-entry = 0">
         <fo:block xsl:use-attribute-sets="strow.stentry__content">
            <xsl:value-of select="$left-cell-count"/>
            <xsl:text>. </xsl:text><xsl:apply-templates/>
         </fo:block>
      </xsl:when>
      <xsl:otherwise>
         <fo:block xsl:use-attribute-sets="strow.stentry__content">
            <xsl:value-of select="$right-cell-count"/>
            <xsl:text>. </xsl:text><xsl:apply-templates/>
         </fo:block>
      </xsl:otherwise>
   </xsl:choose>
</xsl:otherwise>
```

If the number of preceding `<stentry>` elements in the row is zero, it's a left cell. Otherwise, it's a right cell. The `<xsl:text>` statement just adds a period and a space after the number.

5. Save changes, run a PDF build, and check your work.

You might find, if your table's last cell contains no content, that the empty cell is also numbered (this result can vary from renderer to renderer). To eliminate numbering in an empty cell, continue with the next step.

6. Add the following `<xsl:when>` condition to the code:

```
<xsl:otherwise>
   <xsl:choose>
      <xsl:when test="not(text()) and not(node())">
         <fo:block xsl:use-attribute-sets="strow.stentry__content">
            <xsl:apply-templates/>
         </fo:block>
      </xsl:when>
      <xsl:when test="$preceding-entry = 0">
         <fo:block xsl:use-attribute-sets="strow.stentry__content">
            <xsl:value-of select="$left-cell-count"/>
            <xsl:text>. </xsl:text><xsl:apply-templates/>
         </fo:block>
      </xsl:when>
      <xsl:otherwise>
         <fo:block xsl:use-attribute-sets="strow.stentry__content">
            <xsl:value-of select="$right-cell-count"/>
            <xsl:text>. </xsl:text><xsl:apply-templates/>
         </fo:block>
      </xsl:otherwise>
   </xsl:choose>
   </xsl:otherwise>
```

This statement looks to see if the cell contains either text or an element, and if not, it simply renders the non-existent content without numbering, leaving you with nothing in the cell.

There's one more thing you might want to do. Right now, every `<simpletable>` will be numbered. If you don't always want `<simpletable>` to be numbered, you need some way to tell your plugin what to do.

There are two approaches. If you always want numbering on when `<simpletable>` is inside `<fig>`, and you always want numbering off when `<simpletable>` is outside `<fig>`, you can use an `<xsl:choose>` statement to test based on context. If you need more flexibility, your best bet is to use the *outputclass* attribute on `<simpletable>` to identify when to turn numbering on or off.

Exercise: Specify the header for a table

Objective: Determine whether the first row or the first column serves as the table header.

Most of the time, the table header is the first row of a table, and the column names appear at the top of each column, as in the gray area shown here:

But in some cases, you might want the rows to have headings rather than the columns, as in the gray area shown here:

The PDF plugin now includes an easy way to make this distinction: the *rowheader* attribute.

This attribute has three possible values:

- firstcol - specifies that the first column of the table should be the header
- headers - specifies that the first row of the table should be the header
- norowheader - specifies that the table should have no headers at all

Add the *rowheader* attribute to the `<table>` element with the appropriate value, and you're all set.

If you specify "firstcol" for the *rowheader* attribute, use the new *tbody.row.entry__firstcol* attribute, found in **tables-attr.xsl**, to format the header column. If you specify "headers", use the existing *thead.row.entry* attribute, also found in **tables-attr.xsl**.

Exercise: Omit the table heading on subsequent pages

Objective: Omit the table heading on subsequent pages when a table flows beyond one page.

By default, when a table spans more than one page, the table heading row appears on each page. If you want to omit this row, you can make one simple change.

1. Copy the *table.tgroup* attribute set from
   `DITA-OT/plugins/org.dita.pdf2/cfg/fo/attrs/`**tables-attr.xsl** to your copy of **tables-attr.xsl**.
2. Add the following attribute:

```
<xsl:attribute name="table-omit-header-at-break">true</xsl:attribute>
```

3. Save changes, run a PDF build, and check your work.

Exercise: Remove column separators in a table header

Objective: Remove column separators in the header row of a table.

Let's say that you want to format a table so that, while it has vertical and horizontal rules in the body, it doesn't have vertical rules in the header row, like this:

State	Capital	Population
Alabama	Montgomery	374,536
Alaska	Juneau	33,275
Arizona	Phoenix	4,192,887

The most DITA-compliant approach is simply to set `colsep="0"` on each of the `<entry>` elements in the table's `<thead>` element:

```
<thead>
   <row>
      <entry colsep="0">State</entry>
      <entry colsep="0">Capital</entry>
      <entry colsep="0">Population</entry>
   </row>
</thead>
```

However, maybe this design decision is recent and needs to be applied retroactively to hundreds of tables. You could apply XSLT across your content set to apply `colspec="0"` across the board, but if you prefer to keep the formatting entirely plugin-dependent, here's one way to do that.

1. In `DITA-OT/plugins/org.dita.pdf2/xsl/fo/`**tables.xsl**, find the following template:

```
<xsl:template match="*[contains(@class, ' topic/thead ')]/*[contains(@class, ' topic/row ')]/*
►[contains(@class, ' topic/entry ')]">
   <fo:table-cell xsl:use-attribute-sets="thead.row.entry">
      <xsl:call-template name="commonattributes"/>
      <xsl:call-template name="applySpansAttrs"/>
      <xsl:call-template name="applyAlignAttrs"/>
      <xsl:call-template name="generateTableEntryBorder"/>
      <fo:block xsl:use-attribute-sets="thead.row.entry__content">
         <xsl:apply-templates select="." mode="ancestor-start-flag"/>
         <xsl:call-template name="processEntryContent"/>
         <xsl:apply-templates select="." mode="ancestor-end-flag"/>
      </fo:block>
   </fo:table-cell>
</xsl:template>
```

This is the template that processes `<entry>` elements in table header rows (`<thead>`).

2. Now look at the template just below it, that begins with:

```
<xsl:template match="*[contains(@class, ' topic/tbody ')]/*[contains(@class, ' topic/row ')]
►/*[contains(@class, ' topic/entry ')]">
```

This is the template that processes `<entry>` elements in table body rows (`<tbody>`).

Notice that both of these templates in turn call a template named *generateTableEntryBorder*. This means that by default, the PDF plugin applies the same rules to header and body cells. You want to apply different ruling to header and body cells. There are probably many ways to do this, but one straightforward way is to change the first template above so that it calls a template other than *generateTableEntryBorder*.

3. Copy the template shown in step 1 from `DITA-OT/plugins/org.dita.pdf2/xsl/fo/`**tables.xsl** to your copy of **tables.xsl**.

4. Change the line highlighted in step 1 to:

```
<xsl:call-template name="generateTableEntryBorderHeader"/>
```

5. Copy the *generateTableEntryBorder* template from `DITA-OT/plugins/org.dita.pdf2/xsl/fo/`**tables.xsl** to your copy of **tables.xsl**.

6. Change the name of the copy to *generateTableEntryBorderHeader*.

7. Delete the entire contents of *generateTableEntryBorderHeader*, leaving only the following:

```
<xsl:template name="generateTableEntryBorderHeader">
    <!-- do not generate any column separators -->
</xsl:template>
```

By removing all processing for rules for the table header row, you specify that no rules whatsoever will be output. Unless you make other changes to other templates or attribute sets, the horizontal rule above the header row and the left and right vertical rules will come from the table's *frame* attribute. The horizontal rule below the header row will come from the body row ruling. The net visible result, then, is no vertical rules between columns in the header row.

You're all set, but if you want to go a bit further and make the presence or absence of column separators in the header row dependent on *outputclass* (for example), read on.

8. In your copy of the template shown in step 1, use `<xsl:choose>` to call either the original *generateTableEntryBorder* template or your new *generateTableEntryBorderHeader* template, depending on the table's *outputclass*.

For example, to have `outputclass="nocolsep"` trigger the absence of column separators:

```
...
<xsl:call-template name="applyAlignAttrs"/>
<xsl:choose>
  <xsl:when test="ancestor::*[contains(@class, ' topic/table ')][@outputclass = 'nocolsep']">
    <xsl:call-template name="generateTableEntryBorderHeader"/>
  </xsl:when>
  <xsl:otherwise>
    <xsl:call-template name="generateTableEntryBorder"/>
  </xsl:otherwise>
</xsl:choose>
<fo:block xsl:use-attribute-sets="thead.row.entry__content">
...
```

Exercise: Output a table in landscape orientation

Objective: Specify that a table should be output in landscape mode.

If you have a very wide table, you might find that it doesn't fit on a portrait- or vertically-oriented page. In the past, there was no easy way to rotate an entire table fit a landscape- or horizontally-oriented page. The DITA Open Toolkit includes a new attribute, *orient*, that enables you to easily specify that a table should print in landcape orientation.

1. Add the *orient* attribute to the `<table>` element of the table you want to print in landscape mode, with a value of "land".

The rotated table looks like **Figure 38**:

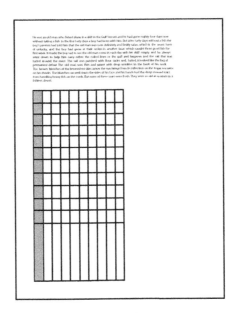

Figure 38: Table in landscape orientation

The two values for *orient* are "land" and "port". The "land" value is defined as "90 degrees counterclockwise from the text flow." Note that if you're outputting on a landscape page to begin with, then setting `orient="land"` could result in a portrait-oriented table. The "port" value is the default and is defined as "the same orientation as the text flow."

After re-orienting the table, you might need to adjust its position. Many thanks to Bob Thomas of Tagsmiths (and a member of the OASIS DITA Technical Committee) for volunteering the following helpful changes.

2. If they are not already present, copy the *table* and *table__container* attribute sets from `DITA-OT/plugins/org.dita.pdf2/cfg/fo/attrs/`**tables-attr.xsl** to your copy of **tables-attr.xsl**.

3. To force the table to a new page, add the *page-break-before* attribute to the *table__container* attribute set:

```
<xsl:attribute-set name="table__container">
    <!-- Tagsmiths: Force rotated tables to start on a new page -->
    <xsl:attribute name="page-break-before" select="if (@orient eq 'land') then 'always'
 ▶ else 'auto'"/>
</xsl:attribute-set>
```

4. To add space between the table and a preceding section title, add the *space-before* attribute set:

```
<xsl:attribute-set name="table__container">
    <!-- Tagsmiths: Ensure that the table has spacing below section titles. Note that this
        setting will introduce vertical space between two contiguous tables. -->
    <xsl:attribute name="space-before">0.6em</xsl:attribute>
</xsl:attribute-set>
```

5. To prevent inappropriate indents, add the *start-indent* attribute to the *table* attribute set, with a value of "0pt":

```
<xsl:attribute-set name="table">
   <!-- Tagsmiths: Break start-indent inheritance to prevent it from being applied
        a second time. The first application of start-indent occurs on this block's
        parent fo:block-container. -->
    <xsl:attribute name="start-indent">0pt</xsl:attribute>
</xsl:attribute-set>
```

Chapter 14

Images

Formatting images is tricky. You can set certain properties for images in your plugin, but the actual rendering of the images takes place in the PDF renderer, so many image-related issues need to be addressed within those applications. However, there are many things you can do with images that are independent of the renderer, and this chapter covers some of those things.

For a list of attributes that affect images, refer to **Common attribute sets** *(p. 406)*.

Image specifications

Here are your specifications for images.

Images

Figure title	12pt Trebuchet bold dark red; above figure
Figure numbering	Includes "Figure" chapter/appendix number, table number (Figure: 1-1). Figure numbering restarts with each chapter/appendix.
Scaling	automatically scale large images to fit on page
Alignment	centered

Files you need

If you have completed other exercises, some of these files might already be present in your plugin.

If not already present, create the following file:

- `DITA-OT/plugins/com.company.pdf/cfg/fo/attrs/`**commons-attr.xsl**

Be sure to add the appropriate `xsl:import` statement to **custom.xsl** (attrs).

If not already present, create the following files in `DITA-OT/plugins/com.company.pdf/cfg/fo/xsl/`:

- **commons.xsl**
- **tables.xsl**

Be sure to add the appropriate `xsl:import` statements to **custom.xsl** (xsl).

For instructions on creating attribute files and XSLT stylesheets in your PDF plugin, refer to **Create an attribute set file in your plugin** *(p. 31)* and **Create an XSLT stylesheet in your plugin** *(p. 32)*.

If not already present, create the following file:

- `DITA-OT/plugins/com.company.pdf/cfg/common/vars/`**en.xml**

For instructions on creating localization variables fles, refer to **Create a localization variables file** *(p. 78)*.

Exercise: Change or delete the label for a figure title

Objective: Change the label text (such as **Figure:**) for a figure title or delete the label altogether.

1. Copy the following variables from `DITA-OT/plugins/org.dita.pdf2/cfg/common/vars/`**en.xml** to the copy of **en.xml** in your plugin:

```
<variable id="Figure.title">Figure <param ref-name="number"/>:
▶ <param ref-name="title"/></variable>
<variable id="Figure Number">Figure <param ref-name="number"/></variable>
```

These two variables are similar, but not quite the same. To understand how they work, take a look at **Including titles and autonumbers separately in table and figure links** *(p. 316)*.

Be aware that changes to these variables also affect the format of links to figures.

2. Do one of the following:

- To change the label, enter the new value in the *Figure.title* variable:

```
<variable id="Figure.title">Image <param ref-name="number"/>:
▶ <param ref-name="title"/></variable>
```

- To delete the label, remove the current value:

```
<variable id="Figure.title"><param ref-name="number"/>: <param ref-name="title"/></variable>
```

3. Repeat this process for any other language-specific variables files you've customized.

4. Save changes, run a PDF build, and check your work.

Exercise: Format figure titles

Objective: Determine how figure titles look in your PDF.

Titles within `<fig>` elements use the *fig.title* attribute set.

1. Copy the *fig.title* attribute set from `DITA-OT/plugins/org.dita.pdf2/cfg/fo/attrs/`**commons-attr.xsl** to your copy of **commons-attr.xsl**.

2. Add the following attributes:

```
<xsl:attribute name="font-size">12pt</xsl:attribute>
<xsl:attribute name="font-family">sans-serif</xsl:attribute>
<xsl:attribute name="color">#990033</xsl:attribute>
```

3. Save changes, run a PDF build, and check your work.

Exercise: Add chapter, appendix or part numbers to figure titles

Objective: Use the *getChapterPrefix* template method to number figure captions.

Out of the box, the PDF plugin doesn't add chapter, part, or appendix numbers to figure titles. Fortunately, it's easy to add them yourself. You will use the *getChapterPrefix* template to add chapter numbers to figure titles (If you've already done the **Add chapter, appendix, or part numbers to TOC page numbers** *(p. 336)* exercise, then you already have this template in your plugin, and you can skip Step 1). If you didn't work through that exercise, refer to it now and add the template.

1. If it is not already present, copy the entire *getChapterPrefix* template into **custom.xsl** (xsl).

You can find this template in the **GetChapterPrefix template** *(p. 481)* appendix.

> **Note** You could copy this template into any of the .xsl files you've added to your plugin, including **commons.xsl**, but it's probably best to copy it into **custom.xsl**. That way you can see right away it's a custom template that didn't come with the PDF plugin.

2. Copy the following template from DITA-OT/plugins/org.dita.pdf2/xsl/fo/**commons.xsl** to your copy of **commons.xsl**:

```
<xsl:template match="*[contains(@class,' topic/fig ')]/*[contains(@class,' topic/title ')]">
    <fo:block xsl:use-attribute-sets="fig.title">
    <xsl:call-template name="commonattributes"/>
    <xsl:call-template name="getVariable">
        <xsl:with-param name="id" select="'Figure.title'"/>
        <xsl:with-param name="params">
            <number>
                <xsl:apply-templates select="." mode="fig.title-number"/>
            </number>
            <title>
                <xsl:apply-templates/>
            </title>
        </xsl:with-param>
    </xsl:call-template>
    </fo:block>
</xsl:template>
```

This is the template that processes figure titles.

3. Add the highlighted code to the `<number>` section as shown below:

```
<number>
    <xsl:call-template name="getChapterPrefix"/>
    <xsl:text>-</xsl:text>
    <xsl:apply-templates select="." mode="fig.title-number"/>
</number>
```

4. Save **commons.xsl** and run a build to check your work.

At this point, figures are numbered consecutively from the first to the last, with no restarting per chapter, appendix, or part. If that's what you want, you can stop here. If you need numbering to restart with each chapter, keep going.

5. Edit the number element in the code above as shown:

```
<number>
    <xsl:call-template name="getChapterPrefix" />
    <xsl:text>-</xsl:text>
    <xsl:choose>
        <xsl:when test="count(ancestor-or-self::*[contains(@class, ' topic/topic')]
        ▶[position()=last()][count(preceding-sibling::*
        ▶[contains(@class, ' topic/topic')]) &gt; 0])">
            <xsl:value-of select="count(./preceding::*
            ▶[contains(@class, ' topic/fig ')]
            ▶[child::*[contains(@class, ' topic/title ')]][ancestor-or-self::*
            ▶[contains(@class, ' topic/topic')][position()=last()]]) -
            ▶count(ancestor-or-self::*[contains(@class, ' topic/topic')]
            ▶[position()=last()]/preceding-sibling::*[contains(@class, ' topic/topic')]
            ▶//*[contains(@class, ' topic/fig ')][child::*
            ▶[contains(@class, ' topic/title ')]])+1"/>
        </xsl:when>
        <xsl:otherwise>
            <xsl:value-of select="count(./preceding::*[contains(@class, ' topic/fig ')]
            ▶[child::*[contains(@class, ' topic/title ')]][ancestor-or-self::
            ▶*[contains(@class, ' topic/topic')][position()=last()]])+1"/>
        </xsl:otherwise>
    </xsl:choose>
</number>
```

(Delete the line `<xsl:apply-templates select="." mode="fig.title-number"/>`.)

> **Note** When you copy the code from this example and see a line break marker (▶), make sure you don't leave any space between the end of the previous line and the character that follows the marker. Yes, those really are very long test and select attribute values.

6. Save changes, run a PDF build, and check your work.

Exercise: Place titles above images

Objective: Place image captions above the images rather than below, which is the default.

By default, the PDF plugin places titles below images. It's simple to change this if you want titles to be above images.

1. Copy the template that begins with the following from
   `DITA-OT/plugins/org.dita.pdf2/xsl/fo/`**commons.xsl** to your copy of **commons.xsl**:

   ```
   <xsl:template match="*[contains(@class,' topic/fig ')]">
   ```

2. Within that template, reverse the order of these two lines

   ```
   <xsl:apply-templates select="*[not(contains(@class,' topic/title '))]"/>
   ```

   ```
   <xsl:apply-templates select="*[contains(@class,' topic/title ')]"/>
   ```

   The line with `*[not(contains...` processes everything within `<fig>` that is not a title. The other line processes the title. By reversing the lines, you output the title first and then everything else.

   There's one more thing you need to do.

3. Copy the *fig.title* attribute set from
   `DITA-OT/plugins/org.dita.pdf2/cfg/fo/attrs/`**commons-attr.xsl** to your copy of
   commons-attr.xsl.

4. Change the *keep-with-previous.within-page* attribute to *keep-with-next.within-page*.

   The value remains "always". By default, the PDF plugin places the title after the image. So it made sense to keep the title with the previous block—the image. However, now that you've reversed the order, the title no longer needs to stay with the previous block, which might be anything. Instead, it needs to stay with the next block—the image.

5. Save changes, run a PDF build, and check your work.

Exercise: Dynamically scale images to the page width

Objective: Automatically resize large images so that they fit on a page in your PDF, without having to specify actual width or height values within the topic.

If you publish PDFs with different page dimensions, you may need to adjust the size of images. An image that's 6 inches wide will fit on an 8.5 x 11 page with 1 inch margins, but it won't fit on a 5.5 x 8.5 page. Of course, you can use *scalefit* on individual images to instruct the DITA Open Toolkit to dynamically scale them, but you have to remember to apply *scalefit* manually to each image. You can take this burden off writers and have the PDF plugin dynamically scale larger images to fit the available width.

Dynamic scaling only affects images whose actual width is greater than the area between the left and right margins. Images whose actual width is less than that area are not scaled down.

> **Note** If you use *scale*, *height*, or *width*, those values take precedence over the dynamic scaling. Unless you need to specify a precise scale for an image, you can delete the *scale*, *width*, and *height* values and let the plugin do the work.

This method uses the *placement* attribute as a trigger for dynamic resizing. Odds are, you don't want to dynamically resize small inline images such as icons. So for them, use `placement="inline"` (or don't use *placement* at all, since the default is "inline"). You probably do want to resize large images, such as diagrams or screen shots. In that case, using `placement="break"` forces the image into its own block and triggers dynamic resizing.

1. Copy the *image__block* and *image__inline* attribute sets from `DITA-OT/plugins/org.dita.pdf2/cfg/fo/attrs/`**commons-attr.xsl** to your copy of **commons-attr.xsl**.

2. Add the following attributes and values to *image__block*:

```
<xsl:attribute-set name="image__block">
    <xsl:attribute name="content-width">scale-to-fit</xsl:attribute>
    <xsl:attribute name="content-height">100%</xsl:attribute>
    <xsl:attribute name="width">100%</xsl:attribute>
    <xsl:attribute name="scaling">uniform</xsl:attribute>
</xsl:attribute-set>
```

Images with `placement="break"` will use this attribute set.

3. Add the following attributes and values to *image__inline*:

```
<xsl:attribute-set name="image__inline">
    <xsl:attribute name="content-width">auto</xsl:attribute>
    <xsl:attribute name="content-height">auto</xsl:attribute>
    <xsl:attribute name="width">auto</xsl:attribute>
    <xsl:attribute name="scaling">uniform</xsl:attribute>
</xsl:attribute-set>
```

Images with `placement="inline"` or no value for *placement* will use this attribute set.

Now you need to make sure your plugin calls the correct attribute set.

4. Copy the template that starts with the following from `DITA-OT/plugins/org.dita.pdf2/xsl/fo/`**commons.xsl** to your copy of **commons.xsl**:

```
<xsl:template match="*" mode="placeImage">
```

This template inserts the image. Specifically, the `<fo:external-graphic>` element does the work. You used this element earlier if you completed the **Add a background image to a page** *(p. 113)* or **Set up headers that include external files** *(p. 132)* exercises.

5. In this template, between the `</xsl:call-template>` line and the `<fo:external-graphic>` element, add the code highlighted below:

```
...
</xsl:call-template>
<xsl:choose>
    <xsl:when>
    </xsl:when>
    <xsl:otherwise>
    </xsl:otherwise>
</xsl:choose>
<fo:external-graphic src="url({$href})" xsl:use-attribute-sets="image">
```

This sets up a test for the value of *placement* so you can use *image__block* or *image__inline* accordingly. You need to test for three possible values for *placement*: "break", "inline", or missing (i.e., the attribute is not used). But you really only have two scenarios, because the value "inline" is the default if placement is missing. The `<xsl:when>` statement covers the case where *placement* is set to "break", and the `<xsl:otherwise>` statement covers the case where *placement* is set to "inline" or is missing.

6. Cut the entire `<fo:external-graphic>` section of code from its current location and paste it inside the `<xsl:when>`.

7. Paste it again inside the `<xsl:otherwise>`.

8. Edit the `<xsl:when>` statement to test for `placement="break"`:

```
<xsl:when test="@placement='break'">
```

9. Set the `<fo:external-graphic>` element in the `<xsl:when>` to use the *image__block* attribute set.

10. Set the `<fo:external-graphic>` element in the `<xsl:otherwise>` to use the *image__inline* set.

11. Save changes, run a PDF build, and check your work.

> **Important:**
>
> Be careful with auto-scaling. The width-to-height ratio of the image cannot exceed the width-to-height ratio of the body area of the page. For example, suppose the body area is 6in wide and 9in high (a ratio of 2:3), and you have an image that's 9in wide and 18in high (a ratio of 1:2). The DITA OT will scale the image down to 6in wide and 12in high.
>
> That still leaves the image too long, even if you put it on a page by itself. Because the image can't fit on a page, the PDF renderer won't render it, and it won't appear in the PDF.
>
> To be on the safe side, figure out what the maximum width-to-height ratio is among all of your images and either set up your pages to accommodate that ratio or resize any images that won't fit after scaling.

A good, simple rule of thumb is: if you're going to scale for width, make sure your images are wider than they are high. If you're going to scale for height, make sure your images are higher than they are wide.

You might be thinking, "To make this work, I still have to add an attribute... *placement* instead of *scalefit*." That's true. If it is your policy (or you can make it your policy) always to put non-inline images in a `<fig>`, then you can change the `<xsl:when>` test from `<xsl:when test="@placement='break'">` to `<xsl:when test="parent::*[contains(@class,' topic/fig ')]">`. Then you don't have to add any attributes for scaling.

Exercise: Change the default alignment for all images

Objective: Determine how all images are aligned (left, right, or center) by default.

By default, the `org.dita.pdf2` plugin aligns block images (that is, images with `placement="break"`) with the left margin of the preceding text, like this:

It's important to configure the Widget correctly.

1. Click Configure.
   The Configuration window opens.

Figure 39: Left-aligned images

You can set the *align* attribute to *right* or *center* on a per-image basis. But what if you want all your block images to be aligned the same way? You don't want to have to set *align* for every single image. Instead, you can change the default alignment.

1. Copy the template that includes the following from
`DITA-OT/plugins/org.dita.pdf2/xsl/fo/`**commons.xsl** to your copy of **commons.xsl**:

```
<xsl:template match="*" mode="placeImage">
    <xsl:param name="imageAlign"/>
    <xsl:param name="href"/>
    <xsl:param name="height" as="xs:string?"/>
    <xsl:param name="width" as="xs:string?"/>
    <xsl:call-template name="processAttrSetReflection">
        <xsl:with-param name="attrSet" select="concat('__align__', $imageAlign)"/>
        <xsl:with-param name="path" select="'../../cfg/fo/attrs/commons-attr.xsl'"/>
    </xsl:call-template>
</xsl:template>
```

If you did the exercise **Dynamically scale images to the page width** *(p. 306)*, this section should look familiar to you. It defines the parameter *imageAlign* and then uses it in the first `<xsl:with-param>` statement.

2. Paste the template into your **commons.xsl**.

Here's how the template works: *imageAlign* takes the value of the *align* attribute on the image. If *align* is not set, the default is "left". I'll use "left" as the value of *imageAlign*. The first `<xsl:with-param>` statement concatenates (using the `concat` function) the text "__align__" with the value of *imageAlign*, giving you "__align__left" as the value of the *attrSet* parameter. The *attrSet* parameter is then used in a call to the template named ***processAttrSetReflection***, which returns the attribute set *__align__left*, which lives in **commons-attr.xsl**.

Why all of this just to call an attribute set? Because the value of *xsl:use-attribute-sets* cannot be a variable. That's why you needed to repeat the `<fo:external-graphic>` block in **Dynamically scale images to the page width** *(p. 306)*, even though the only difference between the two blocks was the *xsl:use-attribute-sets* attribute. However, certain attribute sets, including the four that have the prefix *__align__*, are so frequently used that the Open Toolkit has a template, ***processAttrSetReflection***, to make those cases easier to handle. This template lets you choose one of a select group of attribute sets based on the value of a variable (*attrSet*). Don't think too hard about this. In developing your average plugin, you don't have to deal much with it.

There are a couple of ways to center-align all of your images. You could try to figure out a way to change the default value of *align* from "left" to "center". Don't do that. Changing attribute defaults is not only a bad idea, it's really hard.

Instead you could edit the code above so it always calls the *__align__center* attribute set unless *align* is specified on the image, in which case it continues to call the attribute set that corresponds to the value of *align*. This is a much better idea and here's how.

3. In the code section above, add an `<xsl:choose>` statement like the following:

```
...
<xsl:param name="width" as="xs:string?"/>
<xsl:choose>
    <xsl:when>
    </xsl:when>
    <xsl:otherwise>
    </xsl:otherwise>
</xsl:choose>
<xsl:call-template name="processAttrSetReflection">
...
```

4. Cut the entire `<xsl:call-template>` section and paste it into the `<xsl:when>` statement.

5. Paste it again into the `<xsl:otherwise>` statement.

6. In the `<xsl:call-template>` section that's inside the `<xsl:when>`, change the first `<xsl:with-param>` line so it reads like this:

```
<xsl:with-param name="attrSet" select="concat('__align__', 'center')"/>
```

You could also just use `select="'__align__center'"` instead, since both halves of the concat function are known. Either should work fine.

7. Edit the `<xsl:when>` so that it tests whether the *align* attribute is present:

```
<xsl:when test="not(@align)">
```

The entire section of edited code should look like this:

```
<xsl:template match="*" mode="placeImage">
    <xsl:param name="imageAlign"/>
    <xsl:param name="href"/>
    <xsl:param name="height"/>
    <xsl:param name="width"/>
    <xsl:choose>
        <xsl:when test="not(@align)">
            <xsl:call-template name="processAttrSetReflection">
                <xsl:with-param name="attrSet" select="concat('__align__', 'center')"/>
                <xsl:with-param name="path" select="'../../cfg/fo/attrs/commons-attr.xsl'"/>
            </xsl:call-template>
        </xsl:when>
        <xsl:otherwise>
            <xsl:call-template name="processAttrSetReflection">
                <xsl:with-param name="attrSet" select="concat('__align__', $imageAlign)"/>
                <xsl:with-param name="path" select="'../../cfg/fo/attrs/commons-attr.xsl'"/>
            </xsl:call-template>
        </xsl:otherwise>
    </xsl:choose>
```

8. Save changes, run a PDF build, and check your work.

Chapter 15

Related links, cross-references, and footnotes

For the most part, link generation in the DITA Open Toolkit is automatic based on the way you've authored your topics and set up your map. There are three main kinds of links in DITA:

- cross-references
- family links
- related links

Cross-references are created when you use `<xref>` or some of the linking attributes available in DITA. Family links are created based on your map structure, attributes such as *collection-type*, and the ANT parameter `args.rellinks`. Related links appear based on the presence or absence of a relationship table in your map and the `args.rellinks` parameter.

Most link processing is built into the Open Toolkit and occurs before any plugin processing. This book isn't going to talk about changing that processing. This chapter covers customizing the appearance and format of cross-references, related links, and footnotes.

For a list of attribute sets that affect links, refer to **Link attribute sets** *(p. 430)*.

General link formatting specifications

Here are the specifications for general link formatting.

Links

Related Links section title	Additional information
Related Links section divider	light gray 2pt line, 3in wide
Link text format	10pt Trebuchet normal dark red
Page reference text format	10pt Trebuchet italic black
Page reference text	(p. #-#), includes chapter number
Cross-reference format	bold non-italic dark gray
Footnote callout	9pt Trebuchet bold italic dark red, enclosed in square brackets in body text but not in footnote itself
Footnote text	9pt Trebuchet italic dark gray
Line above footnotes	light gray dotted

Files you need

If you have completed other exercises, some of these files might already be present in your plugin.

If they are not already present, create the following files:

- `DITA-OT/plugins/com.company.pdf/cfg/fo/attrs/`**links-attr.xsl**
- `DITA-OT/plugins/com.company.pdf/cfg/fo/attrs/`**static-content-attr.xsl**

Be sure to add the appropriate `<xsl:import>` statement to **custom.xsl** (attrs).

If it is not already present, create the following file:

- `DITA-OT/plugins/com.company.pdf/cfg/fo/xsl/`**links.xsl**

Be sure to add the appropriate `<xsl:import>` statement to **custom.xsl** (xsl).

For instructions on creating attribute files and XSLT stylesheets in your PDF plugin, refer to **Create an attribute set file in your plugin** *(p. 31)* and **Create an XSLT stylesheet in your plugin** *(p. 32)*.

If it is not already present, create the following file:

- `DITA-OT/plugins/com.company.pdf/cfg/common/vars/`**en.xml**

For instructions on creating localization variables files, refer to **Create a localization variables file** *(p. 78)*.

General links and cross-references

Exercise: Change the format of cross-references

Objective: Determine the appearance of links created using the `<xref>` element.

Cross-references are links created outside of the reltable functionality—usually via `<xref>` elements that you add within topics. Cross-references are formatted separately from related links. According to your template specifications, they should be bold, non-italic, and dark gray. Unless you specify otherwise, the cross-references should have the same font and size as the surrounding text.

1. Copy the *xref* attribute set from `DITA-OT/plugins/org.dita.pdf2/cfg/fo/attrs/`**links-attr.xsl** to your copy of **links-attr.xsl**.

   > **Important:** This attribute set also controls the appearance of the links in the mini-TOC. If you want to format them differently from cross-references, you'll need to create a new attribute set and use it instead of *xref* to format the mini-TOC links. Hint: look in **commons.xsl** to see where *xref* is applied to the mini-TOC and call your new attribute set from there.

2. Add the *color*, *font-style* and *font-weight* attributes as shown:

```
<xsl:attribute-set name="xref" use-attribute-sets="common.link">
    <xsl:attribute name="color">#8A8A8A</xsl:attribute>
    <xsl:attribute name="font-style">normal</xsl:attribute>
    <xsl:attribute name="font-weight">bold</xsl:attribute>
</xsl:attribute-set>
```

3. Save changes, run a PDF build, and check your work.

Exercise: Change the on-page text for related links

Objective: Change the default text (such as "on page") that is automatically generated as part of related links and cross-references.

By default, related links include the boilerplate text "on page." You can easily change or delete this boilerplate text. For this example, you're going to change the text to "(p.)."

1. Copy the variable *On the page* from DITA-OT/plugins/org.dita.pdf2/cfg/common/vars/**en.xml** to your copy of **en.xml**.
2. Change the value as shown:

```
<variable id="On the page"> (p. <param ref-name="pagenum"/>)</variable>
```

3. Save changes, run a PDF build, and check your work.

Including titles and autonumbers separately in table and figure links

Objective: Determine the format of links to tables and figures.

There are now two localization variables that pertain to table links: *Table Number* and *Table.title*; and two that pertain to figure links: *Figure Number* and *Figure.title*.

Table Number and *Figure Number* include only the autonumber, while *Table.title* and *Figure.title* include the label, autonumber, and title:

```
<variable id="Table.title">Table <param ref-name="number"/>: <param ref-name="title"/></variable>
<variable id="Table Number">Table <param ref-name="number"/></variable>
```

```
<variable id="Figure.title">Figure <param ref-name="number"/>: <param ref-name="title"/></variable>
<variable id="Figure Number">Figure <param ref-name="number"/></variable>
```

You can find the declarations for all four variables in the localization variables files in this directory: DITA-OT/plugins/org.dita.pdf2/cfg/common/vars/.

What's the point of these separate variables? Well, for some time, there have been two build parameters, *args.tablelink.style* and *args.figurelink.style*. These handy parameters enable you to determine, at the build level, whether links to figures and tables include just the figure/table number or the figure/table number and title. In the past, they were not available for PDF transforms, but now they are.

Both parameters have three valid values:

- NUMBER - include just the table/figure label and number in the link
- TITLE - include just the table/figure title in the link
- NUMTITLE - include the table/figure label, number, and title in the link (the default for PDFs; supported only for PDFs)

These variables are used in **links.xsl**.

Here's the template that creates links for figures. It covers all three cases for *args.figurelink.style*:

```
<xsl:template match="*[contains(@class, ' topic/fig ')][*[contains(@class, ' topic/title ')]]"
mode="retrieveReferenceTitle">
    <xsl:choose>
        <xsl:when test="$figurelink.style='NUMBER'">
            <xsl:call-template name="getVariable">
                <xsl:with-param name="id" select="'Figure Number'"/>
                <xsl:with-param name="params">
                    <number>
                        <xsl:apply-templates select="*[contains(@class, ' topic/title ')]"
                        ▶ mode="fig.title-number"/>
                    </number>
                </xsl:with-param>
            </xsl:call-template>
        </xsl:when>
        <xsl:when test="$figurelink.style='TITLE'">
            <xsl:apply-templates select="*[contains(@class, ' topic/title ')]"
            ▶ mode="insert-text"/>
        </xsl:when>
        <xsl:otherwise>
            <xsl:call-template name="getVariable">
                <xsl:with-param name="id" select="'Figure.title'"/>
                <xsl:with-param name="params">
                    <number>
                        <xsl:apply-templates select="*[contains(@class, ' topic/title ')]"
                        ▶ mode="fig.title-number"/>
                    </number>
                    <title>
                        <xsl:apply-templates select="*[contains(@class, ' topic/title ')]"
                        ▶ mode="insert-text"/>
                    </title>
                </xsl:with-param>
            </xsl:call-template>
        </xsl:otherwise>
    </xsl:choose>
</xsl:template>
```

If you include the variables with a value of NUMBER, the Open Toolkit uses the *Table Number* and *Figure Number* variables (the first `<xsl:when>` above).

If you include the variables with a value of TITLE, the Open Toolkit uses neither variable, but simply outputs the title of the table or figure (the second `<xsl:when>` above).

If you don't include *args.tablelink.style* and *args.figurelink.style* in your build file, the Open Toolkit assumes a value of NUMTITLE and uses the *Table.title* and *Figure.title* variables (the `<xsl:otherwise>` above).

Related links

Exercise: Include related links in a PDF

Objective: Set up your ANT build to include related links in your PDF.

You can dynamically create a list of related topics at the end of each topic by using a reltable or the *collection-type* attribute. By default, these links do not appear in a PDF. If you want to display them, you need to set the `args.rellinks` property in your ANT build file.

1. Add the following property and appropriate value to your ANT build file:

```
<property name="args.rellinks" value="nofamily"/>
```

The value "nofamily" outputs reltable-based and hardcoded links but not family links (that is, links to the parent topic or child topics and links to previous or next topics). To include these type of links as well, use the value "all".

Related links created when you use the *collection-type* attribute are family links, so none of them appear in your PDF if you use the "nofamily" value with the `args.rellinks` property.

There is no `args.rellinks` value that outputs family links but not reltable-based links.

2. Save changes, run a PDF build, and check your work.

Exercise: Change or delete the "Related Links" label

Objective: Use a label other than the default "Related Links" for the Related Links section.

By default, the plugin creates the Related Links section with the heading "Related Links." You need to change it to "Additional information."

1. Copy the *Related Links* variable from `DITA-OT/plugins/org.dita.pdf2/cfg/common/vars/`**en.xml** to your copy of **en.xml**.
2. Change the value to "Additional information".
3. Repeat this process for any other localization variable files you've customized.
4. Save changes, run a PDF build, and check your work.

Exercise: Add a divider above the Related Links section

Objective: Create a divider line between the main topic text and the Related links section.

To set the Related Links section off from the main topic body, you might want to add a divider of some sort. In this exercise, you'll add a light gray line.

1. Copy the *related-links.title* attribute set from
`DITA-OT/plugins/org.dita.pdf2/cfg/fo/attrs/`**links-attr.xsl** to your copy of **links-attr.xsl**.

2. Add the *border-top* attribute as shown below:

```
<xsl:attribute-set name="related-links.title">
    <xsl:attribute name="font-weight">bold</xsl:attribute>
    <xsl:attribute name="border-top">2pt solid #E6E6E6</xsl:attribute>
</xsl:attribute-set>
```

By default, the line extends from margin to margin. You only want it to be 3in long. There's no attribute that sets border length. However, you can limit the width of the block that contains the line.

You can't set the width of `<fo:block>`, but you can set the width of `<fo:block-container>`. This is the same container you used to position text on the page (**Place cover page information in a specific location** *(p. 183)*) and also to rotate text (**Rotate text** *(p. 228)*).

3. Copy the template that uses the *related-links.title* attribute set from
`DITA-OT/plugins/org.dita.pdf2/xsl/fo/`**links.xsl** to your copy of **links.xsl**

It's the template that begins

```
<xsl:template match="*[contains(@class,' topic/related-links ')]">
```

The `<fo:block>` that uses *related-links.title* is commented out by default.

4. Uncomment the lines between

```
<xsl:if test="exists($includeRelatedLinkRoles)">
```

and

```
<fo:block xsl:use-attribute-sets="related-links">
```

5. Wrap the `<fo:block>` that uses the *related-links.title* attribute set in an `<fo:block-container>` with the width set to 3in:

```
<fo:block-container width="3in">
    <fo:block xsl:use-attribute-sets="related-links.title">
        <xsl:call-template name="getVariable">
            <xsl:with-param name="id" select="'Related Links'"/>
        </xsl:call-template>
    </fo:block>
</fo:block-container>
```

6. Save changes, run a PDF build, and check your work.

> **Note** There is another way you can do this, similar to the method used in **Change the footnote separator line** *(p. 323)*. You can insert a leader before the block for the related links and have it use an attribute set that includes the attribute *leader-length* set to a percentage of the page width.

Exercise: Change the format of related links

Objective: Determine the appearance of links created via a relationship table or via topic family relationships.

By default, links created from a reltable or the *collection-type* attribute have the format "*Change the format of cross-references* on page 2" with the link text in blue italics. You want to change this format. The PDF specs say `<related-link>` text should be Trebuchet 10pt normal (not bold, not italic) dark red. The page reference should be the same, but in italics and black.

I. Copy the *link__content* attribute set from DITA-OT/plugins/org.dita.pdf2/cfg/fo/attrs/**links-attr.xsl** to your copy of **links-attr.xsl**.

This attribute set uses the *common.link* attribute set (found in **common-attr.xsl**), which includes the *color* and *font-style* attributes to specify blue italic text for links. You need to override these attributes in *link__content*.

2. Add the *color* and *font-style* attributes as shown:

```
<xsl:attribute-set name="link__content" use-attribute-sets="common.link">
    <!--<xsl:attribute name="margin-left">8pt</xsl:attribute>-->
    <xsl:attribute name="color">#990033</xsl:attribute>
    <xsl:attribute name="font-style">normal</xsl:attribute>
</xsl:attribute-set>
```

3. In `DITA-OT/plugins/org.dita.pdf2/xsl/fo/`**links.xsl**, search for the text "On the page."

You're searching for this because the plugin uses a localization variable named *On the page* to insert the boilerplate text "on page" in related links. If you find where this variable is called, you can find the section of code that inserts the page reference and figure out how to format it.

You'll find this text in a template that begins:

```
<xsl:template name="insertPageNumberCitation">
```

The text is in this section:

```
<xsl:otherwise>
   <fo:inline>
      <xsl:call-template name="getVariable">
         <xsl:with-param name="id" select="'On the page'"/>
         <xsl:with-param name="params">
            <pagenum>
               <fo:inline>
                  <fo:page-number-citation ref-id="{$destination}"/>
               </fo:inline>
            </pagenum>
         </xsl:with-param>
      </xsl:call-template>
   </fo:inline>
</xsl:otherwise>
```

4. Copy this template into your **links.xsl**.

5. Edit the top-level `<fo:inline>` to call a new attribute set:

```
<fo:inline xsl:use-attribute-sets="link__page">
   <xsl:call-template name="getVariable">
   ...
```

6. Add the new *link__page* attribute set to your **links-attr.xsl**:

```
<xsl:attribute-set name="link__page" use-attribute-sets="base-font">
   <xsl:attribute name="font-style">italic</xsl:attribute>
</xsl:attribute-set>
```

7. Save changes, run a PDF build, and check your work.

Footnotes

Exercise: Change the appearance of footnote text and callouts

Objective: Change the font, size, color, weight, and style of footnote text and numbers.

Out of the box, the PDF plugin creates footnotes, but they're pretty plain. They're at the bottom of the page, indented slightly, with a line above. You can dress them up a little.

1. Copy the *fn__body* and *fn__callout* attribute sets
   `DITA-OT/plugins/org.dita.pdf2/cfg/fo/attrs/`**commons-attr.xsl** from **commons-attr.xsl**.

   The *fn__body* attribute set formats the text of the footnote.

   The *fn__callout* attribute set formats the superscripted number at the reference point in the main text and the superscripted number in the footnote itself. The *fn__body* attribute set also applies to these numbers, so you only need to add attributes to account for any differences between the footnote text and the number. In this case, only the color and font-weight are different.

   ‖ **Note** There are also *fn* and *fn__id* attribute sets. Out of the box, they aren't used at all. ‖

2. Add the following attributes and values to *fn__body*:

   ```
   <xsl:attribute name="font-family">sans-serif</xsl:attribute>
   <xsl:attribute name="font-size">9pt</xsl:attribute>
   <xsl:attribute name="font-style">italic</xsl:attribute>
   <xsl:attribute name="color">#8a8a8a</xsl:attribute>
   ```

3. Add the following attributes and values to *fn__callout*:

   ```
   <xsl:attribute name="font-weight">bold</xsl:attribute>
   <xsl:attribute name="color">#990033</xsl:attribute>
   ```

4. Save changes, run a PDF build, and check your work.

Exercise: Change the footnote separator line

Objective: Determine the appearance of the line that separates footnotes from the main page text.

1. Open `DITA-OT/plugins/org.dita.pdf2/xsl/fo/`**static-content.xsl**.

2. Find the *insertBodyFootnoteSeparator* template.

This is the template that actually inserts the line above the footnote section. It may seem strange that it's not in **commons.xsl** with the rest of the code that processes footnotes, but this line is considered static content, just like page headers or footers, because it occurs in a fixed location on the page. This leader uses the *__body__footnote__separator* attribute set, and you can probably guess which file that's found in.

3. Copy the *__body__footnote__separator* attribute set from `DITA-OT/plugins/org.dita.pdf2/cfg/fo/attrs/`**static-content-attr.xsl** to your copy of **static-content-attr.xsl**.

4. Change attributes as follows:

```
<xsl:attribute name="leader-pattern">dots</xsl:attribute>
<xsl:attribute name="rule-style">dotted</xsl:attribute>
<xsl:attribute name="color">#e6e6e6</xsl:attribute>
```

5. Save changes, run a PDF build, and check your work.

Exercise: Add brackets around footnote callouts

Objective: Add square brackets [ ] around footnote callouts in the text body.

1. Copy the template that starts with the following line from `DITA-OT/plugins/org.dita.pdf2/xsl/fo/`**commons.xsl** to your copy of **commons.xsl**:

```
<xsl:template match="*[contains(@class,' topic/fn ')]">
```

You probably suspect that you need to use `<xsl:text>` to add the square brackets, and you're absolutely right. Where to put the `<xsl:text>` is the slightly tricky part here.

2. Edit the first part of the template as follows:

```
<xsl:template match="*[contains(@class,' topic/fn ')]">
    <xsl:text>[</xsl:text>
    <!--<fo:inline>
        <xsl:call-template name="commonattributes"/>
    </fo:inline>-->
    ...
```

3. Edit the end of the template as follows:

```
        . . .
        </fo:footnote-body>
    </fo:footnote>
    <xsl:text>]</xsl:text>
</xsl:template>
```

At this point, if you ran a build, you'd see the square brackets around the callouts. Now you need to format them the same as the callouts. Fortunately, you already have an attribute set that's doing the job: *fn__callout*. You need to apply it to the <xsl:text> by wrapping the <xsl:text> in <fo:inline> and applying the attribute set to the <fo:inline> tag.

4. Edit the code as shown:

```
<xsl:template match="*[contains(@class,' topic/fn ')]">
    <fo:inline xsl:use-attribute-sets="fn__callout">
        <xsl:text>[</xsl:text>
    </fo:inline>
    . . .
```

```
    . . .
    </fo:footnote>
    <fo:inline xsl:use-attribute-sets="fn__callout">
        <xsl:text>]</xsl:text>
    </fo:inline>
</xsl:template>
```

5. Save changes, run a PDF build, and check your work.

Chapter 16

Table of Contents

The DITA Open Toolkit makes it easy to add a Table of Contents (TOC) to your PDF. Many traditional desktop publishing applications treat the TOC as a separate file that must be generated and updated separately during the publishing process. When you create a PDF using the Open Toolkit, the TOC is not a separate file; it's generated as an integral part of the process.

The creation of a Table of Contents is automated within the DITA Open Toolkit's PDF plugin, but there is a lot you can do to change the look and feel as well as the information included.

The chapter also explains how to format and include or exclude a mini-TOC (section-level TOC).

For a list of attributes that affect TOCs, refer to **Table of Contents attribute sets** *(p. 456)*.

How TOC attribute sets interact

To customize the appearance of TOC entries, you edit the corresponding attribute sets in **toc-attr.xsl**. Making those changes is easy; the tricky part is understanding what attribute sets affect what parts of the TOC entry.

First, let's look at the attribute sets that format titles in general. **Figure 40** illustrates these attribute sets.

```
<bookmap>
   <chapter href="myditamap1.ditamap">
      <map>
         <topicref href="topicA.xml">        ──  __toc__chapter__content
            <topicref href="topicB.xml"/>    ┐
            <topicref href="topicC.xml"/>    ├  __toc__topic__content
         </topicref>                          ┘
      </map>
   <part href="myditamap2.ditamap"/>
      <map>
         <topicref href="topicA.xml">        ──  __toc__part__content
            <topicref href="topicB.xml"/>    ┐
            <topicref href="topicC.xml"/>    ├  __toc__topic__content
         </topicref>                          ┘
      </map>
   <appendix href="myditamap3.ditamap"/>
      <map>
         <topicref href="topicA.xml">        ──  __toc__appendix__content
            <topicref href="topicB.xml"/>    ┐
            <topicref href="topicC.xml"/>    ├  __toc__topic__content
         </topicref>                          ┘
      </map>
</bookmap>
```

Figure 40: TOC entry attribute sets

If you dissect a TOC entry, you see that it can have four pieces:

- the label (for example, "Chapter n:")
- the title itself
- the leader
- the page number

Theoretically, you can format each of these four pieces separately. Here is a rough guide to the TOC attribute sets that format each piece of a TOC entry:

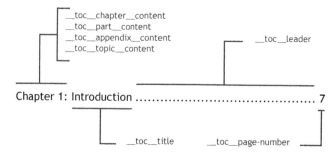

Figure 41: Attribute sets for pieces of TOC entry

This is not a complete discussion of all of the attribute sets that apply to TOCs; it just hits the high points. To fully understand how the TOC attribute sets interact, your best bet is to experiment by assigning an easily visible attribute, such as *color*, to different attribute sets and observing the results.

Title levels in bookmaps and ditamaps

Let's stray outside the scope of this book for a moment and talk about map structure. It's easy to be confused about how the Open Toolkit renders topic titles as levels, so let's discuss discuss it briefly here.

Publishing from bookmaps

In a bookmap, one possibility is that each chapter, part, or appendix is a separate ditamap:

```
<chapter href="myditamap1.ditamap"/>
<part href="myditamap2.ditamap"/>
<appendix href="myditamap3.ditamap"/>
```

And each ditamap has one top-level topic—in this example, **topicA.xml**:

```
<map>
    <topicref href="topicA.xml">
        <topicref href="topicB.xml"/>
        <topicref href="topicC.xml">
            <topicref href="topicD.xml"/>
        </topicref>
    </topicref>
</map>
```

Another possibility is that each chapter, part, or appendix is a nested topic structure:

```
<chapter href="topicA.xml">
    <topicref href="topicB.xml"/>
    <topicref href="topicC.xml">
        <topicref href="topicD.xml"/>
    <topicref href="topicE.xml">
        <topicref href="topicF.xml"/>
    </topicref>
    </topicref>
</chapter>
```

Again, each has one top-level topic—in this example, also **topicA.xml**.

In both instances, the title of topicA is going to appear in the TOC as the chapter title: "Chapter n: Title of topicA." In other words, it will be the level-1 TOC entry. The titles of nested topicrefs appear in the TOC as nested TOC levels corresponding to their levels in the ditamap. Using the example above, topicB and topicC would become level-2 TOC entries and topicD would become a level-3 TOC entry.

Many authors make the mistake of creating a ditamap with no single top-level topic:

```
<map>
   <topicref href="topicA.xml"/>
   <topicref href="topicB.xml"/>
   <topicref href="topicC.xml">
      <topicref href="topicD.xml"/>
   </topicref>
</map>
```

In such a map, the titles of topicA, topicB, and topicC appear in the TOC as chapter titles because they are all at the highest level of the ditamap.

If your ditamap doesn't already include a topic that can naturally act as the top-level topic, you can create a stem topic and nest the other topics inside it. A stem topic contains just a title, which becomes the chapter title. Or you can nest topics inside a single top-level `<topichead>` element.

Of course, if you want several chapter titles generated from one ditamap, then the structure above is fine.

Publishing from ditamaps

If you're publishing from a ditamap, you don't have access to the `<chapter>`, `<appendix>`, or `<part>` elements. Your TOC simply reflects the hierarchy set up in your ditamaps. If you have one primary ditamap that calls several modular ditamaps, and you want each modular ditamap to represent, more or less, a chapter, you need to be sure each modular ditamap has a single top-level topic.

In **Figure 42**, the primary ditamap calls two modular ditamaps. The first ditamap has three topicrefs, all direct children of the map. The title of each of these topics will show up as a chapter in the PDF. The second ditamap has only one top-level topic, and only its title will show up as a chapter in the PDF.

```
                                          <map>
                                             <topicref href="topicA.xml"/>
        <map>                                <topicref href="topicB.xml"/>
           <mapref href="myditamap1.ditamap"/> ─   <topicref href="topicC.xml"/>
           <mapref href="myditamap2.ditamap"/> ─ </map>
        </map>


                                          <map>
                                             <topicref href="topicD.xml">
                                                <topicref href="topicE.xml"/>
                                                <topicref href="topicF.xml"/>
                                             </topicref>
                                          </map>
```

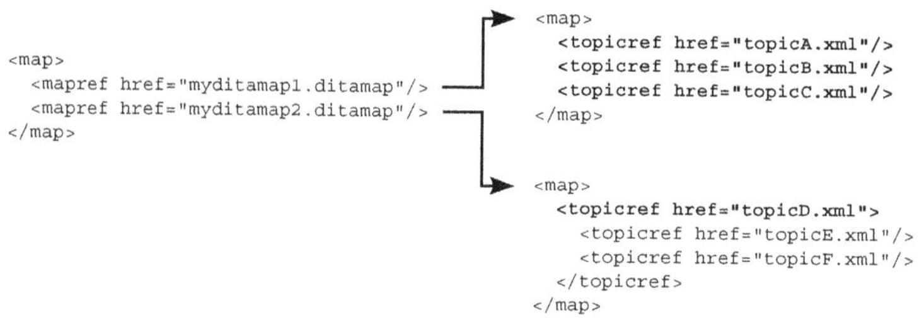

Figure 42: Topics as chapters

Be aware that sub-map titles do not contribute to the TOC title hierarchy. They don't appear at all.

TOC specifications

Here are the specifications for the table of contents.

TOC

TOC title font	28pt Trebuchet normal black
Number of TOC levels	two
TOC level 1 entry font	12pt Trebuchet normal dark red
TOC level 2 entry font	10pt Trebuchet normal dark gray
TOC leader	none
TOC page number	10pt Trebuchet bold dark red; includes chapter, appendix or part number
Level 2 indent	.25in
TOC heading	Table of Contents
Mini-TOC heading	In this section
Mini-TOC links	10pt Trebuchet bold dark red
Mini-TOC link page number	10pt Trebuchet bold black, includes chapter, appendix or part number

Files you need

If you have completed other exercises, some of these files might already be present in your plugin.

If they are not already present, create the following files:

- `DITA-OT/plugins/com.company.pdf/cfg/fo/attrs/`**commons-attr.xsl**
- `DITA-OT/plugins/com.company.pdf/cfg/fo/attrs/`**toc-attr.xsl**
- `DITA-OT/plugins/com.company.pdf/cfg/fo/attrs/`**toc-attr_fop.xsl** (if you're using FOP)

Be sure to add the appropriate `<xsl:import>` statements to **custom.xsl** (attrs).

If they are not already present, create the following files:

- `DITA-OT/plugins/com.company.pdf/cfg/fo/xsl/`**basic-settings.xsl**
- `DITA-OT/plugins/com.company.pdf/cfg/fo/xsl/`**bookmarks.xsl**
- `DITA-OT/plugins/com.company.pdf/cfg/fo/xsl/`**commons.xsl**
- `DITA-OT/plugins/com.company.pdf/cfg/fo/xsl/`**toc.xsl**

Be sure to add the appropriate `<xsl:import>` statements to **custom.xsl** (xsl).

For instructions on creating attribute files and XSLT stylesheets in your PDF plugin, refer to **Create an attribute set file in your plugin** *(p. 31)* and **Create an XSLT stylesheet in your plugin** *(p. 32)*.

If they are not already present, create the following files:

- `DITA-OT/plugins/com.company.pdf/cfg/common/vars/`**en.xml**
- `DITA-OT/plugins/com.company.pdf/cfg/common/vars/`**strings-en-us.xml**

For instructions on creating variables and strings files, refer to **Create a localization variables file** *(p. 78)* and **Create a localization strings file** *(p. 79)*. For instructions on adding the appropriate language and locale mapping statements in **strings.xml**, refer to **Mapping strings files to xml:lang values in the strings.xml file** *(p. 37)*.

Exercise: Add a TOC to a PDF

Objective: Set up your bookmap to generate a Table of Contents for your PDF.

By default, the Open Toolkit automatically creates a TOC if you build your output from a map instead of a bookmap. If you build your PDF from a bookmap, the Open Toolkit only creates a TOC if you explicitly tell it to. Here's how to do that.

1. In your bookmap, add the `<frontmatter>` element if it is not already present.

2. Within `<frontmatter>`, add the `<toc>` element:

```
<frontmatter>
   <booklists>
      <toc/>
   </booklists>
</frontmatter>
```

Exercise: Change the title of the TOC

Objective: Use a heading other than the default "Table of Contents" for your TOC.

1. Copy the *Table of Contents* variable from
`DITA-OT/plugins/org.dita.pdf2/cfg/common/vars/`**en.xml** to your copy of **en.xml**:

2. Change the default value "Contents" to "Table of Contents".

If for some reason you don't want a heading at all, delete the value altogether:

```
<variable id="Table of Contents"></variable>
```

3. Repeat this process for any other language-specific variables files you've customized.

Exercise: Format the TOC title

Objective: Determine the appearance of your Table of Contents title.

1. Copy the *__toc__header* from `DITA-OT/plugins/org.dita.pdf2/cfg/fo/attrs/`**toc-attr.xsl**
to your copy of **toc-attr.xsl**.

2. In the *__toc__header* attribute set, add, or change the following attributes and values:

```
<xsl:attribute-set name="__toc__header" use-attribute-sets="common.title">
    <xsl:attribute name="space-before">0pt</xsl:attribute>
    <xsl:attribute name="space-after">16.8pt</xsl:attribute>
    <xsl:attribute name="font-size">28pt</xsl:attribute>
    <xsl:attribute name="font-weight">normal</xsl:attribute>
    <xsl:attribute name="padding-top">16.8pt</xsl:attribute>
    <xsl:attribute name="color">#000000</xsl:attribute>
</xsl:attribute-set>
```

Note that this attribute set uses the *common.title* attribute set. In the exercise **Specify fonts to use** *(p. 83)* you added the following attribute to the *common.title* attribute set:

```
<xsl:attribute name="font-family">Trebuchet MS, Arial Unicode MS, Helvetica</xsl:attribute>
```

If you did not complete that exercise, add this attribute to the *__toc__header* attribute set now.

3. Save changes, run a PDF build, and check your work.

Exercise: Add or remove entry levels from the TOC

Objective: Limit or expand the number of entry levels (as determined by the map hierarchy) that appear in the Table of Contents.

Remember that you want the TOC to have only two levels: the chapter title and all top-level topics within the chapter. This is a simple, though not intuitive, change to make.

Before beginning this exercise, you should be clear on what the PDF plugin considers to be a TOC level. For details on TOC levels, refer to **Title levels in bookmaps and ditamaps** *(p. 327)*.

1. In DITA-OT/plugins/org.dita.pdf2/xsl/fo/**toc.xsl**, find the template that begins with

```
<xsl:template match="*[contains(@class, ' topic/topic ')]" mode="toc">
```

2. In that template, locate the following line:

```
<xsl:if test="$topicLevel &lt; $tocMaximumLevel">
```

This <xsl:if> statement tests to see if the current topic level is less than the maximum level allowed in the TOC. This maximum level is expressed as a variable, *tocMaximumLevel*.

tocMaximumLevel is defined in **basic-settings.xsl**. By, default, its value is "4".

3. Copy the following parameter from DITA-OT/plugins/org.dita.pdf2/cfg/fo/attrs/**basic-settings.xsl** to your copy of **basic-settings.xsl**:

```
<xsl:param name="tocMaximumLevel" select="4"/>
```

4. Change the value of *select* reflect the number of TOC levels you want.

5. Save changes, run a PDF build, and check your work.

Exercise: Apply different formatting to different TOC entry levels

Objective: Specify that different entry levels of the Table of Contents should have a different appearance.

As a refresher, if you're not clear on how the PDF plugin determines TOC levels, take a look at **Title levels in bookmaps and ditamaps** *(p. 327)*.

1. Copy the *__toc__topic__content* attribute set from `DITA-OT/plugins/org.dita.pdf2/cfg/fo/attrs/`**toc-attr.xsl** to your copy of **toc-attr.xsl**.

   Find the following section in that attribute set:

```
<xsl:attribute-set name="__toc__topic__content">
   <xsl:attribute name="last-line-end-indent">-22pt</xsl:attribute>
   <xsl:attribute name="end-indent">22pt</xsl:attribute>
   <xsl:attribute name="text-indent">-.2in</xsl:attribute>
   <xsl:attribute name="text-align">left</xsl:attribute>
   <xsl:attribute name="text-align-last">justify</xsl:attribute>
   <xsl:attribute name="font-size">
      <xsl:variable name="level" select="count(ancestor-or-self::*
      ►[contains(@class, ' topic/topic ')])"/>
      <xsl:choose>
         <xsl:when test="$level = 1">12pt</xsl:when>
         <xsl:otherwise><xsl:value-of select="$default-font-size"/></xsl:otherwise>
      </xsl:choose>
   </xsl:attribute>
   <xsl:attribute name="font-weight">
      <xsl:variable name="level" select="count(ancestor-or-self::*
      ►[contains(@class, ' topic/topic ')])"/>
         <xsl:choose>
            <xsl:when test="$level = 1">bold</xsl:when>
            <xsl:otherwise>normal</xsl:otherwise>
         </xsl:choose>
   </xsl:attribute>
</xsl:attribute-set>
```

   Notice how the settings for the *font-size* and *font-weight* attributes include `<xsl:choose>` statements that make level-1 headings 12pt bold and all other headings the default font size and normal weight?

   You can apply this same logic to any other attribute. Give this a try by assigning a different color to TOC entry levels.

2. Copy the section highlighted in the code above (the entire *font-size* attribute).

3. Paste the copied text as shown.

```
<xsl:attribute-set name="__toc__topic__content">
   <xsl:attribute name="last-line-end-indent">-22pt</xsl:attribute>
   <xsl:attribute name="end-indent">22pt</xsl:attribute>
   <xsl:attribute name="text-indent">-.2in</xsl:attribute>
   <xsl:attribute name="text-align">left</xsl:attribute>
   <xsl:attribute name="text-align-last">justify</xsl:attribute>
   <xsl:attribute name="font-size">
      <xsl:variable name="level" select="count(ancestor-or-self::*
      ▶[contains(@class, ' topic/topic ')])"/>
      <xsl:choose>
         <xsl:when test="$level = 1">12pt</xsl:when>
         <xsl:otherwise><xsl:value-of select="$default-font-size"/></xsl:otherwise>
      </xsl:choose>
   </xsl:attribute>
   <xsl:attribute name="font-weight">
      <xsl:variable name="level" select="count(ancestor-or-self::*
      ▶[contains(@class, ' topic/topic ')])"/>
         <xsl:choose>
            <xsl:when test="$level = 1">bold</xsl:when>
            <xsl:otherwise>normal</xsl:otherwise>
         </xsl:choose>
   </xsl:attribute>
   <xsl:attribute name="font-size">
      <xsl:variable name="level" select="count(ancestor-or-self::*
      ▶[contains(@class, ' topic/topic ')])"/>
      <xsl:choose>
         <xsl:when test="$level = 1">12pt</xsl:when>
         <xsl:otherwise>10pt</xsl:otherwise>
      </xsl:choose>
   </xsl:attribute>
</xsl:attribute-set>
```

4. Change the attribute name in the pasted section to *color*.

```
<xsl:attribute name="color">
   <xsl:variable name="level" select="count(ancestor-or-self::*
   ▶[contains(@class, ' topic/topic ')])"/>
   <xsl:choose>
      <xsl:when test="$level = 1">12pt</xsl:when>
      <xsl:otherwise>10pt</xsl:otherwise>
   </xsl:choose>
</xsl:attribute>
```

5. You want level-1 TOC headings to be dark red, so the `<xsl:when>` statement, change the value to "#990033".

```
<xsl:when test="$level = 1">#990033</xsl:when>
```

6. You want level-2 TOC headings to be dark gray, so in the `<xsl:otherwise>` statement, change the value to "#8A8A8A".

```
<xsl:otherwise>#8A8A8A</xsl:otherwise>
```

7. While you're at it, you don't want level-1 headings to be bold, so change the *font-weight* attribute to specify "normal":

```
<xsl:attribute name="font-weight">
  <xsl:variable name="level" select="count(ancestor-or-self::*
 ▸[contains(@class, ' topic/topic ')])"/>
    <xsl:choose>
      <xsl:when test="$level = 1">normal</xsl:when>
      <xsl:otherwise>normal</xsl:otherwise>
    </xsl:choose>
</xsl:attribute>
```

8. You also want TOC entries to use the Trebuchet font, so add the *font-family* attribute to this attribute set with a value of "Trebuchet MS".

```
. . .
<xsl:attribute name="color">
    <xsl:variable name="level" select="count(ancestor-or-self::
 ▸*[contains(@class, ' topic/topic ')])"/>
    <xsl:choose>
        <xsl:when test="$level = 1">#990033</xsl:when>
        <xsl:otherwise>#8A8A8A</xsl:otherwise>
    </xsl:choose>
</xsl:attribute>
<xsl:attribute name="font-family">Trebuchet MS, Arial Unicode MS, Helvetica</xsl:attribute>
. . .
```

All TOC entries are Trebuchet, regardless of level, so you don't need an `<xsl:choose>` statement.

9. Finally, you want level-2 TOC entries to be 10pt, so change the `<xsl:otherwise>` value for *font-size* to **10pt**.

10. Save changes, run a PDF build, and check your work.

Reality check

In your TOC, the top-level headings are now dark red and all other headings are now dark gray. If you later add a third level of TOC headings and you wanted them to be black, you could add a second `<xsl:when>` statement after the first, to specify dark gray for level-2 headings and then change the `<xsl:otherwise>` statement to specify black.

Exercise: Format page numbers in the TOC

Objective: Determine the appearance of page numbers in your Table of Contents.

The PDF plugin provides a handy attribute set for formatting TOC page numbers, so it's easy to specify formatting for them. You want page numbers to be 10pt Trebuchet bold dark red.

I. Copy the *__toc__page-number* attribute set from
`DITA-OT/plugins/org.dita.pdf2/cfg/fo/attrs/`**toc-attr.xsl** to your copy of **toc-attr.xsl**.

2. Add the following attributes to that attribute set:

```
<xsl:attribute-set name="__toc__page-number">
    <xsl:attribute name="start-indent">-<xsl:value-of select="$toc.text-indent"/>
    ►</xsl:attribute>
    <xsl:attribute name="keep-together.within-line">always</xsl:attribute>
    <xsl:attribute name="font-size">10pt</xsl:attribute>
    <xsl:attribute name="color">#990033</xsl:attribute>
    <xsl:attribute name="font-weight">bold</xsl:attribute>
</xsl:attribute-set>
```

Even though you want the entries to be Trebuchet MS, you don't need to add the *font-family* attribute. Page numbers are part of the entire TOC entry. The TOC entry as a whole is a block element. The page number is an inline element within the block element. Unless they have their own specification for an attribute, inline elements pick up the formatting of the block element they are part of. In this case, the page number is going to pick up the *font-family* value of the attribute set *__toc__topic__content*, which you set to be Trebuchet MS in the previous exercise.

3. Save changes, run a PDF build, and check your work.

Exercise: Add chapter, appendix, or part numbers to TOC page numbers

Objective: Use a new *getChapterPrefix* template to number TOC entries.

This exercise applies only if you are publishing from a bookmap. Chapter, appendix and part numbering depend on the presence of the `<chapter>`, `<appendix>` and `<part>` elements, which aren't available in a map.

I. Copy the entire *getChapterPrefix* template into **custom.xsl** (xsl).

You can find this template in **GetChapterPrefix template** *(p. 481)*.

> **Note** You could copy this template into any of the .xsl files you've added to your plugin, including **toc.xsl**, but it's just as easy to copy it into **custom.xsl**. You can see right away it's a custom template that didn't come with the PDF plugin.

2. Copy the template that begins with the following from
`DITA-OT/plugins/org.dita.pdf2/xsl/fo/`**toc.xsl** into your copy of **toc.xsl**:

```
<xsl:template match="*[contains(@class, ' topic/topic ')]" mode="toc">
```

Within this template, the `<fo:page-number-citation>` element creates the page number. So you can figure that somehow you need to call the ***getChapterPrefix*** template in conjunction with `<fo:page-number-citation>`.

3. In the template you just copied, find the `<fo:page-number-citation>` element:

```
...
<fo:inline xsl:use-attribute-sets="__toc__page-number">
    <fo:leader xsl:use-attribute-sets="__toc__leader"/>
    <fo:page-number-citation>
    <xsl:attribute name="ref-id">
        <xsl:call-template name="get-id"/>
    </xsl:attribute>
    </fo:page-number-citation>
</fo:inline>
...
```

4. Add lines to call the ***getChapterPrefix*** template and insert a hyphen between the chapter number and the page number:

```
...
<fo:inline xsl:use-attribute-sets="__toc__page-number">
    <fo:leader xsl:use-attribute-sets="__toc__leader"/>
    <xsl:call-template name="getChapterPrefix" />
    <xsl:text>-</xsl:text>
    <fo:page-number-citation>
    <xsl:attribute name="ref-id">
        <xsl:call-template name="get-id"/>
    </xsl:attribute>
    </fo:page-number-citation>
</fo:inline>
...
```

Notice that you add these lines between the line that creates the leader and the code that creates the page number—exactly where the chapter number and hyphen should be in the TOC.

5. Save changes, run a PDF build, and check your work.

Exercise: Change the leader in TOC entries

Objective: Use a different leader format for entries in the Table of Contents.

The default leader in the PDF plugin is dots. If you prefer, you can have underscores or white space as the leader instead.

1. Copy the *__toc__leader* attribute set from
`DITA-OT/plugins/org.dita.pdf2/cfg/fo/attrs/`**toc-attr.xsl**, to your copy of **toc-attr.xsl**:

```
<xsl:attribute-set name="__toc__leader">
    <xsl:attribute name="leader-pattern">dots</xsl:attribute>
</xsl:attribute-set>
```

2. Replace the value "dots" with one of the following:

- space — the leader is white space

- rule — the leader is a series of underscores

These are the only choices available, and how they look in the PDF is a function of the PDF renderer.

3. Save changes, run a PDF build, and check your work.

Exercise: Remove leaders from TOC entries

Objective: Remove the leader between the Table of Contents entry text and the page number.

You don't want any leader at all between the TOC entry title and the page number. That's a simple change.

1. In `DITA-OT/plugins/org.dita.pdf2/xsl/fo/`**toc.xsl**, locate the following line:

```
<fo:leader xsl:use-attribute-sets="__toc__leader"/>
```

There are five of these lines: one for main entries, one for glossary entries, one for entries in the list of tables, one for entries for the list of figures, and one for entries for the list of tables. Right now, you just want to change the first occurrence—the one for main entries.

2. Copy the entire template that contains this line to your copy of **toc.xsl**.

The template begins with these lines:

```
<xsl:template match="*[contains(@class, ' topic/topic ')]" mode="toc">
    <xsl:param name="include"/>
    <xsl:variable name="topicLevel" as="xs:integer">
```

3. Comment out the highlighted line:

```
...
<fo:inline xsl:use-attribute-sets="__toc__page-number">
    <fo:leader xsl:use-attribute-sets="__toc__leader"/>
    <fo:page-number-citation>
    ...
```

There is now nothing between the topic title and the page number—no leader, no space, nothing.

4. Below the line you just commented out, type

```
<xsl:text>  </xsl:text>
```

This line forces two non-breaking spaces between the topic title and the page number. You can put as many spaces as you like. You can also insert a character such as a hyphen (-) or a vertical line (|).

> **Tip:** `<xsl:text>` is a handy tag to use for inserting any kind of hardcoded or boilerplate text within an XSL template.

By default, TOC entries are justified, which forces the page number all the way to the outside margin. Now that there's no leader, it doesn't make sense to leave the entries justified. Let's change that.

5. In your copy of **toc-attr.xsl**, find the *__toc__topic__content* attribute set (which you have probably copied in a previous exercise), and change the value of the *text-align-last* attribute to something other than "justify"... for example, "left".

6. Save changes, run a PDF build, and check your work.

Exercise: Adjust indents for TOC entries

Objective: Determine the amount of space that second- (and third-, etc.) level TOC entries are indented below main entries.

Adjusting TOC indents is one of those thing that's easy to do once you understand how it works, but understanding it is not exactly straightforward. So let's untangle it a little bit.

1. If the *__toc__topic__content* attribute set is not already in your copy of **toc-attr.xsl**, copy it from `DITA-OT/plugins/org.dita.pdf2/cfg/fo/attrs/`.

2. Note the following attribute:

```
<xsl:attribute name="text-indent"><xsl:value-of select="$toc.text-indent"/></xsl:attribute>
```

This attribute is not what you want to use to control indents. It actually shifts the entire TOC to the left or right, depending on the value of the attribute. Consider this snippet of the TOC, with the *text-indent* attribute set to "0pt":

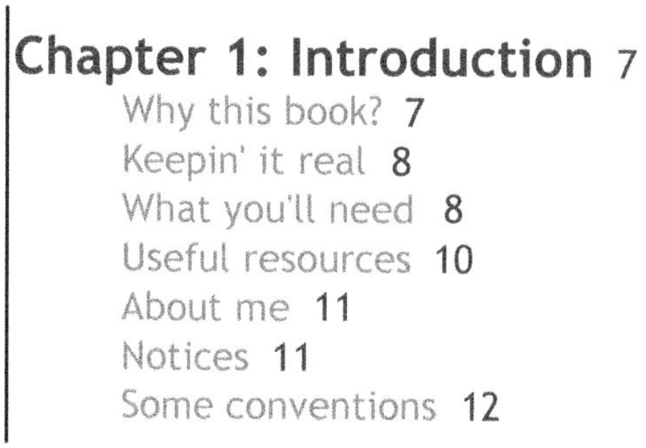

Chapter 1: Introduction 7

Why this book? 7

Keepin' it real 8

What you'll need 8

Useful resources 10

About me 11

Notices 11

Some conventions 12

Figure 43: TOC with text-indent=0

Notice how the level-1 entry is flush left. Now look at this snippet, with the *text-indent* attribute set to the value of the *toc.text-indent* variable, as shown above.

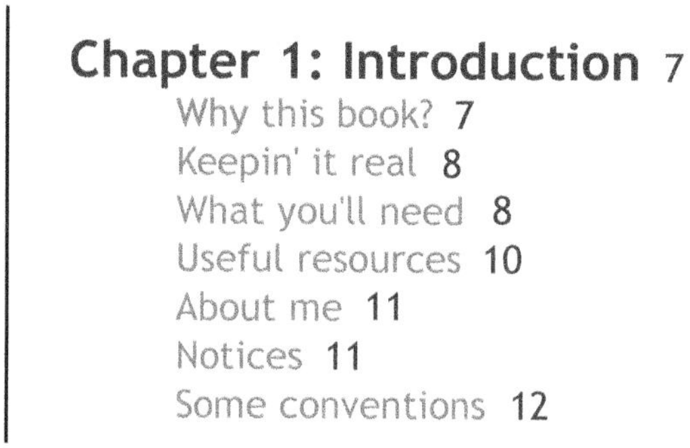

Chapter 1: Introduction 7

Why this book? 7

Keepin' it real 8

What you'll need 8

Useful resources 10

About me 11

Notices 11

Some conventions 12

Figure 44: TOC with text-indent=toc.text-indent

In the second snippet, the level-1 heading is indented whatever the value of the *toc.text-indent* variable is—by default, "14pt".

Now that you understand what *text-indent* does (and doesn't do), take a look at what you actually want to work with.

3. Copy the *__toc__indent* attribute set from
`DITA-OT/plugins/org.dita.pdf2/cfg/fo/attrs/`**toc-attr.xsl** to your copy of **toc-attr.xsl**:

```
<xsl:attribute-set name="__toc__indent">
    <xsl:attribute name="start-indent">
        <xsl:variable name="level" select="count(ancestor-or-self::
        ▸*[contains(@class, ' topic/topic ')])"/>
        <xsl:value-of select="concat($side-col-width, ' + (', string($level - 1), ' * ',
        ▸ $toc.toc-indent, ') + ', $toc.text-indent)"/>
    </xsl:attribute>
</xsl:attribute-set>
```

This is the attribute set that applies a level-based indent to TOC entries. The more nested the entry, the greater the indent. The attribute set is based on a formula and this formula is less than clear at first glance. So here's a breakdown.

First, there's a variable, *level*. This variable gets its value by counting the number of ancestors the TOC entry has. A level-1 entry has no ancestors, so when processing level-1 entries, the value of *level* is 0. A level-2 entry has one ancestor, so when processing level-2 entries, the value of *level* becomes 1. And so on.

Next, the formula itself. This formula uses the values of several other variables:

- *side-col-width* (comes from the variable of the same name in **basic-settings.xsl**; the default is 25pt)
- *level* (calculated as explained above)
- *toc.toc-indent* (comes from the variable of the same name in **toc-attr.xsl**; default is 14pt)
- *toc.text-indent* (comes from the variable of the same name in **toc-attr.xsl**; default is 30pt)

Using the variable defaults, and taking away all the concat stuff, here's the formula rendered a little more plainly:

```
25pt + (0 * 14pt) + 30pt
```

This is the formula for calculating the indent for level-1 headings. In plain English, it says, "multiply 0 by 14. Add 25 to the result of that and then add 30 to the result of that." The result is 55 (pt). Therefore, level-1 TOC entries are indented 55pt. Can you guess the formula for level-2 headings?

```
25pt + (1 * 14pt) + 30pt
```

The result of which is 69 (pt).

If you use the defaults, each TOC entry level is indented 14pt more than the level above it.

Now that you understand the formula, you can see that setting indents is a matter of playing around with different values for the variables the formula uses.

4. Copy the *toc.text-indent* and *toc.toc-indent* variables from
DITA-OT/plugins/org.dita.pdf2/cfg/fo/attrs/**toc-attr.xsl** to your copy of **toc-attr.xsl**

```
<xsl:variable name="toc.text-indent" select="'14pt'"/>
<xsl:variable name="toc.toc-indent" select="'30pt'"/>
```

5. In **toc-attr.xsl**, in the *__toc__topic__content* attribute set, change the value of *text-indent* to "0in".

6. In **toc-attr.xsl**, change the value of *toc.toc-indent* to ".25in".

This represents the .25in indent in the specifications for your PDF.

7. In **toc-attr.xsl**, change the value of *toc.text-indent* to "0in".

With these changes, the formula now becomes:

```
0in + (0 * .25in) + 0in
```

for level-1 headings and

```
0in + (1 * .25in) + 0in
```

for level-2 headings. The result is that level-1 entries are not indented, and level-2 entries are indented .25in. Subsequent entry levels would be indented .5in, .75in, and so on.

8. In **toc-attr.xsl**, change the formula to use *toc.text-indent* instead of *side-col-width*.

```
<xsl:attribute-set name="__toc__indent">
    <xsl:attribute name="start-indent">
        <xsl:variable name="level" select="count(ancestor-or-self::*
        ▶[contains(@class, ' topic/topic ')])"/>
        <xsl:value-of select="concat($toc.text-indent, ' + (', string($level - 1)
        ▶, ' * ', $toc.toc-indent, ') + ', $toc.text-indent)"/>
    </xsl:attribute>
</xsl:attribute-set>
```

You could just change the value of *side-col-width* to 0in, but this change would have effects throughout your PDF. That might be okay, but it also might not, depending on your page layouts. It's better to use a different variable to limit the scope of changes to just what you are trying to do here.

9. Save changes, run a PDF build, and check your work.

Mini-TOC

The mini-TOC is a list of topics that appears on the first page of each chapter (or appendix or part). In other words, it's a chapter-level TOC. (The TOC that appears on the first page of this chapter is an example of a mini-TOC.) Out-of-the-box, the PDF plugin creates a mini-TOC that looks like this:

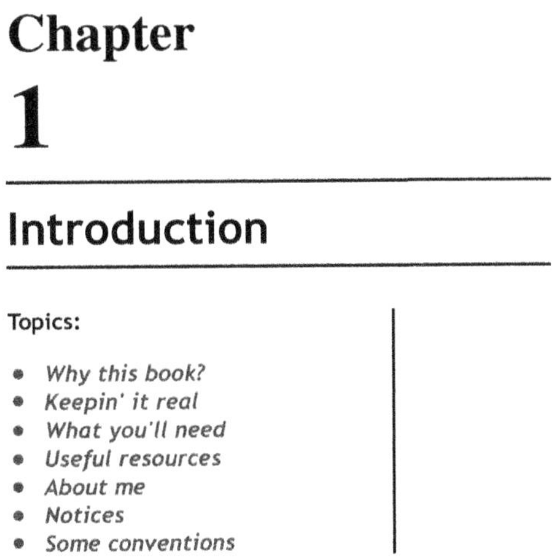

Chapter

1

Introduction

Topics:

- *Why this book?*
- *Keepin' it real*
- *What you'll need*
- *Useful resources*
- *About me*
- *Notices*
- *Some conventions*

Figure 45: Mini-TOC example

Assuming that your chapter is a map with one topic-level topicref and multiple nested topicrefs, the mini-TOC includes links to all level-2 topicrefs. In the example below, the bold topicrefs would be included in the mini-TOC.

```
<map>
   <topicref href="introduction.dita">
      <topicref href="why_this_book.dita"/>
      <topicref href="keepin_it_real.dita"/>
      <topicref href="what_youll_need.dita"/>
      <topicref href="useful_resources.dita"/>
      <topicref href="about_me.dita"/>
      <topicref href="notices.dita"/>
      <topicref href="some_conventions.dita"/>
   </topicref>
</map>
```

Because the mini-TOC is by nature brief, no topics nested below the second level appear in it.

Exercise: Change or delete the title for the mini-TOC

Objective: Use a custom title for the mini-TOC other than the default Topics.

1. Copy the following variable from `DITA-OT/plugins/org.dita.pdf2/cfg/common/vars/`**en.xml** to your copy of **en.xml**:

```
<!-- The heading to put at the top of a chapter when creating a
chapter-level "mini-TOC" -->
<variable id="Mini Toc">Topics: </variable>
```

2. Do one of the following:

   - To change the heading, enter the new value.

   - To have no heading, delete the value (including the colon).

     ```
     <variable id="Mini Toc"></variable>
     ```

3. Repeat this process for any other language-specific variables files you've customized.

Exercise: Format mini-TOC links

Objective: Determine the appearance of links in the mini-TOC.

By default, the plugin uses the *xref* attribute set to format the links in the mini-TOC. This same attribute set is also used to format cross-references (inline links created using the `<xref>` element). If you want the two to have different formatting, you have to assign different attribute sets to them.

1. Copy the following template from `DITA-OT/plugins/org.dita.pdf2/xsl/fo/`**commons.xsl** to your copy of **commons.xsl**.

```
<xsl:template match="*[contains(@class, ' topic/topic ')]" mode="in-this-chapter-list">
 <fo:list-item xsl:use-attribute-sets="ul.li">
  <fo:list-item-label xsl:use-attribute-sets="ul.li__label">
   <fo:block xsl:use-attribute-sets="ul.li__label__content">
    <xsl:call-template name="getVariable">
     <xsl:with-param name="id" select="'Unordered List bullet'"/>
    </xsl:call-template>
   </fo:block>
  </fo:list-item-label>
  <fo:list-item-body xsl:use-attribute-sets="ul.li__body">
   <fo:block xsl:use-attribute-sets="ul.li__content">
    <fo:basic-link internal-destination="{@id}" xsl:use-attribute-sets="xref">
     <xsl:value-of select="*[contains(@class, ' topic/title ')]"/>
    </fo:basic-link>
   </fo:block>
```

```
    </fo:list-item-body>
  </fo:list-item>
</xsl:template>
```

Formatting is applied to mini-TOC links in this section of the template:

```
<fo:basic-link internal-destination="{@id}" xsl:use-attribute-sets="xref">
    <xsl:value-of select="child::*[contains(@class, ' topic/title ')]"/>
</fo:basic-link>
```

2. Change the `<fo:basic-link>` to use attribute set *xref__mini__toc*.

```
<fo:basic-link internal-destination="{@id}" xsl:use-attribute-sets="xref__mini__toc">
```

3. Create the *xref__mini__toc* attribute set in your copy of **commons-attr.xsl** as follows:

```
<xsl:attribute-set name="xref__mini__toc">
    <xsl:attribute name="color">#990033</xsl:attribute>
    <xsl:attribute name="font-style">normal</xsl:attribute>
    <xsl:attribute name="font-weight">bold</xsl:attribute>
</xsl:attribute-set>
```

4. Save changes, run a PDF build, and check your work.

Exercise: Add page numbers to mini-TOC entries

Objective: Add page numbers to the entries in the mini-TOC.

By default, there are no page numbers in the mini-TOC. your template specification calls for them and also calls for them to include the chapter, appendix, or part number.

You're going to use the same ***getChapterPrefix*** template that you used in the **Add chapter, appendix, or part numbers to TOC page numbers** *(p. 336)* exercise to insert the chapter, appendix, or part number in the mini-TOC. If you did not complete that exercise, turn to it now and complete the first step to add the ***getChapterPrefix*** to your plugin.

1. Copy the template that begins with the following line from `DITA-OT/plugins/org.dita.pdf2/xsl/fo/`**commons.xsl** to your copy of **commons.xsl**:

```
<xsl:template match="*[contains(@class, ' topic/topic ')]" mode="in-this-chapter-list">
```

2. In that template, find the following section of code:

```
<fo:list-item-body xsl:use-attribute-sets="ul.li__body">
    <fo:block xsl:use-attribute-sets="ul.li__content">
        <fo:basic-link internal-destination="{@id}" xsl:use-attribute-sets="xref">
            <xsl:value-of select="child::*[contains(@class, ' topic/title ')]"/>
        </fo:basic-link>
    </fo:block>
</fo:list-item-body>
```

3. Edit the code as follows:

```
<fo:basic-link internal-destination="{@id}" xsl:use-attribute-sets="xref">
    <xsl:value-of select="*[contains(@class, ' topic/title ')]"/>
    <fo:inline>
        <xsl:text>  </xsl:text>
        <xsl:call-template name="getChapterPrefix" />
        <xsl:text>-</xsl:text>
        <fo:page-number-citation ref-id="{concat('_OPENTOPIC_TOC_PROCESSING_'
        ▸, generate-id())}"/>
    </fo:inline>
</fo:basic-link>
```

The `<xsl:text>` that inserts two non-breaking spaces, the `<xsl:call-template>`, and the `<xsl:text>` that inserts a hyphen should all be familiar to you. The `<fo:page-number-citation>` is a little bit new. Even though it's used in other places, this is the first time you've inserted it yourself. It's used in the regular TOC to insert page numbers, and you're reusing it here.

4. Save changes, run a PDF build, and check your work.

Exercise: Remove the mini-TOC from a table

Objective: Output the mini-TOC as an unordered list within the main flow of text rather than as a table to the side of the main flow of text.

By default, the mini-TOC is placed in a table at the beginning of each chapter. It looks like **Mini-TOC** *(p. 343)*. You can remove it from the table so that it appears inline with the rest of your text.

1. Copy the template that begins with the following from **commons.xsl** to your copy of **commons.xsl**:

```
<xsl:template match="*" mode="createMiniToc">
```

> **Note** It's a little odd that the template that processes the mini-TOC isn't in **toc.xsl**, especially because the attributes that format the mini-TOC are in **toc-attr.xsl**, but this is just another one of those things that make the DITA Open Toolkit so entertaining.

2. Comment out the text in bold, as shown:

```
<xsl:template match="*" mode="createMiniToc">
    <!--<fo:table xsl:use-attribute-sets="__toc__mini__table">
        <fo:table-column xsl:use-attribute-sets="__toc__mini__table__column_1"/>
        <fo:table-column xsl:use-attribute-sets="__toc__mini__table__column_2"/>
        <fo:table-body xsl:use-attribute-sets="__toc__mini__table__body">
            <fo:table-row>
                <fo:table-cell>-->
                    <fo:block xsl:use-attribute-sets="__toc__mini">
                        <xsl:if test="*[contains(@class, ' topic/topic ')]">
                            <fo:block xsl:use-attribute-sets="__toc__mini__header">
                                <xsl:call-template name="getVariable">
                                    <xsl:with-param name="id" select="'Mini Toc'"/>
                                </xsl:call-template>
                            </fo:block>
                            <fo:list-block xsl:use-attribute-sets="__toc__mini__list">
                                <xsl:apply-templates select="*[contains(@class, ' topic/topic ')]"
                                ▶mode="in-this-chapter-list"/>
                            </fo:list-block>
                        </xsl:if>
                    </fo:block>
                <!--</fo:table-cell>-->
                <!--<fo:table-cell xsl:use-attribute-sets="__toc__mini__summary">-->
                <!--Really, it would be better to just apply-templates, but the attribute sets
                for shortdesc, body and abstract might indent the text.  Here, the topic body
                is in a table cell, and should not be indented, so each element is handled
                specially.-->
                    <fo:block>
                        <xsl:apply-templates select="*[contains(@class,' topic/titlealts ')]"/>
                        <xsl:if test="*[contains(@class,' topic/shortdesc ')
                                    or contains(@class, ' topic/abstract ')]/node()">
                            <fo:block xsl:use-attribute-sets="p">
                                <xsl:apply-templates select="*[contains(@class,' topic/shortdesc ')
                                    or contains(@class, ' topic/abstract ')]/node()"/>
                            </fo:block>
                        </xsl:if>
                        <xsl:apply-templates select="*[contains(@class,' topic/body ')]/*"/>
                        <!-- Added with RFE 2976463 to fix dropped links in topics with a
                            mini-TOC. -->
                        <xsl:if test="*[contains(@class,' topic/related-links ')]//
                        ▶*[contains(@class,' topic/link ')][not(@role) or @role!='child']">
                            <xsl:apply-templates select="*[contains(@class,'
                            ▶topic/related-links ')]"/>
                        </xsl:if>
                    </fo:block>
                <!--</fo:table-cell>-->
            </fo:table-row>
        </fo:table-body>
    </fo:table>-->
</xsl:template>
```

3. Save changes, run a PDF build, and check your work.

Reality check

You're left with two blocks of text. The first one contains the heading of the mini-TOC (notice that heading is determined by the variable *Mini Toc*, which comes from **en.xml**). The second is an unordered list of links to the topics in the chapter. The attribute sets that format these blocks are in **toc-attr.xsl**.

Exercise: Insert a page break after the mini-TOC

Objective: Add a page break after the mini-TOC rather than flowing into the first topic of the chapter on the same page.

If you prefer to have the mini-TOC on a page by itself to introduce the chapter, that's easy to do.

1. Copy the *__toc__mini__list* attribute set from `DITA-OT/plugins/org.dita.pdf2/cfg/fo/attrs/`**toc-attr.xsl** to your copy of **toc-attr.xsl**.
2. Add the following attribute and value:

```
<xsl:attribute name="page-break-after">always</xsl:attribute>
```

3. Save changes, run a PDF build, and check your work.

Exercise: Eliminate the mini-TOC

Objective: Set up your ANT build to remove the mini-TOC from the beginning of each chapter.

By default, the Open Toolkit produces a mini-TOC at the beginning of each chapter, appendix, or part. As handy as this can be for navigation, you can eliminate it if you don't want it.

• Add the following line to your ANT build file:

```
<property name="args.chapter.layout" value="BASIC"/>
```

> **Note** Unlike many other ANT build properties, *args.chapter.layout* has only one value, **BASIC**. If you include this property in your ANT build file, you can easily toggle between having a mini-TOC and not having one by commenting the property in or out.

Other things you can do

Here are a few additional customizations you might want to make to your plugin. These customizations aren't part of the PDF specifications you're following, but they can be useful (and cool).

Exercise: Change or delete chapter/appendix/part prefixes in TOC entries

Objective: Delete or define different values for Table of Contents entries.

By default, the TOC includes the prefixes "Chapter", "Appendix", and "Part" before chapters, appendices and parts. You might want different prefixes or no prefixes at all.

1. Copy the following variables from `DITA-OT/plugins/org.dita.pdf2/cfg/common/vars/`**en.xml** to your copy of **en.xml**:

```
<!-- The string used to label a chapter within the table of contents. -->
<variable id="Table of Contents Chapter">Chapter <param ref-name="number"/>: </variable>

<!-- The string used to label an appendix within the table of contents. -->
<variable id="Table of Contents Appendix">Appendix 
►<param ref-name="number"/>: </variable>

<!-- The string used to label a part within the table of contents. -->
<variable id="Table of Contents Part">Part <param ref-name="number"/>: </variable>

<!-- The string used to label a preface within the table of contents. -->
<variable id="Table of Contents Preface">Preface: </variable>

<!-- The string used to label a notice within the table of contents. -->
<variable id="Table of Contents Notices"></variable>
```

2. Do one of the following:

- To change the prefix, enter the new value for the appropriate variable(s).
- To have no prefix, delete the value (including the non-breaking spaces ` ` and the colon) for the appropriate variable(s).

> **Note** There's a more complicated way to do this; edit the code that calls the variable in **toc.xsl**. But why make things harder than they have to be? Deleting the value essentially outputs a prefix that contains nothing, which 99.99% of the time is fine.

3. Repeat this process for any other language-specific variables files you've customized.
4. Save changes, run a PDF build, and check your work.

Exercise: Omit page numbers for specific TOC entry levels

Objective: Leave out page numbers for specific entry levels in the Table of Contents, such as chapter title entries.

If you want to leave out page numbers for level-1 TOC entries (or any other level), it's easy to do.

1. Copy the template that begins with the following from `DITA-OT/plugins/org.dita.pdf2/xsl/fo/`**toc.xsl**, to your copy of **toc.xsl**:

```
<xsl:template match="*[contains(@class, ' topic/topic ')]" mode="toc">
    <xsl:param name="include"/>
    <xsl:variable name="topicLevel" as="xs:integer">
    ...
```

2. In that template, find the following section:

```
<fo:inline xsl:use-attribute-sets="__toc__page-number">
    <fo:leader xsl:use-attribute-sets="__toc__leader"/>
    <fo:page-number-citation>
        <xsl:attribute name="ref-id">
            <xsl:call-template name="generate-toc-id"/>
        </xsl:attribute>
    </fo:page-number-citation>
</fo:inline>
```

The `<fo:page-number-citation>` element creates the page number for the TOC entry. You want to add a condition to test for the TOC entry level and not output the page number for level-1 entries.

3. Enclose the `<fo:page-number-citation>` element in an `<xsl:if>` test as follows:

```
<fo:inline xsl:use-attribute-sets="__toc__page-number">
    <fo:leader xsl:use-attribute-sets="__toc__leader"/>
    <xsl:if test="$topicLevel !=1">
      <fo:page-number-citation>
        <xsl:attribute name="ref-id">
            <xsl:call-template name="generate-toc-id"/>
        </xsl:attribute>
      </fo:page-number-citation>
    </xsl:if>
</fo:inline>
```

> **Note** If you've made other changes to this code based on previous exercises in this chapter, you might need to give an extra thought or two to where to put the `<xsl:if>` test. Hint: you want to include everything that has to do with the page numbering inside `<xsl:if>`—spaces, leaders, etc.

This `<xsl:if>` tests the current topic level. In this example, the statement is testing to see if the current topic level is not 1—in other words, not a top-level heading. If the statement is true (and the current topic level is not 1), the page number is generated in the TOC entry. If the statement is false (and the current topic level is 1), no page number is generated.

In place of the value "1", you should put whatever level you want to test for. If there is only one level you do not want page numbers for, you should use the != (not equal to) operator. If there is only one level you want page numbers for, you should use the = (equal to) operator.

If you want to test for several levels at once, you can use the < (less than) or > (greater than) operators or you can use an <xsl:choose> statement rather than an <xsl:if> statement.

4. Save changes, run a PDF build, and check your work.

Exercise: Change TOC numbering formats

Objective: Change the way chapters, appendices, and parts are numbered in the Table of Contents.

By default, the PDF plugin numbers chapters as 1, 2, 3, etc.; appendices as A, B, C, etc.; and parts as I, II, III, etc. You can change these numbering formats.

1. Copy the following templates from DITA-OT/plugins/org.dita.pdf2/xsl/fo/**commons.xsl** to your copy of **commons.xsl**, depending on what you want to change:

```
<xsl:template match="*[contains(@class, ' bookmap/chapter ')] |
            opentopic:map/*[contains(@class, ' map/topicref ')]" mode="topicTitleNumber"
        ▸ priority="-1">
    <xsl:variable name="chapters">
        <xsl:document>
            <xsl:for-each select="$map/descendant::*[contains(@class, ' bookmap/chapter ')]">
                <xsl:sequence select="."/>
            </xsl:for-each>
        </xsl:document>
    </xsl:variable>
    <xsl:for-each select="$chapters/*[current()/@id = @id]">
        <xsl:number format="1" count="*[contains(@class, ' bookmap/chapter ')]"/>
    </xsl:for-each>
</xsl:template>

<xsl:template match="*[contains(@class, ' bookmap/appendix ')]" mode="topicTitleNumber">
    <xsl:number format="A" count="*[contains(@class, ' bookmap/appendix ')]"/>
</xsl:template>

<xsl:template match="*[contains(@class, ' bookmap/part ')]" mode="topicTitleNumber">
    <xsl:number format="I" count="*[contains(@class, ' bookmap/part ')]"/>
</xsl:template>
```

2. Change the value of *format* (in bold above) to the appropriate numbering format.

Valid formats are: 1, 01, A, a, I, i.

3. Save changes, run a PDF build, and check your work.

Exercise: Allow line breaks in TOC entries

Objective: Specify that entries within the Table of Contents can wrap to the next line if necessary.

By default, TOC entries do not wrap. A best practice is to avoid extremely long topic titles, but if you use one, it will run right off the page. So, set up TOC entries so they can wrap when necessary.

1. Copy the *__toc__title* attribute set from `DITA-OT/plugins/org.dita.pdf2/cfg/fo/attrs/`**toc-attr.xsl** to your copy of **toc-attr.xsl**.

```
<xsl:attribute-set name="__toc__title">
   <xsl:attribute name="end-indent"><xsl:value-of select="$toc.text-indent"/></xsl:attribute>
</xsl:attribute-set>
```

2. Add the following attribute and value:

```
<xsl:attribute name="keep-together.within-line">auto</xsl:attribute>
```

3. If you want to allow TOC entries to hyphenate, add the following attribute and value to *__toc__title*:

```
<xsl:attribute name="hyphenate">true</xsl:attribute>
```

It's not strictly necessary to add this attribute, but by default, hyphenation is off. If you have any long words in your titles, line breaks might occur in awkward places. By allowing hyphenation, you reduce the changes of strange-looking line breaks. Your choice.

4. Save changes, run a PDF build, and check your work.

Exercise: Specify the numbering format for TOC page numbers

Objective: Use a numbering format other than the default of Roman numerals for page numbers in the Table of Contents.

By default, the DITA Open Toolkit uses lowercase Roman numerals for Table of Contents page numbers. Lots of folks maintain that Roman numerals are a moldy relic of the past. If you agree, here's how to change that default.

1. In `DITA-OT/plugins/org.dita.pdf2/xsl/fo/`**toc.xsl**, find the template that begins with

```
<xsl:template name="createToc">
    <xsl:if test="$generate-toc">
        <xsl:variable name="toc">
            <xsl:choose>
    ...
```

2. Within that template, find the following:

```
<xsl:if test="count($toc/*) > 0">
    <fo:page-sequence master-reference="toc-sequence" xsl:use-attribute-sets="page-sequence.toc">
```

You see that the page numbering comes from an attribute set named *page-sequence.toc*.

3. Copy the *page-sequence.toc* attribute set from `DITA-OT/plugins/org.dita.pdf2/cfg/fo/attrs/`**commons-attr.xsl** to your copy of **commons-attr.xsl**.

4. Add the following attribute and value:

```
<xsl:attribute name="format">__</xsl:attribute>
```

For __, substitute the format you want. Valid formats are 1, 01, A, a, I, i.

5. If you want the TOC to start on page 1, also add the following:

```
<xsl:attribute name="initial-page-number">1</xsl:attribute>
```

By default, the PDF plugin starts numbering at the cover page, so the first page of the TOC in a double-sided PDF is page 3 unless you add this attribute.

6. Save changes, run a PDF build, and check your work.

Exercise: Eliminate the TOC (map-based PDF)

Objective: Create a PDF with no TOC when publishing from a map (as opposed to a bookmap).

By default, the Open Toolkit always creates a Table of Contents when you publish a PDF from a map, so if you want to eliminate it, you need to do a little work.

I. Find the following template in DITA-OT/plugins/org.dita.pdf2/xsl/fo/**root-processing.xsl**:

```
<xsl:template match="*[contains(@class, ' map/map ')]" mode="generatePageSequences">
    <xsl:call-template name="createFrontMatter"/>
    <xsl:call-template name="createToc"/>
    <xsl:choose>
        <xsl:when test="$map-based-page-sequence-generation">
            <fo:page-sequence master-reference="ditamap-body-sequence"
            ▶ xsl:use-attribute-sets="page-sequence.body">
                <xsl:call-template name="startPageNumbering"/>
                <xsl:call-template name="insertBodyStaticContents"/>
                <fo:flow flow-name="xsl-region-body">
                    <xsl:for-each select="opentopic:map/*[contains(@class, ' map/topicref ')
                    ▶]">
                        <xsl:for-each select="key('topic-id', @id)">
                            <xsl:apply-templates select="." mode="processTopic"/>
                        </xsl:for-each>
                    </xsl:for-each>
                </fo:flow>
            </fo:page-sequence>
            <xsl:call-template name="createIndex"/>
        </xsl:when>
        <!-- legacy topic based page-sequence generation -->
        <xsl:otherwise>
            <xsl:apply-templates/>
        </xsl:otherwise>
    </xsl:choose>
    <xsl:call-template name="createBackCover"/>
</xsl:template>
```

The line `<xsl:call-template name="createToc"/>` is what triggers the creation of a TOC. You could just comment it out but you would also eliminate the TOC when publishing a PDF from a bookmap because the test at the beginning of this template, shown below, is also true of bookmarks because their class is '- map/map bookmap/bookmap ':

```
<xsl:template match="*[contains(@class, ' map/map ')]
```

You could try to conditionalize this line so that it applies only when you're publishing a bookmap, but it's not that easy. The more straightforward approach is to conditionalize the *createTOC* template so that it applies only when you're publishing a bookmap.

2. Copy the following template from DITA-OT/plugins/org.dita.pdf2/xsl/fo/**toc.xsl** to your copy of **toc.xsl**:

```
<xsl:template name="createToc">
    <xsl:if test="$generate-toc">
        <xsl:variable name="toc">
            <xsl:choose>
                <xsl:when test="$map//*[contains(@class,' bookmap/toc ')][@href]"/>
                <xsl:when test="$map//*[contains(@class,' bookmap/toc ')]">
                    <xsl:apply-templates select="/" mode="toc"/>
                </xsl:when>
                <xsl:when test="/*[contains(@class,' map/map ')][not(contains(@class,
                ▶' bookmap/bookmap '))]">
                    <xsl:apply-templates select="/" mode="toc"/>
                    <xsl:call-template name="toc.index"/>
                </xsl:when>
```

```
            </xsl:choose>
        </xsl:variable>
        <xsl:if test="count($toc/*) > 0">
            <fo:page-sequence master-reference="toc-sequence" xsl:use-attribute-sets=
            ▶"page-sequence.toc">
                <xsl:call-template name="insertTocStaticContents"/>
                <fo:flow flow-name="xsl-region-body">
                    <xsl:call-template name="createTocHeader"/>
                    <fo:block>
                        <fo:marker marker-class-name="current-header">
                            <xsl:call-template name="getVariable">
                                <xsl:with-param name="id" select="'Table of Contents'"/>
                            </xsl:call-template>
                        </fo:marker>
                        <xsl:copy-of select="$toc"/>
                    </fo:block>
                </fo:flow>
            </fo:page-sequence>
        </xsl:if>
    </xsl:if>
</xsl:template>
```

3. Wrap the entire contents of the template in an `<xsl:if>`:

```
<xsl:template name="createToc">
  <xsl:if test="/*[contains(@class,' bookmap/bookmap ')]">
    <xsl:if test="$generate-toc">
        <xsl:variable name="toc">
            <xsl:choose>
                <xsl:when test="$map//*[contains(@class,' bookmap/toc ')][@href]"/>
                <xsl:when test="$map//*[contains(@class,' bookmap/toc ')]">
                    <xsl:apply-templates select="/" mode="toc"/>
                </xsl:when>
                <xsl:when test="/*[contains(@class,' map/map ')][not(contains(@class,
                ▶' bookmap/bookmap '))]">
                    <xsl:apply-templates select="/" mode="toc"/>
                    <xsl:call-template name="toc.index"/>
                </xsl:when>
            </xsl:choose>
        </xsl:variable>
        <xsl:if test="count($toc/*) > 0">
            <fo:page-sequence master-reference="toc-sequence" xsl:use-attribute-sets=
            ▶"page-sequence.toc">
                <xsl:call-template name="insertTocStaticContents"/>
                <fo:flow flow-name="xsl-region-body">
                    <xsl:call-template name="createTocHeader"/>
                    <fo:block>
                        <fo:marker marker-class-name="current-header">
                            <xsl:call-template name="getVariable">
                                <xsl:with-param name="id" select="'Table of Contents'"/>
                            </xsl:call-template>
                        </fo:marker>
                        <xsl:copy-of select="$toc"/>
                    </fo:block>
                </fo:flow>
            </fo:page-sequence>
        </xsl:if>
    </xsl:if>
  </xsl:if>
</xsl:template>
```

If the `<xsl:if>` test is true (you are publishing a bookmap) then the entire template is applied. If the `<xsl:if>` is false (you are publishing a map) then the template is still applied but it becomes essentially an empty template that has no effect on the output—it does not create a TOC.

4. Save your changes and run a PDF build using a map to check your work.

You'll find there is still a TOC bookmark. You probably want to eliminate that too, so let's take a look at how. (When publishing from a bookmap, there's a TOC bookmark only if there is actually a TOC, so this scenario is a problem only for maps.)

Exercise: Eliminate the TOC bookmark (map-based PDF)

Objective: Omit the TOC PDF bookmark when your map-based PDF does not include a TOC.

If you set up your plugin to not generate a TOC when publishing from a map (as opposed to a bookmap), you'll find the PDF still has a Contents bookmark that now links to nothing. You need to eliminate it.

1. Copy the template that begins with the following from DITA-OT/plugins/org.dita.pdf2/xsl/fo/**bookmarks.xsl** to your copy of **bookmarks.xsl**.

```
<xsl:template name="createBookmarks">
    <xsl:variable name="bookmarks" as="element()*">
    <xsl:choose>
        <xsl:when test="$retain-bookmap-order">
            <xsl:apply-templates select="/" mode="bookmark"/>
        </xsl:when>
    ...
```

2. Find this section in that template:

```
<xsl:choose>
    <xsl:when test="$map//*[contains(@class,' bookmap/toc ')][@href]"/>
        <xsl:when test="$map//*[contains(@class,' bookmap/toc ')]
    ► | /*[contains(@class,' map/map ')][not(contains(@class,' bookmap/bookmap '))]">
            <fo:bookmark internal-destination="{$id.toc}">
                <fo:bookmark-title>
                    <xsl:call-template name="getVariable">
                        <xsl:with-param name="id" select="'Table of Contents'"/>
                    </xsl:call-template>
                </fo:bookmark-title>
            </fo:bookmark>
    </xsl:when>
</xsl:choose>
```

The two `<xsl:when>` statements are saying, "If there's a `<toc>` element in the bookmap that points to a file via *href*, or if there's a `<toc>` element in the bookmap, period, *or if you're publishing from a map*, create a TOC bookmark."

The first two "or" conditions are okay. That last "or" condition—the highlighted section—is the part that's creating a TOC bookmark for your map-based PDF even though you have eliminated the TOC when publishing from maps.

3. To eliminate the TOC bookmark, delete the "or" part of the second `<xsl:when>` statement to leave the following:

```
<xsl:choose>
    <xsl:when test="$map//*[contains(@class,' bookmap/toc ')][@href]"/>
        <xsl:when test="$map//*[contains(@class,' bookmap/toc ')]">
            <fo:bookmark internal-destination="{$id.toc}">
            ...
```

4. Save changes, run a PDF build, and check your work.

Chapter 17

Index

An index is an important part of any longer printed publication. It adds tremendously to the usability of the document. For the most part, the DITA Open Toolkit automates index creation so that you don't have to worry about much more than adding `<indexterm>` elements as appropriate.

This chapter explains how to take the basic out-of-the-box index and give it your look and feel.

For a list of attributes that affect the index, refer to **Index attribute sets** *(p. 428)*.

This chapter is different...

Unlike all the other exercises in this book, which are written assuming you are using the FOP PDF renderer, the exercises in this chapter are written to accommodate all three PDF renderers. This is because all three work differently with respect to certain aspects of index generation.

Topics that apply to specific renderers include those renderer names in their titles. If a title does not include renderer names, it applies to all three.

There are additional differences in index generation between the three. Not all of those differences are documented here; I only cover the differences that are directly related to these exercises. If you need additional information about the three renderers, check the documentation for Antenna House, XEP and FOP or use Beyond Compare or a similar application to compare files in `org.dita.pdf2.axf`, `org.dita.pdf2.fop` and `org.dita.pdf2.xep`.

Index entry levels

Working with index entry indents is an extra-special treat. If you worked through the **Adjust indents for TOC entries** *(p. 339)* exercise, you might be thinking it's just another formula. But you'd be wrong.

Many of the principles that you've learned so far go out the window when it comes to indexes. There is a reason for this, sort of. Unlike TOC entries, which are based on the map hierarchy and therefore are easy to characterize by level, index entries don't have any kind of leveling. This makes it harder to treat entries on a "main" or "sub" basis. But it can be done with a little brute force and awkwardness.

The index is set up as an `<xsl:choice>`. In the `<xsl:when>` option, there's a table. **Main** index entries go in the table header. The table has two `<table-body>` elements. It's a little hard to see exactly what's happening here, but for this exercise, it doesn't matter. In the `<xsl:otherwise>` option, there's an `<fo:block>`. The **sub** index entries go in this block. Here is the structure:

```
<xsl:choice>
    <xsl:when>
        <fo:table>
            <fo:table-header>
                [main entry]
            </fo:table-header>
            <fo:table-body>
                [some other stuff]
            </fo:table-body>
            <fo:table-body>
                [some other stuff]
            </fo:table-body>
        </fo:table>
    </xsl:when>
    <xsl:otherwise>
        <fo:block>
            [sub-entries]
        </fo:block>
    </xsl:otherwise>
</xsl:choice>
```

Figure 46: Structure of index table

The problem with the way things are set up out of the box is that both the main entry and the sub-entries use the attribute set *index-indents*, so whatever changes you make to this attribute set affect both levels. You need to tease the main and sub-entries apart.

The curious case of page numbers on main index entries

If you're using **index_xep.xsl**, then when you start testing your index, you'll notice that top-level index entries with nested entries list *all* the page numbers of *all* the sub-entries. This isn't so bad if there are only a few sub-entries, but if the main entry has, say, 20 sub-entries, the list gets long and ugly.

You probably want to fix this. The easiest thing to do is to not include **index_xep.xsl**, because the problem is limited to that file. However, there's a tradeoff, **index_xep.xsl** produces a more attractive index, and **index.xsl** has some formatting issues of its own. For example, sub-entries are randomly squished and page numbers arbitrarily flow to the next line even when there's plenty of space. Unfortunately, the page number issue has been around for several years, and it probably won't be fixed soon.

Index specifications

Let's review the specifications for your index.

Index

Index title font	28pt Trebuchet normal black
Index level 1 entry font	11pt Book Antiqua normal dark gray
Index level 2 entry font	11pt Book Antiqua normal dark gray
Index page number	10pt Book Antiqua normal italic dark red
Index letter group headings	12pt Trebuchet bold dark red
Index pages	two-column
Level 2 indent	.125in
Index heading	Index

Files you need

If you have completed other exercises, some of these files might already be present in your plugin.

If they are not already present, create the following files:

- `DITA-OT/plugins/com.company.pdf/cfg/fo/attrs/`**index-attr.xsl**
- `DITA-OT/plugins/com.company.pdf/cfg/fo/attrs/`**index-attr_axf.xsl** (for Antenna House)
- `DITA-OT/plugins/com.company.pdf/cfg/fo/attrs/`**index-attr_xep.xsl** (for XEP)
- `DITA-OT/plugins/com.company.pdf/cfg/fo/attrs/`**layout-masters-attr.xsl**

Be sure to add the appropriate `<xsl:import>` statements to **custom.xsl** (attrs).

If they are not already present, create the following files:

- `DITA-OT/plugins/com.company.pdf/cfg/fo/xsl/`**bookmarks.xsl**
- `DITA-OT/plugins/com.company.pdf/cfg/fo/xsl/`**index.xsl**
- `DITA-OT/plugins/com.company.pdf/cfg/fo/xsl/`**root-processing.xsl**

Depending on the PDF renderer you are using, copy the appropriate file to `DITA-OT/plugins/com.company.pdf/cfg/fo/xsl/`:

- `DITA-OT/plugins/org.dita.pdf2.fop/xsl/fo/`**index_fop.xsl** (for FOP)
- `DITA-OT/plugins/org.dita.pdf2.axf/xsl/fo/`**index_axf.xsl** (for Antenna House)
- `DITA-OT/plugins/org.dita.pdf2.xep/xsl/fo/`**index_xep.xsl** (for XEP)

Be sure to add the appropriate `<xsl:import>` statements to **custom.xsl** (xsl).

For instructions on creating attribute files and XSLT stylesheets in your PDF plugin, refer to **Create an attribute set file in your plugin** *(p. 31)* and **Create an XSLT stylesheet in your plugin** *(p. 32)*.

If they are not already part of your plugin, create the following files:

- `DITA-OT/plugins/com.company.pdf/cfg/common/vars/`**en.xml**
- `DITA-OT/plugins/com.company.pdf/cfg/common/vars/`**strings-en-us.xml**

For instructions on creating localization variables and strings files, refer to **Create a localization variables file** *(p. 78)* and **Create a localization strings file** *(p. 79)*.

Be sure to add the appropriate language and locale mapping statements in **strings.xml**. You can find instructions in **Mapping strings files to xml:lang values in the strings.xml file** *(p. 37)*.

Exercise: Add an index to your PDF

Objective: Set up your bookmap to generate an index.

When you publish from a bookmap, you can control whether or not there's an index simply by including or excluding the `<indexlist>` element from the back matter or front matter of the bookmap. But when you publish from a map, all renderers automatically produce an index—as long as your content includes some `<indexterm>` elements.

These steps assume you're using a bookmap. If you're using a map, you can skip this exercise because you'll get the index by default. If you don't want it, refer to **Eliminate the index (map-based PDF)** *(p. 372)* to learn how to remove it.

1. In your bookmap, add the `<backmatter>` element if it is not already present.

   This step assumes you want the index at the end of the PDF. If you want it at the beginning, use the `<frontmatter>` element instead. Unfortunately, if you're publishing from a map, you can't control where the index is placed; it's at the end.

2. Add the `<indexlist>` element to `<backmatter>`.

   Your bookmap structure should resemble the following:

```
<backmatter>
    <booklists>
        <indexlist/>
    </booklists>
</backmatter>
```

3. Run a build and verify there's an index.

   Remeber, you must have some `<indexterm>` elements in your content. If you don't, you won't get an index. So, if you don't get an index, first make sure you have some `<indexterm>`s.

Exercise: Change the title of the index

Objective: Use a title other than the default "Index" for your index.

1. Copy the following string from DITA-OT/xsl/common/**strings-en-us.xml** to your copy of **strings-en-us.xml**:

```
<str name="Index">Index</str>
```

2. Change the value to the heading you want.
3. Repeat this process for any other language-specific strings files you've customized.
4. Save changes, run a PDF build, and check your work.

Exercise: Format the index title

Objective: Determine the appearance of your index's title.

1. Copy the *__index__label* attribute set from
DITA-OT/plugins/org.dita.pdf2/cfg/fo/attrs/**index-attr.xsl** to your copy of **index-attr.xsl**.

If you are using Antenna House, make the following change to the *__index__label* attribute set in your copy of **index-attr_axf.xsl** instead of in **index-attr.xsl**.

2. Add or change the following attributes and values:

```
<xsl:attribute-set name="__index__label">
    <xsl:attribute name="space-before">20pt</xsl:attribute>
    <xsl:attribute name="space-after">20pt</xsl:attribute>
    <xsl:attribute name="space-after.conditionality">retain</xsl:attribute>
    <xsl:attribute name="font-size">28pt</xsl:attribute>
    <xsl:attribute name="font-weight">normal</xsl:attribute>
    <xsl:attribute name="keep-with-next.within-column">always</xsl:attribute>
    <xsl:attribute name="span">all</xsl:attribute>
    <xsl:attribute name="color">#000000</xsl:attribute>
    <xsl:attribute name="font-family">Trebuchet MS, Arial Unicode MS, Helvetica</xsl:attribute>
</xsl:attribute-set>
```

> **Note** There are several other attributes in this attribute set that you can use to adjust the space above and below the title.

3. Save changes, run a PDF build, and check your work.

Exercise: Format index letter headings

Objective: Determine the appearance of the letter headings above each alphabetical section of your index.

1. Copy the *__index__letter-group* attribute set from
DITA-OT/plugins/org.dita.pdf2/cfg/fo/attrs/**index-attr.xsl** to your copy of **index-attr.xsl**.

If you are using Antenna House, make the following change to the *__index__letter-group* attribute set in your copy of **index-attr_axf.xsl** instead of in **index-attr.xsl**.

2. Add or change the following attributes and values:

```
<xsl:attribute-set name="__index__letter-group">
    <xsl:attribute name="font-size">12pt</xsl:attribute>
    <xsl:attribute name="font-weight">bold</xsl:attribute>
    <xsl:attribute name="space-after">7pt</xsl:attribute>
```

```
    <xsl:attribute name="keep-with-next.within-column">always</xsl:attribute>
    <xsl:attribute name="color">#990033</xsl:attribute>
    <xsl:attribute name="font-family">Trebuchet MS, Arial Unicode MS, Helvetica</xsl:attribute>
</xsl:attribute-set>
```

3. Save changes, run a PDF build, and check your work.

Exercise: Format index entries

Objective: Determine the appearance of entries in your index.

Because of the way **index.xsl** is written, it's a little tricky to figure out which attribute sets affect what. Some experimentation shows that the *index.entry* attribute set formats main index entries, while the *index.entry__content* attribute set formats sub-entries.

That much makes sense, but some attributes of *index.entry*, such as *font-size*, apply only to main entries without sub-entries. Main entries with sub-entries and the sub-entries pick up their font size from *index.entry__content*. That doesn't make a lot of sense, but it's one of the things you have to live with in the PDF plugin.

1. Copy the *index.entry* attribute set from
DITA-OT/plugins/org.dita.pdf2/cfg/fo/attrs/**index-attr.xsl** to your copy of **index-attr.xsl**.

This attribute set formats main entries.

2. Add or change the following attributes and values:

```
<xsl:attribute-set name="index.entry">
    <xsl:attribute name="space-after">14pt</xsl:attribute>
    <xsl:attribute name="font-size">11pt</xsl:attribute>
    <xsl:attribute name="color">#8A8A8A</xsl:attribute>
</xsl:attribute-set>
```

3. Copy the *index.entry__content* attribute set from
DITA-OT/plugins/org.dita.pdf2/cfg/fo/attrs/**index-attr.xsl** to your copy of **index-attr.xsl**.

This attribute set formats sub-entries.

4. Add or change the following attributes and values:

```
<xsl:attribute-set name="index.entry__content">
    <xsl:attribute name="start-indent">18pt</xsl:attribute>
    <xsl:attribute name="font-size">11pt</xsl:attribute>
    <xsl:attribute name="color">#8A8A8A</xsl:attribute>
</xsl:attribute-set>
```

5. Save changes, run a PDF build, and check your work.

Exercise: Format index page numbers (FOP)

Objective: Determine the appearance of page numbers in your index if you are using FOP.

From working with the index attribute sets, you might think that the *__index__page__link* attribute set formats page numbers in index entries, right? Actually, that's wrong, but it's not your fault. It should logically work this way, but it doesn't.

Here's why. If you are using FOP, take a look at this template in `DITA-OT/plugins/org.dita.pdf2.fop/xsl/fo/`**index_fop.xsl**:

```
<xsl:template match="opentopic-index:refID" mode="make-index-links">
   <xsl:variable name="value" as="xs:string" select="opentopic-func:get-unique-refid-value(.)"/>

   <fo:basic-link internal-destination="{$value}" xsl:use-attribute-sets="common.link">
      <fo:page-number-citation ref-id="{$value}"/>
      <xsl:variable name="start-range-value" as="attribute(value)?"
      ▶ select="ancestor-or-self::opentopic-index:index.entry[@start-range]/@value"/>
      <xsl:apply-templates mode="make-page-number-citation" select="key('refid-by-end-range-value'

      ▶, $start-range-value)"/>
   </fo:basic-link>
</xsl:template>
```

The `<fo:basic-link>` element uses the *common.link* attribute set, not the *__index__page__link* attribute set. This is easy to change.

1. Copy the template shown above to your copy of **index_fop.xsl.**, then edit it to use the *__index__page__link* attribute set instead of the *common.link* attribute set.

```
<xsl:template match="opentopic-index:refID" mode="make-index-links">
   <xsl:variable name="value" as="xs:string" select="opentopic-func:get-unique-refid-value
   ▶(.)"/>
   <fo:basic-link internal-destination="{$value}" xsl:use-attribute-sets=
   ▶"__index__page__link">
      <fo:page-number-citation ref-id="{$value}"/>
      <xsl:variable name="start-range-value" as="attribute(value)?"
      ▶ select="ancestor-or-self::opentopic-index:index.entry[@start-range]/@value"/>
      <xsl:apply-templates mode="make-page-number-citation" select=
      ▶"key('refid-by-end-range-value', $start-range-value)"/>
   </fo:basic-link>
</xsl:template>
```

2. Copy the *__index__page__link* attribute set from `DITA-OT/plugins/org.dita.pdf2/cfg/fo/attrs/`**index-attr.xsl** to your copy of **index-attr.xsl**.

3. Add or change the following attributes and values:

```
<xsl:attribute name="color">#990033</xsl:attribute>
<xsl:attribute name="font-style">italic</xsl:attribute>
```

```
<xsl:attribute name="font-weight">normal</xsl:attribute>
<xsl:attribute name="font-size">10pt</xsl:attribute>
```

4. Save changes, run a PDF build, and check your work.

Exercise: Format index page numbers (Antenna House, XEP)

Objective: Determine the appearance of page numbers if you are using Antenna House or XEP.

From working with the index attribute sets, you might think that the *__index__page__link* attribute set formats page numbers in index entries, right? Actually, that's wrong, but it's not your fault. It should work this way, but it doesn't.

Here's why. In DITA-OT/plugins/org.dita.pdf2/xsl/fo/**index.xsl**, find this section:

```
<xsl:if test="not($no-page)">
    <xsl:if test="$idxs">
        <xsl:copy-of select="$index.separator"/>
            <fo:index-page-citation-list>
                <xsl:for-each select="$idxs">
                    <fo:index-key-reference ref-index-key="{@value}" xsl:use-attribute-sets=
                    ▶"__index__page__link"/>
                </xsl:for-each>
            </fo:index-page-citation-list>
    </xsl:if>
</xsl:if>
```

The `<fo:index-key-reference>` element uses the *__index__page__link* attribute set. You're all set; you just need to edit the *__index__page__link* itself.

1. Copy the *__index__page__link* attribute set from DITA-OT/plugins/org.dita.pdf2/cfg/fo/attrs/**index-attr.xsl** to your copy of **index-attr.xsl**.

2. In your copy of **index-attr.xsl**, add or edit the following attributes in the *__index__page__link* attribute set.

```
<xsl:attribute name="color">#990033</xsl:attribute>
<xsl:attribute name="font-style">italic</xsl:attribute>
<xsl:attribute name="font-weight">normal</xsl:attribute>
<xsl:attribute name="font-size">10pt</xsl:attribute>
```

3. Save changes, run a PDF build, and check your work.

Exercise: Adjust indents for index entries

Objective: Determine the amount of space that second- (and third-, etc.) level index entries are indented below main entries.

If you worked through the Table of Contents chapter and set indents there (in **Adjust indents for TOC entries** *(p. 339)*, you might be thinking that index indents use a formula as well, but happily they don't. They simply use an attribute value.

1. If it's not already present, copy the *index.entry__content* attribute set from
   `DITA-OT/plugins/org.dita.pdf2/cfg/fo/attrs/` to your copy of **index-attr.xsl**.
2. Change the value of the *start-indent* attribute, as shown:

```
<xsl:attribute-set name="index.entry__content">
   <xsl:attribute name="start-indent">.125in</xsl:attribute>
</xsl:attribute-set>
```

3. Save changes, run a PDF build, and check your work.

Other things you can do

Here are a few additional customizations you might want to make to your plugin. These customizations aren't part of the PDF specifications you're following, but they can be useful (and cool).

Exercise: Change the column count on index pages (XEP)

Objective: Determine the number of columns on index pages if you are using XEP.

By default, the Open Toolkit processes each index page with two columns. If you want more columns on a page, you can specify the column count you want.

1. Find the following template in your copy of **index_xep.xsl**:

```
<xsl:template match="/" mode="index-postprocess">
   <fo:block xsl:use-attribute-sets="__index__label" id="{$id.index}">
      <xsl:call-template name="getVariable">
         <xsl:with-param name="id" select="'Index'"/>
      </xsl:call-template>
   </fo:block>
   <rx:flow-section column-count="2">
```

```
      <xsl:apply-templates select="//opentopic-index:index.groups" mode="index-postprocess"/>
   </rx:flow-section>
</xsl:template>
```

This section defines the default two-column layout of index pages using XEP. The `rx:` namespace on the `<rx:flow-section>` element means that this element is specific to XEP.

2. Change the default value "2" to the number of columns you want.

3. Save changes, run a PDF build, and check your work.

Exercise: Change the column count on index pages (FOP, Antenna House)

Objective: Determine the number of columns on index pages if you are using FOP or Antenna House.

By default, the Open Toolkit processes each index page with two columns. If you want more columns on a page, you can specify the column count you want. Unlike XEP, when you use FOP or Antenna House, the number of columns on an index page isn't part of index processing. It's part of your page master definitions.

1. Copy the following attribute sets from
`DITA-OT/plugins/org.dita.pdf2/cfg/fo/`**layout-masters-attr.xsl** to your copy of
layout-masters-attr.xsl:

```
<xsl:attribute-set name="region-body__index.odd" use-attribute-sets="region-body.odd">
   <xsl:attribute name="column-count">2</xsl:attribute>
</xsl:attribute-set>

<xsl:attribute-set name="region-body__index.even" use-attribute-sets="region-body.even">
   <xsl:attribute name="column-count">2</xsl:attribute>
</xsl:attribute-set>
```

2. Change the default values "2" to the number of columns you want.

3. Save changes, run a PDF build, and check your work.

Exercise: Add chapter, appendix, or part numbers to index entries

Objective: Use a new *getChapterPrefixForIndex* template to add chapter, appendix, or part numbers to index entries.

This exercise applies only if you are publishing from a bookmap. Chapter, appendix and part numbering depend on the presence of the `<chapter>`, `<appendix>` and `<part>` elements, which are not available in a map.

Before we plunge into the steps, here is some additional helpful information from Eliot Kimber:

> A feature in FO 1.1 is support for "folios," that is, page number definitions at the page sequence level that include additional prefix or suffix values to use with the page number, such as the chapter number.
>
> Because the folio prefix and suffix are defined at the page sequence level, all page number references automatically reflect the full folio. Otherwise you need to manually construct the full page number in your XSLT and there is no way to do it for index entries when you are using FO-provided index page number generation.
>
> See **http://www.w3.org/TR/xsl/#fo_folio-prefix**
>
> In some cases, generating a page number reference for each occurrence of `<indexterm>` results in two references to the same page in this case. There's no way, in the general case, to avoid that at the XSLT level because you can't know if two close index entries for the same term do or do not fall on the same page.
>
> In documents where all the indexing is done in the topic prologs this problem is less likely to occur but for print, in particular, indexing really needs to be done inline, especially for topics that span multiple pages, and thus the case is more likely to occur.
>
> The OT's solution for indexing with FOP, which doesn't implement the FO indexing features, exhibits this behavior because it can't avoid it.
>
> Ken Holman has developed a technique for generating complete indexes by generating the index information in the PDF and then doing a sort of final pass to collapse duplicate page number references. The code is available from his Web site at **http://www.cranesoftwrights.com**.
>
> If the solution presented here is not sufficient for you, then consider investing in Antenna House or XEP or have a look at Ken Hohlman's solution.

1. Copy the entire *getChapterPrefixForIndex* template into **custom.xsl** (xsl).

You can find this template in **GetChapterPrefixForIndex template** *(p. 483)*.

> **Note** You could copy this template into any of the **.xsl** files you've added to your plugin, including **index.xsl** or **index_xep.xsl**, but it's just as easy to copy it into **custom.xsl**. You can see right away it's a custom template that didn't come with the PDF plugin.

Remember that in **index.xsl**, the `<fo:index-page-citation-list>` element creates the page number. So you can figure that somehow you need to call the *getChapterPrefixForIndex* template in conjunction with `<fo:index-page-citation-list>`.

However, as its name suggests, `<fo:index-page-citation-list>` creates a list of page numbers. Anything you do to this element affects the whole list. If you want to add the chapter number to each individual page, which you do, you'll have to change a template.

2. Open either `DITA-OT/plugins/org.dita.pdf2/xsl/fo/`**index.xsl** (for FOP or Antenna House) or `DITA-OT/plugins/org.dita.pdf2.xep/xsl/fo/`**index_xep.xsl** (for XEP) and copy the template that begins with the following to your copy of **index.xsl** or **index_xep.xsl**:

```
<xsl:template match="*" mode="make-index-ref">
```

If you're using FOP or Antenna House, make these changes in **index.xsl**. If you're using XEP, make these changes in **index_xep.xsl**. (`<fo:index-page-citation-list>` is not used in **index_axf.xsl** or **index_fop.xsl**.)

3. In the template you just copied, find this section:

```
<xsl:if test="not($no-page)">
    <xsl:if test="$idxs">
        <xsl:copy-of select="$index.separator"/>
        <fo:index-page-citation-list>
            <xsl:for-each select="$idxs">
                <fo:index-key-reference ref-index-key="{@value}" xsl:use-attribute-sets=
                ▸"__index__page__link"/>
            </xsl:for-each>
        </fo:index-page-citation-list>
    </xsl:if>
</xsl:if>
```

4. Replace the entire section of code with the following:

```
<xsl:if test="not($no-page)">
    <xsl:if test="$idxs">
        <xsl:copy-of select="$index.separator"/>
        <fo:inline xsl:use-attribute-sets="__index__page__link">
            <xsl:for-each select="$idxs">
                <xsl:for-each select="//opentopic-index:refID[@value = $idxs/@value]
                ▶[not(ancestor::opentopic-index:index.groups)]">
                    <fo:basic-link internal-destination="{./ancestor::*[@id][1]/@id}">
                        <xsl:call-template name="getChapterPrefixForIndex">
                            <xsl:with-param name="currentNode"
                            ▶select="./ancestor::*[@id][1]/@id"/>
                        </xsl:call-template>
                        <xsl:text>-</xsl:text>
                        <fo:page-number-citation ref-id="{./ancestor::*[@id][1]/@id}"/>
                    </fo:basic-link>
                    <xsl:if test="following::opentopic-index:refID[@value = $idxs/@value]
                    ▶[not(ancestor::opentopic-index:index.groups)]">
                        <xsl:text>, </xsl:text>
                    </xsl:if>
                </xsl:for-each>
            </xsl:for-each>
        </fo:inline>
    </xsl:if>
</xsl:if>
```

There are a few critical parts of this new code. You've called the ***getChapterPrefixForIndex*** template, which inserts the chapter number before the page number. And you've added the dash between the chapter and page number. Most critically, you've switched the `<fo:index-page-citation-list>` element to `<fo:page-number-citation>`.

5. Save changes, run a PDF build, and check your work.

Exercise: Eliminate the index (map-based PDF)

Objective: Create a PDF with no index when publishing from a map (as opposed to a bookmap).

By default, the Open Toolkit always creates an index when you publish a PDF from a map, so if you want to eliminate it, you need to do a little work.

1. Find the following template in `DITA-OT/plugins/org.dita.pdf2/xsl/fo/`**root-processing.xsl**:

```
<xsl:template match="*[contains(@class, ' map/map ')]" mode="generatePageSequences">
    <xsl:call-template name="createFrontMatter"/>
    <xsl:call-template name="createToc"/>
    <xsl:choose>
        <xsl:when test="$map-based-page-sequence-generation">
            <fo:page-sequence master-reference="ditamap-body-sequence" xsl:use-attribute-sets
            ▶="page-sequence.body">
                <xsl:call-template name="startPageNumbering"/>
                <xsl:call-template name="insertBodyStaticContents"/>
                <fo:flow flow-name="xsl-region-body">
                    <xsl:for-each select="opentopic:map/*[contains(@class, ' map/topicref ')]">
```

```
                      <xsl:for-each select="key('topic-id', @id)">
                          <xsl:apply-templates select="." mode="processTopic"/>
                      </xsl:for-each>
                  </xsl:for-each>
              </fo:flow>
          </fo:page-sequence>
          <xsl:call-template name="createIndex"/>
      </xsl:when>
      <!-- legacy topic based page-sequence generation -->
      <xsl:otherwise>
          <xsl:apply-templates/>
      </xsl:otherwise>
  </xsl:choose>
  <xsl:call-template name="createBackCover"/>
</xsl:template>
```

The line `<xsl:call-template name="createIndex"/>` triggers the creation of an index. You
could comment it out, but that would also eliminate the index when you publish from a bookmap,
because the match `<xsl:template match="*[contains(@class, ' map/map ')]` is also true
of bookmaps, because their class is `'- map/map bookmap/bookmap '`. You could try to
conditionalize this line so that it only applies when you're publishing a bookmap, but it's not that
easy. The more straightforward approach is to conditionalize the *createIndex* template so it only
applies when you're publishing a bookmap.

2. Copy the following template from `DITA-OT/plugins/org.dita.pdf2/xsl/fo/`**index.xsl** to your
copy of **index.xsl**:

```
<xsl:template name="createIndex">
    <xsl:if test="(//opentopic-index:index.groups//opentopic-index:index.entry) and
  ▶ (count($index-entries//opentopic-index:index.entry) &gt; 0)">
        <xsl:variable name="index">
            <xsl:choose>
                <xsl:when test="$map//*[contains(@class,' bookmap/indexlist ')][@href]"/>
                <xsl:when test="$map//*[contains(@class,' bookmap/indexlist ')]">
                    <xsl:apply-templates select="/" mode="index-postprocess"/>
                </xsl:when>
                <xsl:when test="/*[contains(@class,' map/map ')]
                ▶[not(contains(@class,' bookmap/bookmap '))]">
                    <xsl:apply-templates select="/" mode="index-postprocess"/>
                </xsl:when>
            </xsl:choose>
        </xsl:variable>
        <xsl:if test="count($index/*) > 0">
            <fo:page-sequence master-reference="index-sequence" xsl:use-attribute-sets
            ▶="page-sequence.index">
                <xsl:call-template name="insertIndexStaticContents"/>
                <fo:flow flow-name="xsl-region-body">
                    <fo:marker marker-class-name="current-header">
                        <xsl:call-template name="getVariable">
                            <xsl:with-param name="id" select="'Index'"/>
                        </xsl:call-template>
                    </fo:marker>
                    <xsl:copy-of select="$index"/>
                </fo:flow>
            </fo:page-sequence>
        </xsl:if>
    </xsl:if>
</xsl:template>
```

3. Wrap the entire contents of the template in an `<xsl:if>`:

```
<xsl:template name="createIndex">
    <xsl:if test="/*[contains(@class,' bookmap/bookmap ')]">
        <xsl:if test="(//opentopic-index:index.groups//opentopic-index:index.entry) and
        ▸ (count($index-entries//opentopic-index:index.entry) &gt; 0)">
            <xsl:variable name="index">
                <xsl:choose>
                    <xsl:when test="$map//*[contains(@class,' bookmap/indexlist ')][@href]"/>
                    <xsl:when test="$map//*[contains(@class,' bookmap/indexlist ')]">
                        <xsl:apply-templates select="/" mode="index-postprocess"/>
                    </xsl:when>
                    <xsl:when test="/*[contains(@class,' map/map ')]
                    ▸ [not(contains(@class,' bookmap/bookmap '))]">
                        <xsl:apply-templates select="/" mode="index-postprocess"/>
                    </xsl:when>
                </xsl:choose>
            </xsl:variable>
            <xsl:if test="count($index/*) > 0">
                <fo:page-sequence master-reference="index-sequence" xsl:use-attribute-sets
                ▸="page-sequence.index">
                    <xsl:call-template name="insertIndexStaticContents"/>
                    <fo:flow flow-name="xsl-region-body">
                        <fo:marker marker-class-name="current-header">
                            <xsl:call-template name="getVariable">
                                <xsl:with-param name="id" select="'Index'"/>
                            </xsl:call-template>
                        </fo:marker>
                        <xsl:copy-of select="$index"/>
                    </fo:flow>
                </fo:page-sequence>
            </xsl:if>
        </xsl:if>
    </xsl:if>
</xsl:template>
```

If the `<xsl:if>` test is true (you are publishing a bookmap) then the entire template is applied. If the `<xsl:if>` is false (you are publishing a map) then the template is still applied but it becomes essentially an empty template that has no effect on the output—it does not create an index.

4. Save changes, run a PDF build, and check your work.

You'll find there is still an Index bookmark. You probably want to eliminate that too, so let's take a look at how. (When publishing from a bookmap, there's an Index bookmark only if there is actually an index, so this scenario is a problem only for maps.)

Exercise: Eliminate the Index bookmark (map-based PDF)

Objective: Omit the Index PDF bookmark when your map-based PDF does not include an index.

If you set up your plugin to not generate an index when publishing from a map (as opposed to a bookmap), you'll find the PDF still has an Index bookmark that now links to nothing. You need to eliminate it.

1. Copy the following template from `DITA-OT/plugins/org.dita.pdf2/xsl/fo/`**bookmarks.xsl** to your copy of **bookmarks.xsl**.

```
<xsl:template match="*" mode="bookmark-index">
  <xsl:if test="//opentopic-index:index.groups//opentopic-index:index.entry">
    <xsl:choose>
      <xsl:when test="$map//*[contains(@class,' bookmap/indexlist ')][@href]"/>
      <xsl:when test="$map//*[contains(@class,' bookmap/indexlist ')]
          ▶ | /*[contains(@class,' map/map ')][not(contains(@class,' bookmap/bookmap '))]">

          <fo:bookmark internal-destination="{$id.index}" starting-state="hide">
            <fo:bookmark-title>
              <xsl:call-template name="getVariable">
                <xsl:with-param name="id" select="'Index'"/>
              </xsl:call-template>
            </fo:bookmark-title>
            <xsl:if test="$bookmarks.index-group-size !=0 and
                count(//opentopic-index:index.groups//opentopic-index:index.entry)
                ▶ &gt; $bookmarks.index-group-size">
              <xsl:apply-templates select="//opentopic-index:index.groups"
                ▶ mode="bookmark-index"/>
            </xsl:if>
          </fo:bookmark>
        </xsl:when>
    </xsl:choose>
  </xsl:if>
</xsl:template>
```

2. Find this section in that template:

```
<xsl:choose>
    <xsl:when test="$map//*[contains(@class,' bookmap/indexlist ')][@href]"/>
    <xsl:when test="$map//*[contains(@class,' bookmap/indexlist ')]
        | /*[contains(@class,' map/map ')][not(contains(@class,' bookmap/bookmap '))]">
        <fo:bookmark internal-destination="{$id.index}" starting-state="hide">
        ...
```

The two `<xsl:when>` statements are saying, "If there's an `<indexlist>` element in the bookmap that points to a file via *href*, or if there's an `<indexlist>` element in the bookmap, period, *or if you're publishing from a map*, create an Index bookmark." The first two "or" conditions are okay. That last "or" condition—the highlighted section—creates an Index bookmark for your map-based PDF, even though you have eliminated the index when publishing from maps.

3. To eliminate the Index bookmark, delete the "or" part of the second `<xsl:when>` leaving this:

```
...
<xsl:choose>
    <xsl:when test="$map//*[contains(@class,' bookmap/indexlist ')][@href]"/>
    <xsl:when test="$map//*[contains(@class,' bookmap/indexlist ')]">
        <fo:bookmark internal-destination="{$id.index}" starting-state="hide">
  ...
```

4. Save changes, run a PDF build, and check your work.

Chapter 18

List of Tables and List of Figures

If your content includes a lot of images, diagrams, or tables, it's helpful to your readers to include a list of these items so they can turn quickly to them. Sometimes, they need the diagram or the table much more than the surrounding text! Fortunately, more recent versions of the DITA Open Toolkit automate the inclusion of a List of Tables or List of Figures—if you are publishing from a bookmap. If you're publishing from a map, the process is more difficult and beyond the scope of this book.

This chapter explains how to add these lists to your PDF and how to format them for your look and feel.

For a list of attributes that affect lists of tables and figures, refer to **Lot-Lof attribute sets** *(p. 436)*.

List of Tables and List of Figures specifications

Here are your specifications for the List of Tables (LOT) and List of Figures (LOF).

List of Tables

LOT title font	28pt Trebuchet normal black
LOT level 1 entry font	12pt Trebuchet normal black
LOT leader	none
LOT page number	12pt Trebuchet bold black, includes chapter number
LOT heading	Tables
LOT entries	include chapter number-table number (Table B-4)

List of Figures

LOF title font	28pt Trebuchet normal black
LOF level 1 entry font	12pt Trebuchet normal black
LOF leader	none
LOF page number	12pt Trebuchet bold black, includes chapter number
LOF heading	Figures
LOF entries	include chapter number-figure number (Figure 3-6)

Files you need

If you have completed other exercises, some of these files might already be present in your plugin.

If they are not already present, create the following files in `DITA-OT/plugins/com.company.pdf/cfg/fo/attrs/`:

- **lot-lof-attr.xsl**
- **toc-attr.xsl**

Be sure to add the appropriate `<xsl:import>` statements to **custom.xsl** (attrs).

If they are not already present, create the following files in `DITA-OT/plugins/com.company.pdf/cfg/fo/xsl/`:

- **lot-lof.xsl**
- **toc.xsl**

Be sure to add the appropriate `<xsl:import>` statements to **custom.xsl** (xsl).

For instructions on creating attribute files and XSLT stylesheets in your PDF plugin, refer to **Create an attribute set file in your plugin** *(p. 31)* and **Create an XSLT stylesheet in your plugin** *(p. 32).*

If it is not already part of your plugin, create the following file:

- `DITA-OT/plugins/com.company.pdf/cfg/common/vars/`**en.xml**

For instructions on creating localization variables files, refer to **Create a localization variables file** *(p. 78).*

Exercise: Add a List of Tables or List of Figures

Objective: Set up your bookmap to automatically create a List of Tables or a List of Figures.

It's common to have a List of Tables or List of Figures in a PDF. It's pretty easy to do, too, which is nice. Unfortunately, the functionality isn't available if you're publishing from a map.

1. Add a List of Tables or a List of Figures to either the frontmatter or the backmatter of your bookmap:

- To add a List of Tables, add the `<tablelist/>` element.
- To add a List of Figures, add the `<figurelist/>` element.

```
<frontmatter>
    <booklists>
        <toc/>
        <tablelist/>
        <figurelist/>
    </booklists>
</frontmatter>


<backmatter>
    <booklists>
        <tablelist/>
        <figurelist/>
    </booklists>
</backmatter>
```

2. Save changes, run a PDF build, and check your work.

Exercise: Format the List of Tables and List of Figures titles

Objective: Determine the appearance of the title in the List of Tables or List of Figures.

By default, both the LOT and LOF headings are formatted by the *__lotf__heading* attribute set which in turn uses the *__toc__header* attribute set in **toc-attr.xsl**.

If you do want different formatting, you can override the attributes in *__toc__header* by adding the same attributes but with different values in *__lotf__heading*.

It's worth pointing out that the formatting of the List of Tables and List of Figures is largely based on the formatting for the Table of Contents. If you want a completely different look and feel for them, you'll probably have to create and call some new attribute sets.

For a refresher on that topic, look at **Create a new attribute set** *(p. 65)* and **Call a new attribute set** *(p. 66)*.

Exercise: Change the title of the List of Tables and List of Figures

Objective: Use a custom title for the lists of tables and figures.

1. Copy the following variables from `DITA-OT/plugins/org.dita.pdf2/cfg/common/vars/`**en.xml** to your copy of **en.xml**:

```
<!-- The heading string to put at the top of the List of Tables -->
<variable id="List of Tables">List of Tables</variable>

<!-- The heading string to put at the top of the List of Figures -->
<variable id="List of Figures">List of Figures</variable>
```

2. Change the default value **List of Tables** to **Tables**.
3. Change the default value **List of Figures** to **Figures**.
4. Repeat this process for any other language-specific variables files you've customized.
5. Save changes, run a PDF build, and check your work.

Exercise: Format page numbers in the List of Tables and Figures

Objective: Determine the appearance of page numbers in the List of Tables or List of Figures.

If you worked through the **Table of Contents** *(p. 325)* chapter, you removed leaders for TOC entries (**Remove leaders from TOC entries** *(p. 338)*), and you also added chapter and appendix numbers to page numbers in TOC entries (**Add chapter, appendix, or part numbers to TOC page numbers** *(p. 336)*). If you didn't work through that chapter, you should review those two exercises now, so you understand what to do here.

You want to make the TOC entries for the List of Tables and List of Figures match the other TOC entries, so remove leaders and add chapter or appendix numbers for those entries as well.

> **Tip:** The formatting for the TOC entries for the List of Table and List of Figures is based on the styling for other topic entries in the TOC. For example, the entries use the attribute set __lotf__title, which calls attribute set __lotf__content, which calls attribute set __toc__topic__content__booklist, which calls attribute set __toc__topic__content. So if you want the styling of the TOC entries for the List of Tables and List of Figures to differ from that of the other TOC entries, you're going to have to override the attributes in the toc-related attribute sets with different values in the lotf-related attribute sets.

1. If it is not already present, copy the entire *getChapterPrefix* template into **custom.xsl** (xsl).

   You can find this template in **GetChapterPrefix template** *(p. 481)*.

   > **Note** You could copy this template into any of the .xsl files you've added to your plugin, including **lot-lof.xsl**, but it's just as easy to copy it into **custom.xsl**. You can see right away it's a custom template that didn't come with the PDF plugin.

2. Open `DITA-OT/plugins/org.dita.pdf2/xsl/fo/`**lot-lof.xsl** and copy the two templates that begin with the following lines to your copy of **lot-lof.xsl**:

```
<xsl:template match="*[contains (@class, ' topic/table ')][child::*[contains
►(@class, ' topic/title ' )]]" mode="list.of.tables">


<xsl:template match="*[contains (@class, ' topic/fig ')][child::*
►[contains(@class, ' topic/title ' )]]" mode="list.of.figures">
```

3. Find this code in both templates:

```
<fo:inline xsl:use-attribute-sets="__lotf__page-number">
   <fo:leader xsl:use-attribute-sets="__lotf__leader"/>
   <fo:page-number-citation>
      <xsl:attribute name="ref-id">
         <xsl:call-template name="get-id"/>
      </xsl:attribute>
   </fo:page-number-citation>
</fo:inline>
```

4. Comment out the line `<fo:leader xsl:use-attribute-sets="__lotf__leader"/>`.

5. Just after the commented-out line, add two non-breaking spaces using the `<xsl:text>` element:

```
<xsl:text>  </xsl:text>
```

6. After that line, call the ***getChapterPrefix*** template to add the chapter or appendix number and a dash before the page number.

When you've made all these changes, the sections should look like this:

```
...
<fo:inline xsl:use-attribute-sets="__lotf__page-number">
    <!--<fo:leader xsl:use-attribute-sets="__lotf__leader"/>-->
    <xsl:text>  </xsl:text>
    <xsl:call-template name="getChapterPrefix" />
    <xsl:text>-</xsl:text>
    <fo:page-number-citation>
        <xsl:attribute name="ref-id">
            <xsl:call-template name="get-id"/>
        </xsl:attribute>
    </fo:page-number-citation>
</fo:inline>
...
```

7. Save changes, run a PDF build, and check your work.

Depending on other exercises you've done, the entries might be very strangely spaced. This is because they are justified. In the next exercise, you'll see how to fix that.

Exercise: Format entries in the List of Tables and List of Figures

Objective: Determine the appearance of entries in the List of Tables or List of Figures.

By default, the styling of the entries in the List of Tables and List of Figures is based on the styling for entries in the Table of Contents. For example, the entries use the attribute set *__lotf__title*, which calls attribute set *__lotf__content*, which calls attribute set *__toc__topic__content__booklist*, which calls attribute set *__toc__topic__content*. If you want the styling of the entries for the List of Tables and List of Figures to differ from that of the other TOC entries, you have to override the TOC entry styling with specific attributes in the lotf attribute sets.

If you formatted the TOC according the exercises in this book, the entries in the List of Tables and List of Figures should be dark red or dark gray, because they pick up the value of the *color* attribute from the *__toc__topic__content* attribute set. Let's change the entries to black, following the template specification, and make the page numbers bold.

1. Copy the *__lotf__content* and *__lotf__page-number* attribute sets from DITA-OT/plugins/org.dita.pdf2/cfg/fo/attrs/**lot-lof-attr.xsl** to your copy of **lot-lof-attr.xsl**.

2. Add the following attributes and values to the *__lotf__content* attribute set.

```
<xsl:attribute name="color">#000000</xsl:attribute>
<xsl:attribute name="text-align-last">left</xsl:attribute>
```

All of the entries in the LOT and LOF are formatted using the *__toc__topic__content* attribute set by default. They are level-1 entries. If you completed the exercises in **Table of Contents** *(p. 325)*, level-1 entries are already formatted at 12pt Trebuchet, so you don't need to change anything in the *__lotf__content* attribute set. Go to step 4. If you did not complete the exercises, continue to step 3.

3. Add the following attributes and values to the *__lotf__content* attribute set:

```
<xsl:attribute name="font-size">12pt</xsl:attribute>
<xsl:attribute name="font-family">Trebuchet MS, Arial Unicode MS, Helvetica</xsl:attribute>
```

4. Add the following attribute and value to the *__lotf__page-number* attribute set:

```
<xsl:attribute name="font-weight">bold</xsl:attribute>
```

5. Save changes, run a PDF build, and check your work.

Exercise: Add chapter, appendix, or part numbers to table or figure numbers in the List of Tables or List of Figures

Objective: Use the *getChapterPrefix* template to number table or figure captions in the lists of tables and figures.

This exercise applies only if you are publishing from a bookmap. Chapter, appendix and part numbering depend on the presence of the `<chapter>`, `<appendix>` and `<part>` elements, which are not available in a map. The List of Tables and List of Figures are also available only in a bookmap.

You've already set up table and figure numbering to include the chapter, appendix, or part number (**Add chapter, appendix or part numbers to table titles** *(p. 288)*) and (**Add chapter, appendix or part numbers to figure titles** *(p. 304)*). So it makes sense to have the table or figure numbers to include chapter, appendix, or part numbers in the List of Tables or List of Figures as well, so they match. Here's how.

1. If they are not already present, copy the two templates that begin with the following lines from DITA-OT/plugins/org.dita.pdf2/xsl/fo/**lot-lof.xsl** to your copy of **lot-lof.xsl**:

```
<xsl:template match="*[contains (@class, ' topic/table ')]
►[child::*[contains(@class, ' topic/title ' )]]" mode="list.of.tables">
```

```
<xsl:template match="*[contains (@class, ' topic/fig ')]
►[child::*[contains(@class, ' topic/title ' )]]" mode="list.of.figures">
```

2. In the first template, find this section:

```
<xsl:call-template name="getVariable">
    <xsl:with-param name="id" select="'Table.title'"/>
    <xsl:with-param name="params">
        <number>
            <xsl:variable name="id">
                <xsl:call-template name="get-id"/>
            </xsl:variable>
            <xsl:variable name="tableNumber">
                <xsl:number format="1" value="count($tableset/*[@id = $id]/
                    ►preceding-sibling::*) + 1" />
            </xsl:variable>
            <xsl:value-of select="$tableNumber"/>
        </number>
    <title>
        <xsl:apply-templates select="./*[contains(@class, ' topic/title ')]"
        ► mode="insert-text"/>
    </title>
    </xsl:with-param>
</xsl:call-template>
```

You can see that in this section, the text "Table:" is inserted (via the *Table.title* localization variable), and then the table number is generated.

As you've done before, you're going to call the ***getChapterPrefix*** template to insert the chapter or appendix number. And you need to call that template from within the `<number>` element.

3. Add the template call as shown here:

```
...
<number>
    <xsl:variable name="id">
        <xsl:call-template name="get-id"/>
    </xsl:variable>
    <xsl:variable name="tableNumber">
        <xsl:number format="1" value="count($tableset/*[@id = $id]/preceding-sibling::*)
        ► + 1" />
    </xsl:variable>
    <xsl:call-template name="getChapterPrefix"/><xsl:text>-</xsl:text>
    <xsl:value-of select="$tableNumber"/>
</number>
...
```

4. Make the same change in the second template.

5. Save changes, run a PDF build, and check your work.

Chapter 19

Bookmarks

While there's not a lot you can do to customize PDF bookmarks, you can do a few things. This chapter explains how to control whether or not your PDF has bookmarks and how to make a few basic changes.

Just to be clear, when we talk about PDF bookmarks, we're talking about the bookmarks that appear in the left panel when you view a PDF in Acrobat Reader, Foxit Reader, or any other PDF reader. While the PDF plugin generates the bookmarks, it's largely up to the PDF reader to render them however it sees fit.

Files you need

If you have completed other exercises, some of these files might already be present in your plugin.

If it is not already present, create the following file:

- `DITA-OT/plugins/com.company.pdf/cfg/fo/xsl/`**bookmarks.xsl**

Be sure to add the appropriate `<xsl:import>` statement to **custom.xsl** (xsl).

There are no attributes that you can customize for bookmarks; the PDF viewer controls the appearance of bookmarks.

> **Note** That's not completely true. If you add something like
>
> ```
> <xsl:attribute name="color">purple</xsl:attribute>
> ```
>
> to the ___fo__root_ attribute set, then your bookmarks will be purple. You don't want to go adding attributes to ___fo__root_ just to change the appearance of bookmarks, though, because the ___fo__root_ attribute set affects everything in the PDF. It's far too broad a brush to use.

For instructions on creating attribute files and XSLT stylesheets in your PDF plugin, refer to **Create an attribute set file in your plugin** *(p. 31)* and **Create an XSLT stylesheet in your plugin** *(p. 32)*.

Exercise: Add chapter, appendix or part numbers to bookmarks

Objective: Use a new *getChapterPrefix* template to number TOC entries.

If you use different numbering schemes for chapters, appendices, and parts, you might want to reflect those numbers in your PDF bookmarks. Here's how.

This exercise applies only if you are publishing from a bookmap. Chapter, appendix and part numbering depend on the `<chapter>`, `<appendix>` and `<part>` elements, which are not available in a map.

1. If it's not already present, copy the entire *getChapterPrefix* template into **custom.xsl** (xsl).

You can find this template in **GetChapterPrefix template** *(p. 481)*.

> **Note** You could copy this template into any of the .xsl files you've added to your plugin, including **toc.xsl**, but it's just as easy to copy it into **custom.xsl**. You can see right away it's a custom template that didn't come with the PDF plugin.

2. Open `DITA-OT/plugins/org.dita.pdf2/xsl/fo/`**bookmarks.xsl** and copy the template that begins with the following line to your copy of **bookmarks.xsl**.

```
<xsl:template match="*[contains(@class, ' topic/topic ')]" mode="bookmark">
```

3. In that template, find this section:

```
...
<fo:bookmark-title>
    <xsl:value-of select="normalize-space($topicTitle)"/>
</fo:bookmark-title>
...
```

You're looking for the `<fo:bookmark-title>` element and the following line:

```
<xsl:value-of select="normalize-space($topicTitle)"/>
```

This line creates the bookmark entry text. You want to add the chapter number here, using the same *getChapterPrefix* template that you used to add chapter/appendix/part numbers to TOC entries (if you completed that exercise).

However, you don't want to add the chapter number to all bookmarks. You just want to add it to the bookmarks that are chapter titles. In your maps, the topics whose titles become chapter titles are the highest-level topicrefs, meaning they don't have any ancestor topics. You can use this hierarchy to limit the application of the *getChapterPrefix* template to only those top-level topicrefs.

4. Add an `<xsl:if>` test and the template call as shown here:

```
...
<fo:bookmark-title>
    <xsl:if test="not(ancestor::*[contains(@class, ' topic/topic ')])">
        <xsl:call-template name="getChapterPrefix"/><xsl:text> </xsl:text>
    </xsl:if>
    <xsl:value-of select="normalize-space($topicTitle)"/>
</fo:bookmark-title>
...
```

This `<xsl:if>` applies the template only to topics that do not have an ancestor topic ... that is, only to the top-level topics in a map.

5. Save changes, run a PDF build, and check your work.

About TOC and index bookmarks

Bookmap-based PDFs

If you include a `<toc>` element in your bookmap, you get a TOC bookmark as well. If you don't include a `<toc>` element in your bookmap, then of course, you don't get a bookmark for it.

The same is true of the `<indexlist>` element: no `<indexlist>`, no bookmark. (Of course, you must have some `<indexterm>` elements in your content.)

And the same is true of the List of Tables and List of Figures; you must include a `<tablelist>` and `<figurelist>` element in your bookmap to get them along with a corresponding bookmark.

Easy enough. But what if you want a TOC or index or List of Tables/Figures, but you don't want the corresponding bookmarks? Find out how to omit the bookmarks in the following exercises:

- **Eliminate the Table of Contents bookmark (bookmap-based PDF)** *(p. 388)*
- **Eliminate the Index bookmark (bookmap-based PDF)** *(p. 389)*
- **Eliminate the List of Tables or List of Figures bookmarks (bookmap-based PDF)** *(p. 389)*

Map-based PDFs

On the other hand, if you're publishing from a map, you always get a TOC and an index—again, if your content includes any `<indexterm>` elements. (Out of the box, you can't add a List of Tables or List of Figures to a map.) It's possible to make some code changes that omit the TOC and index, and those changes are explained in **Eliminate the TOC (map-based PDF)** *(p. 353)* and **Eliminate the index (map-based PDF)** *(p. 372)*. But even after eliminating those items, you still get the corresponding bookmark, which links to nothing.

You need to eliminate the TOC or index and the corresponding bookmark separately. Find out how to do that in these exercises:

Eliminate the TOC bookmark (map-based PDF) *(p. 356)*

Eliminate the Index bookmark (map-based PDF) *(p. 374)*

Exercise: Eliminate the Table of Contents bookmark (bookmap-based PDF)

Objective: Omit the Table of Contents PDF bookmark even when your bookmap-based PDF includes a Table of Contents.

If you do not include a Table of Contents in your bookmap, you do not get a Contents bookmark. However, you might want to include a TOC but not include a bookmark. Here's how.

1. Open `DITA-OT/plugins/org.dita.pdf2/xsl/fo/`**bookmarks.xsl** and copy the template that begins with the following line to your copy of **bookmarks.xsl**:

```
<xsl:template name="createBookmarks">
```

2. Within that template, comment out the following section:

```
<!--<xsl:choose>
    <xsl:when test="$map//*[contains(@class,' bookmap/toc ')][@href]"/>
    <xsl:when test="$map//*[contains(@class,' bookmap/toc ')]
    ► | /*[contains(@class,' map/map ')][not(contains(@class,' bookmap/bookmap '))]">
        <fo:bookmark internal-destination="{$id.toc}">
            <fo:bookmark-title>
                <xsl:call-template name="getVariable">
                    <xsl:with-param name="id" select="'Table of Contents'"/>
                </xsl:call-template>
            </fo:bookmark-title>
        </fo:bookmark>
    </xsl:when>
</xsl:choose>-->
```

3. Save your changes, run a PDF build and check your work.

Exercise: Eliminate the Index bookmark (bookmap-based PDF)

Objective: Omit the Index PDF bookmark even when your bookmap-based PDF includes an index.

If you do not include an index in your bookmap, you do not get an Index bookmark. However, you might want to include an index but not include a bookmark. Here's how.

1. Open `DITA-OT/plugins/org.dita.pdf2/xsl/fo/` and find the template that begins with the following line (this template creates the Index bookmark):

   ```
   <xsl:template match="*" mode="bookmark-index">
   ```

2. Copy just the line `<xsl:template match="*" mode="bookmark-index">` to your copy of **bookmarks.xsl**.

3. Edit that line to add a closing slash: `<xsl:template match="*" mode="bookmark-index"/>`.

   This has the same effect as copying the entire template to your copy of **bookmarks.xsl**, then commenting out its contents. Keep this handy trick in mind for any time you want to negate the effects of an entire template.

4. Save your changes, run a PDF build and check your work.

Exercise: Eliminate the List of Tables or List of Figures bookmarks (bookmap-based PDF)

Objective: Omit the List of Tables or List of Figures PDF bookmarks even when your bookmap-based PDF includes those items.

If you do not include a List of Tables or List of Figures in your bookmap, you do not get corresponding bookmarks. However, you might want to include a LOT or LOF but not include a corresponding bookmark.

1. In `DITA-OT/plugins/org.dita.pdf2/xsl/fo/`**bookmarks.xsl**, find the template that begins:

   ```
   <xsl:template match="ot-placeholder:tablelist" mode="bookmark">
   ```

   This template processes the List of Tables bookmark.

2. In `DITA-OT/plugins/org.dita.pdf2/xsl/fo/`**bookmarks.xsl**, find the template that begins:

```
<xsl:template match="ot-placeholder:figurelist" mode="bookmark">
```

This template processes the List of Figures bookmark.

3. To delete the List of Tables bookmark, copy the following line to your copy of **bookmarks.xsl**:

```
<xsl:template match="ot-placeholder:tablelist" mode="bookmark">
```

4. Edit that line to add a closing slash:

```
<xsl:template match="ot-placeholder:tablelist" mode="bookmark"/>
```

This has the same effect as copying the entire template, then commenting out its contents.

5. To delete the List of Figures bookmark, copy the following line to your copy of **bookmarks.xsl**:

```
<xsl:template match="ot-placeholder:figurelist" mode="bookmark">
```

6. Edit that line to add a closing slash:

```
<xsl:template match="ot-placeholder:figurelist" mode="bookmark"/>
```

7. Save your changes, run a PDF build and check your work.

Other things you can do

Here are a few additional customizations you might want to make to your plugin. These customizations aren't part of the PDF specifications you're following, but they can be useful (and cool).

Exercise: Generate bookmarks for index letter headings

Objective: Determine whether or not to generate bookmarks for each letter heading in the index.

If you have a very long index, it might be helpful to include bookmarks for the individual letter headings so that users reading the PDF can quickly go to the specific index section they need rather than having to scroll through page after page. The PDF plugin now includes an easy way to create those letter headings.

1. Copy the following from `DITA-OT/plugins/org.dita.pdf2/xsl/fo/`**bookmarks.xsl** to your copy of **bookmarks.xsl**:

```
<!-- Determines whether letter headings in an index generate bookmarks.
     0 = no bookmarks.
     Any other number = if total # of terms exceeds $bookmarks.index-group-size, generate
     headers.
     To always generate headers, set to 1. -->
<xsl:param name="bookmarks.index-group-size" as="xs:integer">100</xsl:param>
```

As the helpful comments explain, this parameter determines whether or not to create the letter headings. The default value of "100" indicates that if there are more than 100 index entries for a letter, the letter heading will be created. A setting of "0" specifies never to create the letter headings. A setting of "1" specifies to always create the letter headings, regardless of how many entries there are for a letter.

2. Do one of the following.

   - Change the value to "1" to always create the letter headings.
   - Change the value to something other than "1" or "0" to specify the point at which the plugin should create the letter headings.

3. Save changes, run a PDF build, and check your work.

Appendix A

Specifications used in these exercises

Your mission: create a plugin to produce a PDF that includes the following:

- front cover page
- table of contents
- list of tables and figures
- numbered chapters (1, 2, 3, etc.)
- numbered appendices (A, B, C, etc.)
- index
- back cover page

Front cover page

The front cover page includes:

Specification	Details
Product name	14pt Trebuchet black left-aligned
Product version	14pt Trebuchet black left-aligned
Book title	36pt Trebuchet bold dark red left-aligned
Product logo	left-aligned
Link to company website	12pt Trebuchet black left-aligned, aligned to bottom page margin

Table 1: Front cover page

Back cover page

The back cover page includes:

Product logo	Support contact info
Book item number	Legal disclaimer
Publication date	Copyright info

Page layout

Here are the requirements for the body pages:

Page size	8.5in width by 11in height
Page layout	double-sided
Top margin	1in
Bottom margin	1in
Inside margin	1in
Outside margin	.75in
Header	text top-aligned .5in from top of page
Footer	text bottom-aligned .5in from bottom of page

The front and back cover pages have different requirements:

Top margin	1in
Bottom margin	1in
Inside margin	1in
Outside margin	1in
Header	none
Footer	none

General text

Chapter, appendix titles	28pt Trebuchet normal black, no borders. Autonumber is 16pt Trebuchet regular black.
Level 1 heading font	20pt Trebuchet normal black, no borders
Level 2 heading font	18pt Trebuchet normal dark red
Level 3 heading font	16pt Trebuchet normal dark red, .5in indent
Level 4+ heading font	14pt Trebuchet italic dark red, .5in indent
Section headings	12pt Trebuchet bold dark red
Regular body font	11pt Book Antiqua normal dark gray. Used for all body elements except where noted below.
Small body font	10pt Book Antiqua normal dark gray. Used for *note*, *info*, *stepxmp*, *stepresult*, *choice*.
Code samples, system messages	9pt Consolas normal black, light gray background
Line height (leading)	120%

Headers and footers

All footers are justified across the entire footer region width.

Header font	9pt Trebuchet normal black
Footer font	9pt Trebuchet normal black
Footer page number	9pt Trebuchet bold dark red
Header ruling	1pt solid black line below
Footer ruling	1pt solid black line above
Page number	includes chapter, appendix or part number and restarts at 1 in each chapter, appendix or part

Header - body first Header - toc first Header - index first	product logo
Header - body odd	running header | product logo
Header - body even Header - body last Header - toc even Header - toc last Header - index even Header - index last	book title
Header - toc odd	Table of Contents | product logo
Header - index odd	Index | product logo

| Footer - body first
Footer - body odd
Footer - toc first
Footer - toc odd
Footer - index first
Footer - index odd | publication date \| Copyright ©copyright info \| page # |
| Footer - body even
Footer - body last
Footer - toc even
Footer - toc last
Footer - index even
Footer - index last | page # \| product name version |

TOC

TOC title font	28pt Trebuchet normal black
Number of TOC levels	two
TOC level 1 entry font	12pt Trebuchet normal dark red
TOC level 2 entry font	10pt Trebuchet normal dark gray
TOC leader	none
TOC page number	10pt Trebuchet bold dark red; includes chapter, appendix or part number
Level 2 indent	.25in
TOC heading	Table of Contents
Mini-TOC heading	In this section
Mini-TOC links	10pt Trebuchet bold dark red

Mini-TOC link page number	10pt Trebuchet bold black, includes chapter, appendix or part number

List of Tables

LOT title font	28pt Trebuchet normal black
LOT level 1 entry font	12pt Trebuchet normal black
LOT leader	none
LOT page number	12pt Trebuchet bold black, includes chapter number
LOT heading	Tables
LOT entries	include chapter number-table number (Table B-4)

List of Figures

LOF title font	28pt Trebuchet normal black
LOF level 1 entry font	12pt Trebuchet normal black
LOF leader	none
LOF page number	12pt Trebuchet bold black, includes chapter number
LOF heading	Figures
LOF entries	include chapter number-figure number (Figure 3-6)

Index

Index title font	28pt Trebuchet normal black
Index level 1 entry font	11pt Book Antiqua normal dark gray

Index level 2 entry font	11pt Book Antiqua normal dark gray
Index page number	10pt Book Antiqua normal italic dark red
Index letter group headings	12pt Trebuchet bold dark red
Index pages	two-column
Level 2 indent	.125in
Index heading	Index

Colors

dark red	990033
dark gray	8A8A8A
light gray	E6E6E6
black	000000
white	FFFFFF

Tables

Table text font	10pt Book Antiqua normal dark gray
Table cell text padding	6pt all around
Table header row background color	dark red
Table header row font	11pt Trebuchet bold white
Table title	12pt Trebuchet bold dark red, below table

Table frame	2pt dark red
Table rules	2pt dark red under the heading row; 1pt dark gray all other rows and columns
Table title numbering	Includes "Table" chapter/appendix number, table number (Table: 1-1). Table numbering restarts with each chapter/appendix.
Table column widths	respect when specified

Bulleted and numbered lists

Top-level ordered list numbering	1, 2, 3...
Second-level ordered list numbering	i, ii, iii...
Top-level bulleted list bullet	■ black square (■ or ■)
Second-level bulleted list bullet	— em-dash (Ᾱ or —)
Checklist (custom list type)	☐ empty checkbox (□ or □)

Images

Figure title	12pt Trebuchet bold dark red; above figure
Figure numbering	Includes "Figure" chapter/appendix number, table number (Figure: 1-1). Figure numbering restarts with each chapter/appendix.
Scaling	automatically scale large images to fit on page
Alignment	centered

Notes

Note label	dark red, bold. Not used on note type="note".
Note borders	left and right, 2pt dark red
Note indents	.5in left and right
Note font	Small body font, justified
Note image	note.png
Custom note label	Best Practice
Label for fastpath notes	Quick access

Links

Related Links section title	Additional information
Related Links section divider	light gray 2pt line, 3in wide
Link text format	10pt Trebuchet normal dark red
Page reference text format	10pt Trebuchet italic black
Page reference text	(p. #-#), includes chapter number
Cross-reference format	bold non-italic dark gray
Footnote callout	9pt Trebuchet bold italic dark red, enclosed in square brackets in body text but not in footnote itself
Footnote text	9pt Trebuchet italic dark gray
Line above footnotes	light gray dotted

Tasks

Optional step label	(Optional)
Step section labels	12pt Trebuchet bold dark red prereq: "Before you start" context: "About this task" steps: "Procedure" steps-unordered: "Procedure" result: ""Reality check"" example: "Show me" tasktroubleshooting: "Help me fix it" postreq: "What to do next"
Prereq, context, result, example, postreq, tasktroubleshooting text format	11pt Book Antiqua normal dark gray
Info, stepxmp, stepresult, choices, substeps text format	10 pt Book Antiqua normal dark gray
Step numbers	11pt Trebuchet bold dark red. Arabic number with a parenthesis after: 1)
Substep numbers	10pt Trebuchet bold dark red. Roman numeral with a period after: i.

Appendix B

Attribute set lists and descriptions

This section includes a separate list of all of the attribute sets in each of the attribute set files found in the DITA Open Toolkit.

Attribute set file list

Attribute set	Description/purpose
basic-settings.xsl	Contains variables that define default page master dimensions (height, width, margins, sidehead).
commons-attr.xsl	This is the largest attribute set file and is a bit of a catch-all for attribute sets that don't fit neatly into the other files. Among the elements whose attributes you define here are `<title>`, `<note>`, `<shortdesc>`, `<image>`, and many of the prolog metadata elements.
commons-attr_fop.xsl	Contains alternative common attribute sets specific to the FOP PDF renderer. No longer found in `org.dita.pdf2`; has been moved to `org.dita.pdf2.fop`.
commons-attr_xep.xsl	Contains alternative common attribute sets specific to the XEP PDF renderer. No longer found in `org.dita.pdf2`; has been moved to `org.dita.pdf2.xep`.
front-matter-attr.xsl	Contains attribute sets for formatting items on the title page, such as the `<mainbooktitle>`, `<subtitle>`, `<bookowner>`, etc. Typically this information is drawn from the bookmap.

Attribute set	Description/purpose
glossary-attr.xsl	Contains attribute sets for formatting items in glossaries, such as the term, definition, letter-group label, etc.
hi-domain.xsl	Contains attribute sets for formatting elements in the highlight domain, such as `<b>`, `<i>`, `<u>`, etc.
index-attr.xsl	Contains attribute sets for formatting items in the index, such as the entry text, page number, letter-group heading, etc.
index-attr_axf.xsl	Contains alternative index-related attribute sets specific to the Antenna House PDF renderer. No longer found in `org.dita.pdf2`; has been moved to `org.dita.pdf2.axf`.
index-attr_xep.xsl	Contains alternative index-related attribute sets specific to the XEP PDF renderer. Found in `org.dita.pdf2.xep`.
layout-masters-attr.xsl	Contains attribute sets for formatting page masters, such as page dimensions and page region dimensions.
layout-masters-attr_xep.xsl	Contains alternative layout master attribute sets specific to the XEP PDF renderer. No longer found in `org.dita.pdf2`; has been moved to `org.dita.pdf2.xep`.
learning-elements-attr.xsl	Contains attribute sets for formatting elements from the learning specialization.
links-attr.xsl	Contains attribute sets for formatting link-type elements, such as `<xref>`, `<link>`, `<linkinfo>`, `<linklist>`, `<related-links>`, etc.
lists-attr.xsl	Contains attribute sets for formatting elements in lists, such as `<ol>`, `<ul>`, `<li>`, etc. Some attribute sets found here in previous versions of the OT are now in **task-elements-attr.xsl**.
lot-lof_xsl	Contains attribute sets for formatting elements in the List of Tables and List of Figures.

Attribute set	Description/purpose
map-elements-attr.xsl	Contains attribute sets for formatting relationship tables. The attribute sets in this file were found in other attribute set files in previous versions of the Open Toolkit.
markup-domain.xsl	Contains attribute sets for formatting elements in the highlight domain, which currently consists only of `<markupname>`.
pr-domain-attr.xsl	Contains attribute sets for formatting elements in the pr domain, such as `<codeblock>`, `<option>`, etc.
reference-elements-attr.xsl	Contains attribute sets for formating the properties table as well as the `<reference>`, `<refbody>`, and `<refsyn>` elements. The attribute sets in this file were found in other attribute set files in previous versions of the Open Toolkit.
static-content-attr.xsl	Contains attribute sets for formatting items in page headers and footers, such as running header/footer text, page numbers, chapter numbers, etc.
sw-domain-attr.xsl	Contains attribute sets for formatting elements in the software domain, such as `<cmdname>`, `<msgblock>`, etc.
tables-attr.xsl	Contains attribute sets for formatting heading, row and cell elements in tables, including `<table>`, `<simpletable>`, `<properties>`, and `<reltable>`.
tables-attr_axf.xsl	Contains alternative table-related attribute sets specific to the Antenna House PDF renderer. No longer found in `org.dita.pdf2`; has been moved to `org.dita.pdf2.axf`.
tables-attr_fop.xsl	Contains alternative table-related attribute sets specific to the FOP PDF renderer. No longer found in `org.dita.pdf2`; has been moved to `org.dita.pdf2.fop`.
task-elements-attr.xsl	Contains attribute sets for formatting elements within tasks, such as `<cmd>`, `<info>`, `<stepxmp>`, `<result>`, `<prereq>`, etc. The attribute sets in this file were found in other attribute set files in previous versions of the Open Toolkit.

Attribute set	Description/purpose
toc-attr.xsl	Contains attribute sets for formatting items in the table of contents, such as the entry text, page number, leader, etc.
toc-attr_axf.xsl	Contains alternative TOC-related attribute sets specific to the Antenna House PDF renderer. No longer found in `org.dita.pdf2`; has been moved to `org.dita.pdf2.axf`.
toc-attr_fop.xsl	Contains alternative TOC-related attribute sets specific to the FOP PDF renderer. No longer found in `org.dita.pdf2`; has been moved to org.dita.pdf2.fop.
ui-domain.xsl	Contains attribute sets for formatting elements in the user interface domain, such as `<userinput>`, `<wintitle>`, `<menucascade>`, etc.
xml-domain.xsl	Contains attribute sets for formatting elements found in the xml domain, including `<xmlelement>`, `<xmlatt>`, `<textentity>`, etc.

Common attribute sets

The attribute sets you use to format the appearance of elements that are widely used or that aren't easily categorized are found in **commons-attr.xsl**, **commons-attr_fop.xsl**, and **commons-attr_xep.xsl**. Here is a list of those attributes along with their purpose and what XSLT stylesheets they are called from.

Commons-attr.xsl is by far the largest attribute set file. To make it a little easier to work with, I separated attribute sets that have to do with titles, graphics, notes, and general text into separate tables. These categories are my own and not necessarily comprehensive.

> **Note** Some task-related attribute sets previously found in this file are now found in **task-elements-attr.xsl**. **Commons-attr.xsl**, as delivered in the Open Toolkit, has two instances of both the *searchtitle* attribute set and the *searchtitle__label* attribute set. *Searchtitle* has no attributes in either instance and *searchtitle__label* defines only *font-weight* in both instances. If you copy this file into your PDF plugin, you can delete the duplicate instance of each.

Attribute set	Purpose	Called from
common.title	Establishes some common title formatting used by other attribute sets such as __frontmatter__title_, _table.title_, __toc__header_.	front-matter-attr.xsl, tables-attr.xsl, toc-attr.xsl
example.title	Formats content in the `<title>` element within an `<example>` element.	commons.xsl , task-elements.xsl
image.artlabel	Formats the contents of the `<title>` element in images.	commons.xsl
lq_title	Formats the value of the _reftitle_ attribute on an `<lq>` element.	commons.xsl
section.title	Formats content in the `<title>` element within a `<section>` element.	commons.xsl , task-elements.xsl
topic.title	Formats content in the highest-level `<title>` elements.	attr-set-reflection.xsl, commons.xsl, preface.xsl
topic.title__content	Formats content in the highest-level `<title>` elements, excluding any auto-generated label.	attr-set-reflection.xsl
topic.topic.title	Formats content in the second-highest-level `<title>` elements.	attr-set-reflection.xsl
topic.topic.title__content	Formats content in the second-highest-level `<title>` elements, excluding any auto-generated label.	attr-set-reflection.xsl
topic.topic.topic.title	Formats content in the third-highest-level `<title>` elements.	attr-set-reflection.xsl

Attribute set	Purpose	Called from
topic.topic.topic.title__content	Formats content in the third-highest-level `<title>` elements, excluding any auto-generated label.	attr-set-reflection.xsl
topic.topic.topic.topic.title	Formats content in the fourth-highest-level `<title>` elements.	attr-set-reflection.xsl
topic.topic.topic.topic .title__content	Formats content in the fourth-highest-level `<title>` elements, excluding any auto-generated label.	attr-set-reflection.xsl
topic.topic.topic.topic. topic.title	Formats content in the fifth-highest-level `<title>` elements.	attr-set-reflection.xsl
topic.topic.topic.topic. topic.title__content	Formats content in the fifth-highest-level `<title>` elements, excluding any auto-generated label.	attr-set-reflection.xsl
topic.topic.topic.topic. topic.topic.title	Formats content in the sixth-highest-level `<title>` elements.	attr-set-reflection.xsl
topic.topic.topic.topic. topic.topic.title__content	Formats content in the sixth-highest-level `<title>` elements, excluding any auto-generated label.	attr-set-reflection.xsl
navtitle	Formats content in the `<navtitle>` element	commons.xsl
navtitle__label	Formats the auto-generated label of the `<navtitle>` element	commons.xsl
searchtitle	Formats content in the `<searchtitle>` element. Multiple instances in **commons-attr.xsl**.	commons.xsl

Attribute set	Purpose	Called from
searchtitle__label	Formats the auto-generated label of the `<searchtitle>` element. Multiple instances in **commons-attr.xsl**.	commons.xsl
titlealts	Formats content in the `<titlealts>` element.	commons.xsl

Table 1: Title-related common attribute sets

Attribute set	Purpose	Called from
note	Formats content in the `<note>` element.	commons.xsl
note__image__column	Formats the column used to display the image when notes are rendered as a 2-column table.	commons.xsl
note__image__entry	Formats the table cell that contains the note icon in the default two-column table used to display the `<note>` element.	commons.xsl
note__label	Formats the auto-generated label of the `<note>` element. Can be overridden by specific note types using attributes below.	commons.xsl
note__label__attention	Formats the auto-generated label of the `<note>` element where *type*="attention".	commons.xsl
note__label__caution	Formats the auto-generated label of the `<note>` element where *type*="caution".	commons.xsl
note__label__danger	Formats the auto-generated label of the `<note>` element where *type*="danger".	commons.xsl
note__label__fastpath	Formats the auto-generated label of the `<note>` element where *type*="fastpath".	commons.xsl

Attribute set	Purpose	Called from
note__label__important	Formats the auto-generated label of the <note> element where *type*="important".	commons.xsl
note__label__note	Formats the auto-generated label of the <note> element where *type*="note".	commons.xsl
note__label__notice	Formats the auto-generated label of the <note> element where *type*="notice".	commons.xsl
note__label__other	Formats the auto-generated label of the <note> element where *type*="other".	commons.xsl
note__label__remember	Formats the auto-generated label of the <note> element where *type*="remember".	commons.xsl
note__label__restriction	Formats the auto-generated label of the <note> element where *type*="restriction".	commons.xsl
note__label__tip	Formats the auto-generated label of the <note> element where *type*="tip".	commons.xsl
note__label__trouble	Formats the auto-generated label of the <note> element where *type*="trouble".	commons.xsl
note__label__warning	Formats the auto-generated label of the <note> element where *type*="warning".	commons.xsl
note__table	Formats the two-column table used by default to display the note icon and text. Also in commons-attr_fop.xsl.	commons.xsl
note__text__column	Formats the column used to display the text when notes are rendered as a 2-column table. Also in commons-attr_fop.xsl	commons.xsl

Attribute set	Purpose	Called from
note__text__entry	Formats the table cell that contains the note text in the default two-column table used to display the `<note>` element.	commons.xsl

Table 2: Note-related common attribute sets

Attribute set	Purpose	Called from
alt	Formats content in the `<alt>` element	commons.xsl
fig	Formats content in the `<fig>` element.	commons.xsl
fig.title	Formats content in the `<title>` element within a `<fig>` element.	commons.xsl
figgroup	Formats content in the `<figgroup>` element.	commons.xsl
image	Formats the contents of an `<image>` element.	commons.xsl
image__block	Formats the positioning of an `<image>` element where *placement* is not "inline".	commons.xsl
image__float	Formats the positioning of an `<image>` element where *placement* is not "inline".	commons.xsl
image__inline	Formats the contents of an <image> element where *placement*="inline".	commons.xsl

Table 3: Graphics-related common attribute sets

Attribute set	Purpose	Called from
abstract	Formats content in the `<abstract>` element.	commons.xsl
base-font	Establishes some common font formatting used by many other attribute sets.	hi-domain-attr.xsl links-attr.xsl, lot-lof-attr.xsl, hi-domain-attr.xsl, pr-domain-attr.xsl, sw-domain-attr.xsl, tables-attr.xsl, tables-attr_axf.xsl, ui-domain-attr.xsl
body	Formats content in the `<body>` element of top-level topics.	commons.xsl
body__secondLevel	Formats content in the `<body>` element of third-level (double-nested) topics.	commons.xsl
body__toplevel	Formats content in the `<body>` element of second-level (single-nested) topics.	commons.xsl, reference-elements.xsl, preface.xsl
boolean	Formats content in the `<boolean>` element.	commons.xsl
conbody	Formats content in the `<conbody>` element.	commons.xsl
cite	Formats content in the `<cite>` element.	commons.xsl
concept	Formats content in the `<concept>` element.	commons.xsl
common.block	Establishes some common block-level formatting used by many other attribute sets.	lists-attr.xsl
common.link	Establishes some common link formatting used by other attribute sets such as `<link__content>` and `<xref>`.	index-attr.xsl, links-attr.xsl, index_fop.xsl

Attribute set	Purpose	Called from
desc	Formats content in the `<desc>` element.	commons.xsl
div	Formats content in the `<div>` element.	commons.xsl
draft-comment	Formats content in the `<draft-comment>` element	commons.xsl
draft-comment__label	Formats the auto-generated label of the `<draft-comment>` element	commons.xsl
example	Formats content in the `<example>` element.	commons.xsl
example__content	Formats content in the `<example>` element.	task-elements.xsl
fn	Formats content in the `<fn>` element. Not used. Replaced by `<fn__body>`.	none
fn__body	Formats content in the `<fn>` element.	commons.xsl
fn__callout	Formats the value of *callout* in an `<fn>` element.	commons.xsl, links.xsl
fn__id	Formats the value of *id* in an `<fn>` element.	commons.xsl
keyword	Formats content in the `<keyword>` element.	commons.xsl
keywords	Formats content in the `<keywords>` element.	commons.xsl
lines	Formats content in the `<lines>` element.	commons.xsl
lq	Formats content in the `<lq>` element when there is no *href* or *reftitle* attribute.	commons.xsl, attr-set-reflection.xsl
lq_link	Formats the value of *href* on an `<lq>` element.	commons.xsl
lq_simple	Formats content in the `<lq>` element when there is no *href* or *reftitle* attribute.	attr-set-reflection.xsl

Attribute set	Purpose	Called from
object	Formats content in the `<object>` element.	commons.xsl
p	Formats content in the `<p>` element	commons.xsl
param	Formats content in the `<param>` element.	commons.xsl
ph	Formats content in the `<ph>` element	commons.xsl
pre	Formats content in the `<pre>` element	pr-domain.xsl, sw-domain.xsl
prolog	Formats content in the `<prolog>` element	commons.xsl
required-cleanup	Formats content in the `<required-cleanup>` element.	commons.xsl
required-cleanup__label	Formats the auto-generated label of the `<required-cleanup>` element.	commons.xsl
q	Formats content in the `<q>` element.	commons.xsl
section	Formats content in the `<section>` element.	commons.xsl, task-elements-attr.xsl
section__content	Formats content in the `<section>` element.	task-elements-attr.xsl
series	Formats content in the `<series>` prolog element	commons.xsl
shortdesc	Formats content in the `<shortdesc>` element and renders it inline when `<shortdesc>` has fewer than the maximum number of characters. The maximum number is language-specific and determined in **commons.xsl**.	commons.xsl
state	Formats content in the `<state>` element.	commons.xsl
term	Formats content in the `<term>` element.	commons.xsl

Attribute set	Purpose	Called from
tm	Formats content in the `<tm>` element.	commons.xsl
tm__content	Formats the auto-generated symbol in the `<tm>` element except when *tmtype*="service".	commons.xsl
tm__content__service	Formats the auto-generated symbol in the `<tm>` element when *tmtype*="service".	commons.xsl
topic	Formats content in the `<topic>` element.	commons.xsl, preface.xsl
topic__shortdesc	Formats content in the `<shortdesc>` element and renders it as a block when the shortdesc has more than the maximum number of characters. The maximum number is language-specific and determined in **commons.xsl**.	commons.xsl

Table 4: Text-related common attribute sets

Attribute set	Purpose	Called from
__align__center	Aligns content center-justified for all elements where *align*="center".	attr-set-reflection.xsl
__align__justify	Aligns content fully-justified for all elements where *align*="justify".	attr-set-reflection.xsl
__align__left	Aligns content left-justified for all elements where *align*="left".	attr-set-reflection.xsl
__align__right	Aligns content right-justified for all elements where *align*="right".	attr-set-reflection.xsl
__border__all	Creates a border on all sides of elements where *frame*="all".	commons.xsl

Attribute set	Purpose	Called from
__border__bot	Creates a border on the bottom of elements where *frame*="bottom".	commons.xsl
__border__left	Used by the *__border__sides* attribute set to create a left border on elements where *frame*="sides".	commons-attr.xsl
__border__right	Used by the *__border__sides* attribute set to create a right border on elements where *frame*="sides".	commons-attr.xsl
__border__sides	Creates a border on both sides of elements where *frame*="sides".	commons.xsl
__border__top	Creates a border on top of elements where *frame*="top".	commons.xsl
__border__topbot	Creates a border on the top and bottom of elements where *frame*="topbot".	commons.xsl
common.border	Establishes common border formatting used by many other attribute sets.	commons-attr.xsl
common.border__bottom	Establishes common border formatting used by many other attribute sets.	tables-attr.xsl
common.border__left	Establishes common border formatting used by many other attribute sets.	tables-attr.xsl, toc-attr.xsl
common.border__right	Establishes common border formatting used by many other attribute sets.	tables-attr.xsl
common.border__top	Establishes common border formatting used by many other attribute sets.	tables-attr.xsl

Table 5: Border- and alignment-related attribute sets

Attribute set	Purpose	Called from
audience	Formats content in the `<audience>` prolog element	commons.xsl
author	Formats content in the `<author>` prolog element	commons.xsl, front-matter.xsl
brand	Formats content in the `<brand>` prolog element	commons.xsl
category	Formats content in the `<category>` prolog element	commons.xsl
component	Formats content in the `<component>` prolog element	commons.xsl
copyrholder	Formats content in the `<copyrholder>` prolog element	commons.xsl, front-matter.xsl
copyright	Formats content in the `<copyright>` prolog element	commons.xsl, front-matter.xsl
copyryear	Formats content in the `<copyryear>` prolog element	commons.xsl, front-matter.xsl
created	Formats content in the created `<prolog>` element	commons.xsl
critdates	Formats content in the `<critdates>` prolog element	commons.xsl
featnum	Formats content in the `<featnum>` prolog element	commons.xsl
metadata	Formats content in the `<metadata>` prolog element	commons.xsl
othermeta	Formats content in the `<othermeta>` prolog element	commons.xsl
permissions	Formats content in the `<permissions>` prolog element	commons.xsl
platform	Formats content in the `<platform>` prolog element	commons.xsl
prodinfo	Formats content in the `<prodinfo>` prolog element	commons.xsl
prodname	Formats content in the `<prodname>` prolog element	commons.xsl

Attribute set	Purpose	Called from
prognum	Formats content in the `<prognum>` prolog element	commons.xsl
publisher	Formats content in the `<publisher>` prolog element	commons.xsl , front-matter.xsl
resourceid	Formats content in the `<resourceid>` prolog element.	commons.xsl
revised	Formats content in the `<revised>` prolog element.	commons.xsl
source	Formats content in the `<source>` prolog element.	commons.xsl
vrm	Formats content in the `<vrm>` element	commons.xsl
vrmlist	Formats content in the `<vrmlist>` element	commons.xsl

Table 6: Metadata-related attribute sets

Attribute set	Purpose	Called from
topicgroup	Formats content in the `<topicgroup>` element.	commons.xsl
topichead	Formats content in the `<topichead>` element. Not used.	none
topicmeta	Formats content in the `<topicmeta>` element.	map-elements.xsl

Table 7: Map-related attribute sets

Attribute set	Purpose	Called from
__fo__root	Defines very general format properties for an entire publication at the topmost level. Also in commons-attr_xep.xsl.	root-processing.xsl, root-processing_fop.xsl
__force__page__count	Determines whether the PDF pagination is single-sided (no forced pages) or double-sided (force empty even page if necessary). By default, the pagination is double-sided if the input is a bookmap; otherwise, is single-sided.	commons-attr.xsl
page-sequence.appendix	Formats various aspects of appendix pages, such as page numbering format.	commons.xsl
page.sequence.body	Formats various aspects of body pages, such as page numbering format.	commons.xsl
page.sequence.cover	Formats various aspects of cover pages, such as page numbering format.	commons.xsl
page.sequence.glossary	Formats various aspects of glossary pages, such as page numbering format.	commons.xsl
page.sequence.index	Formats various aspects of index pages, such as page numbering format.	commons.xsl
page.sequence.lof	Formats various aspects of LOF pages, such as page numbering format.	commons.xsl
page.sequence.lot	Formats various aspects of LOT pages, such as page numbering format.	commons.xsl
page.sequence.notice	Formats various aspects of notice pages, such as page numbering format.	commons.xsl
page.sequence.part	Formats various aspects of part pages, such as page numbering format.	commons.xsl

Attribute set	Purpose	Called from
page.sequence.preface	Formats various aspects of preface pages, such as page numbering format.	commons.xsl
page.sequence.toc	Formats various aspects of TOC pages, such as page numbering format.	commons.xsl

Table 8: Output-related attribute sets

Attribute set	Purpose	Called from
__spectitle	Formats the value of *spectitle* on elements where specified	commons.xsl
__unresolved__conref	Formats unresolved conrefs	commons.xsl
flag.image	Formats the image used for ditaval flagging. not used.	none
indextermref	Formats content in the `<indextermref>` element.	commons.xsl

Table 9: Miscellaneous attribute sets

Domain attribute sets

Highlight domain

The attribute sets you use to format the appearance of the highlight domain elements are found in **hi-domain-attr.xsl**. Here is a list of those attributes along with their purpose and what XSLT stylesheets they are called from.

Attribute set	Purpose	Called from
b	Formats text in the `<b>` element.	hi-domain.xsl
i	Formats text in the `<i>` element.	hi-domain.xsl
line-through	Formats text in the `<linethrough>` element.	hi-domain.xsl
overline	Formats text in the overline element.	hi-domain.xsl
sub	Formats text in the `<sub>` element.	hi-domain.xsl
sup	Formats text in the `<sup>` element.	hi-domain.xsl
tt	Formats text in the `<tt>` element.	hi-domain.xsl
u	Formats text in the `<u>` element.	hi-domain.xsl

Markup domain

The attribute set you use to format the appearance of the markup domain element is found in **markup-domain-attr.xsl**. Here is that attribute along with its purpose and what XSLT stylesheet it is called from.

Attribute set	Purpose	Called from
markupname	Formats text in the `<markupname>` element.	markup-domain.xsl

Processing domain

The attribute sets you use to format the appearance of the processing domain elements are found in **pr-domain-attr.xsl**. Here is a list of those attributes along with their purpose and what XSLT stylesheets they are called from.

Attribute set	Purpose	Called from
apiname	Formats text in the `<apiname>` element	pr-domain.xsl
codeblock	Formats text in the `<codeblock>` element	pr-domain.xsl
codeblock__bottom	Formats the block following the `<codeblock>`, often used to add a line below the block	pr-domain.xsl
codeblock.line-number	Formats line numbers in a `<codeblock>` element	pr-domain.xsl
codeblock__top	Formats the block preceding the `<codeblock>`, often used to add a line above the block	pr-domain.xsl
codeph	Formats text in the `<codeph>` element	pr-domain.xsl
delim	Formats text in the `<delim>` element	pr-domain.xsl
fragment	Formats text in the `<fragment>` element	pr-domain.xsl
fragment.group	Formats text in the `<groupcomp>`, `<groupseq>`, and `<groupchoice>` elements within fragment	pr-domain.xsl
fragment.title	Formats the `<title>` element within a `<fragment>` element	pr-domain.xsl
fragref	Formats text in the `<fragref>` element	pr-domain.xsl
kwd	Formats text in the `<kwd>` element	pr-domain.xsl
kwd__default	Formats text in `<kwd>` element where *importance*="default"	pr-domain.xsl
oper	Formats text in the `<oper>` element	pr-domain.xsl

Attribute set	Purpose	Called from
option	Formats text in the `<option>` element	pr-domain.xsl
parml	Formats text in the `<parml>` element	pr-domain.xsl
parmname	Formats text in the `<parmname>` element	pr-domain.xsl
pd	Formats text in the `<pd>` element	pr-domain.xsl
plentry	Formats text in the `<plentry>` element	pr-domain.xsl
pt	Formats text in the `<pt>` element	pr-domain.xsl
pt__content	Formats text contained in child elements of `<pt>`	pr-domain.xsl
sep	Formats text in the `<sep>` element	pr-domain.xsl
synblk	Formats text in the `<synblk>` element	pr-domain.xsl
synnote	Formats text in the `<synnote>` element	pr-domain.xsl
synnoteref	Formats text in the `<synnoteref>` element	pr-domain.xsl
synph	Formats text in the `<synph>` element	pr-domain.xsl
syntaxdiagram	Formats text in the `<syntaxdiagram>` element	pr-domain.xsl
syntaxdiagram.group	Formats text in the `<groupcomp>`, `<groupseq>`, and `<groupchoice>` elements within a `<syntaxdiagram>` element	pr-domain.xsl
syntaxdiagram.title	Formats the `<title>` element within a `<syntaxdiagram>` element	pr-domain.xsl
var	Formats text in the `<var>` element	pr-domain.xsl, sw-domain.xsl

Software domain

The attribute sets you use to format the appearance of the software domain elements are found in *sw-domain-attr.xsl*. Here is a list of those attributes along with their purpose and what XSLT stylesheets they are called from.

Attribute set	Purpose	Called from
cmdname	Formats text in the `<cmdname>` element.	sw-domain.xsl
filepath	Formats text in the `<filepath>` element.	sw-domain.xsl
msgblock	Formats text in the `<msgblock>` element.	sw-domain.xsl
msgnum	Formats text in the `<msgnum>` element.	sw-domain.xsl
msgph	Formats text in the `<msgph>` element.	sw-domain.xsl
systemoutput	Formats text in the `<systemoutput>` element.	sw-domain.xsl
userinput	Formats text in the `<userinput>` element.	sw-domain.xsl
varname	Formats text in the `<varname>` element.	sw-domain.xsl

User interface domain

The attribute sets you use to format the appearance of the user interface domain elements are found in *ui-domain-attr.xsl*. Here is a list of those attributes along with their purpose and what XSLT stylesheets they are called from.

Attribute set	Purpose	Called from
menucascade	Formats text in the `<menucascade>` element.	ui-domain.xsl
screen	Formats text in the `<screen>` element.	ui-domain.xsl
screen__bottom	Formats the block following the `<screen>`; often used to add a line	ui-domain.xsl

Attribute set	Purpose	Called from
screen__top	Formats the block preceding the `<screen>`; often used to add a line	ui-domain.xsl
shortcut	Formats text in the `<shortcut>` element.	ui-domain.xsl
uicontrol	Formats text in the `<uicontrol>` element.	ui-domain.xsl
wintitle	Formats text in the `<wintitle>` element.	ui-domain.xsl

XML domain

The attribute sets you use to format the appearance of the XML domain elements are found in *xml-domain-attr.xsl*. Here is a list of those attributes along with their purpose and what XSLT stylesheets they are called from.

Attribute set	Purpose	Called from
numcharref	Formats text in the `<numcharref>` element.	xml-domain.xsl
parameterentity	Formats text in the `<parameterentity>` element.	xml-domain.xsl
textentity	Formats text in the `<textentity>` element.	xml-domain.xsl
xmlatt	Formats text in the `<xmlatt>` element.	xml-domain.xsl
xmlelement	Formats text in the `<xmlelement>` element.	xml-domain.xsl
xmlnsname	Formats text in the `<xmlnsname>` element.	xml-domain.xsl
xmlpi	Formats text in the `<xmlpi>` element.	xml-domain.xsl

Frontmatter attribute sets

The attribute sets that you use to format the appearance of lists are found in **front-matter-attr.xsl**. Here is a list of those attributes along with their purpose and what XSLT stylesheets they are called from.

> **Note** While frontmatter does not refer only to title pages, the attributes in the default version of **front-matter-attr.xsl** all pertain to formatting elements on the title page. Therefore, I have included this attribute set in the title page information.

Attribute set	Purpose	Called from
back-cover	Formats the back cover	front-matter.xsl
__back-cover	Formats the contents of the back cover	front-matter.xsl
bookmap.summary	Formats the contents of the `<summary>` element.	front-matter.xsl
__frontmatter	Formats all items on the title page; can be overridden by settings for lower-level elements like `<title>`	front-matter.xsl
__frontmatter__booklibrary	Formats the contents of the `<booklibrary>` element.	front-matter.xsl
__frontmatter__mainbooktitle	Formats the contents of the `<mainbooktitle>` element.	front-matter.xsl
__frontmatter__owner	If the input is a bookmap, formats the contents of the `<bookowner>` element. If the input is a map, formats the contents of the `<topicmeta>` element.	front-matter.xsl
__frontmatter__owner__container	If the input is a bookmap, formats the contents of the `<bookowner>` element. If the input is a map, formats the contents of the `<topicmeta>` element.	front-matter.xsl

Attribute set	Purpose	Called from
__frontmatter__owner__ container_content	Not used.	none
__frontmatter__subtitle	Formats the contents of the `<booktitlealt>` element.	front-matter.xsl
__frontmatter__title	If the input is a bookmap, formats the `<title>` element. If the input is a map, formats the value of the *title* attribute. Otherwise, formats the value of the `<title>` element of the first topic.	front-matter.xsl

Glossary attribute sets

The attribute sets that you use to format the appearance of glossaries are found in **glossary-attr.xsl**. Here is a list of those attributes along with their purpose and what XSLT stylesheets they are called from.

Attribute set	Purpose	Called from
__glossary__label	Formats the heading of the glossary	glossary.xsl
__glossary__term	Formats content in the `<term>` element within a `<glossentry>`	glossary.xsl
__glossary__def	Formats content in the `<def>` element within a `<glossentry>`	glossary.xsl

Index attribute sets

The attribute sets you use to format the appearance of the index are found in **index-attr.xsl** and **index-attr_axf.xsl**. Here is a list of those attributes along with their purpose and what XSLT stylesheets they are called from.

Attribute set	Purpose	Called from
__index__label	Formats the title of the index. Also in **index-attr_axf.xsl**.	index.xsl, index_xep.xsl
__index__letter-group	Formats the letters the index entries are grouped under. Also in **index-attr_axf.xsl**.	index.xsl
__index__page__link	Formats the page numbers for each index entry	index.xsl, index_xep.xsl
index-indents	Controls the indents of entries and sub-entries	index.xsl, index_xep.xsl
index.entry	Formats the main index entries.	index.xsl
index.entry__content	Formats the index sub-entries	index.xsl, index_xep.xsl
index.see.label	Formats the text inserted by the <index-see> element and *Index See String* variable.	index.xsl
index.see-also- entry__content	Formats the text inserted by the <index-see-also> element. Not used.	none
index.see-also.label	Formats the text inserted by the <index-see-also> element and *Index See Also String* variable.	index.xsl
index.term	Formats the text of index entries.	index_xep.xsl, index_fop.xsl

Layout masters attribute sets

The attribute sets used to format page masters are found in **layout-masters-attr.xsl** and **layout-masters-attr_xep.xsl**. Here is a list of those attributes along with their purpose and what XSLT stylesheets they are called from.

Attribute set	Purpose	Called from
simple-page-master	Formats the height and width of page masters.	layout-masters.xsl
region-body	Establishes common formatting for body regions. Called by attribute sets such as *region-body.odd* and *region-body.even*.	layout-masters.xsl
region-body__index	Establishes common formatting for index page body regions. Called by attribute sets such as *region-body__index.odd* and *region-body__index.even*. Also in **layout-masters-attr_xep.xsl**.	layout-masters.xsl
region-before	Formats the header region of page masters.	layout-masters.xsl
region-after	Formats the footer region of page masters.	layout-masters.xsl
region-body.odd	Formats the body region of the odd body page master.	layout-masters.xsl
region-body.even	Formats the body region of even body page master.	layout-masters.xsl
region-body__frontmatter. odd	Formats the body region of the odd frontmatter page master.	layout-masters.xsl
region-body__frontmatter. even	Formats the body region of the even frontmatter page master.	layout-masters.xsl
region-body__backcover.odd	Formats the body region of the odd back cover page master.	layout-masters.xsl

Attribute set	Purpose	Called from
region-body__backcover.even	Formats the body region of the even back cover page master.	layout-masters.xsl
region-body__index.odd	Formats the body region of the odd index page master.	layout-masters.xsl
region-body__index.even	Formats the body region of the even index page master.	layout-masters.xsl

Table 10: Layout master attribute sets

Link attribute sets

The attribute sets that you use to format the appearance of links are found in **links-attr.xsl**. Here is a list of those attributes along with their purpose and what XSLT stylesheets they are called from.

> **Note** As delivered in the Open Toolkit, **links-attr.xsl** has two instances of the *linkpool* attribute set. Neither set has any attributes in it, and the duplication appears to be a simple oversight. If you copy this file into your PDF plugin, you can delete one instance.

Attribute set	Purpose	Called from
link	Formats text in the `<link>` element. Text of link (excluding page reference) can be overridden by the *linktext* attribute set.	links.xsl
link__content	Formats text of link (excluding page reference) and text in `<linkinfo>`. Can be overridden by more specific attribute sets.	links.xsl
link__shortdesc	Formats text of short description associated with topic referenced by link. Does not work out of the box; specifications for the *shortdesc* attribute set in **commons-attr.xsl** override this attribute set.	links.xsl

Attribute set	Purpose	Called from
linkinfo	Formats text in the `<linkinfo>` element. Repeated multiple times in **links-attr.xsl**. Only one instance needs to be edited. The others can be deleted.	links.xsl
linklist	Formats text in the `<linklist>` element. Can be overridden by attribute sets specific to child elements of `<linklist>`.	links.xsl
linkpool	Formats text in the `<linkpool>` element. Can be overridden by attribute sets specific to child elements of `<linkpool>`. Repeated multiple times in **links-attr.xsl**. Only one instance needs to be edited. The others can be deleted.	links.xsl
linktext	Formats text in the `<linktext>` element.	links.xsl
related-links	Formats the text in the `<related-links>` elements, including the "Related links" heading. Can be overridden by attribute sets specific to child elements of `<related-links>`.	links.xsl
related-links.ol	Formats the contents of the `<related-links>` element when the links are presented as an ordered list.	links.xsl
related-links.ol.li	Formats a link within a `<related-links>` element when the links are presented as an ordered list.	links.xsl
related-links.ol.li__body	Formats the text of a link—excluding the bullet—within a `<related-links>` element when the links are presented as an ordered list.	links.xsl
related-links.ol.li__content	Formats all content in a link except the label within a `<related-links>` element when the links are presented as an ordered list.	links.xsl

Attribute set	Purpose	Called from
related-links.ol.li__label	Formats the entire label block of a link within a `<related-links>` element when the links are presented as an ordered list. Generally used to control keep with next and indent properties, also to create space between label and text.	links.xsl
related-links.ol.li__label __content	Formats a link's bullet within a `<related-links>` element when the links are presented as an ordered list.	links.xsl
related-links.title	Formats the title in the `<related-links>` element	links.xsl
related-links.ul	Formats the contents of the `<related-links>` element when the links are presented as an unordered list.	links.xsl
related-links.ul.li	Formats a link within a `<related-links>` element when the links are presented as an unordered list.	links.xsl
related-links.ul.li__body	Formats the text of a link—excluding the bullet—within a `<related-links>` element when the links are presented as an unordered list.	links.xsl
related-links.ul.li__content	Formats all content in a link except the label within a `<related-links>` element when the links are presented as an unordered list.	links.xsl
related-links.ul.li__label	Formats the entire label block of a link within a `<related-links>` element when the links are presented as an unordered list. Generally used to control keep with next and indent properties, also to create space between label and text.	links.xsl

Attribute set	Purpose	Called from
related-links.ul.li__label __content	Formats a link's bullet within a `<related-links>` element when the links are presented as an unordered list.	links.xsl
related-links__content	Formats the text in the `<related-links>` elements, excluding the "Related links" heading. Can be overridden by attribute sets specific to child elements of `<related-links>`.	links.xsl
xref	Formats text in `<xref>` element. By default, also formats links in the mini-TOC.	commons.xsl, links.xsl

List attribute sets

The attribute sets that you use to format the appearance of lists are found in **lists-attr.xsl**. Here is a list of those attributes along with their purpose and what XSLT stylesheets they are called from.

> **Note** In previous versions of the Open Toolkit, this attribute set file also included some attribute sets pertaining to list-type elements found in tasks. All attribute sets pertaining to task elements are now in task-elements-attr.xsl.

Attribute set	Purpose	Called from
li.itemgroup	Formats elements within an `<itemgroup>` element. `<itemgroup>` is reserved for specialization.	lists.xsl
linklist.title	Formats the `<title>` element within a `<linklist>` element	lists.xsl
ol	Formats the entire `<ol>` block; generally used to control spacing above and below block	ut-domain.xsl, lists.xsl
ol.li	Formats an entire `<li>` block within an `<ol>`; generally used to control spacing above and below block	ut-domain.xsl, lists.xsl

Attribute set	Purpose	Called from
ol.li__body	Formats the text of an `<li>`—excluding the numbering—within an `<ol>`; overridden by attributes in *ol.li__content*	ut-domain.xsl, lists.xsl
ol.li__content	Formats all content within an ordered list (`<ol>`) `<li>` element except labels; can be overridden by properties of lower-level element such as `<p>`	ut-domain.xsl, lists.xsl
ol.li__label	Formats the entire label block in an ordered list item; generally used to control keep with next and indent properties, also to create space between label and text	ut-domain.xsl, lists.xsl
ol.li__label__content	Formats the ordered list item's number/letter	ut-domain.xsl, lists.xsl
sl	Formats the entire `<sl>` block; generally used to control spacing above and below block	lists.xsl
sl.sli	Formats an entire `<sli>` block within an `<sl>`; generally used to control spacing above and below block	lists.xsl
sl.sli__body	Formats the text of an `<sli>`—excluding the label—within an `<sl>`; overridden by attributes in *sl.sli__content*	lists.xsl
sl.sli__content	Formats all content within a simple list (`<sl>`) `<sli>` element except labels; can be overridden by properties of lower-level element such as `<p>`	lists.xsl
sl.sli__label	Formats the entire label block in a simple list item; generally used to control keep with next and indent properties, also to create space between label and text	lists.xsl

Attribute set	Purpose	Called from
sl.sli__label__content	Formats the simple list item's number/letter/bullet	lists.xsl
ul	Formats the entire `<ul>` block; generally used to control spacing above and below block	commons.xsl, lists.xsl
ul.li	Formats an entire `<li>` block within an `<ul>`; generally used to control spacing above and below block	commons.xsl, lists.xsl
ul.li__body	Formats the text of an `<li>`—excluding the bullet—within an `<ul>`; overridden by attributes in *ul.li__content*	commons.xsl, lists.xsl
ul.li__content	Formats all content within an unordered list (`<ul>`) `<li>` element except labels; can be overridden by properties of lower-level element such as `<p>`	commons.xsl, lists.xsl
ul.li__label	Formats the entire label block in an unordered list item; generally used to control keep with next and indent properties, also to create space between label and text	commons.xsl, lists.xsl
ul.li__label__content	Formats the unordered list item's bullet	commons.xsl, lists.xsl

Lot-Lof attribute sets

The attribute sets that you use to format the appearance of List of Tables and Lists of Figures are found in **lot-lof-attr.xsl**. Here is a list of those attributes along with their purpose and what XSLT stylesheets they are called from.

Attribute set	Purpose	Called from
__lotf__heading	Formats the heading of the List of Tables or List of Figures.	lot-lof.xsl
__lotf__indent	Formats the level indents in the List of Tables or List of Figures.	lot-lof.xsl
__lotf__content	Formats the entries in the List of Tables or List of Figures.	lot-lof.xsl
__lotf__leader	Formats leaders in the List of Tables or List of Figures.	lot-lof.xsl
__lotf__title	Formats the entries in the List of Tables or List of Figures.	lot-lof.xsl
__lotf__page-number	Formats page numbers in the List of Tables or List of Figures.	lot-lof.xsl

Map attribute sets

The attribute sets that you use to format the appearance of relationship tables are found in **map-elements-attr.xsl**. Here is a list of those attributes along with their purpose and what XSLT stylesheets they are called from.

> **Note** In previous versions of the DITA Open Toolkit, these attribute sets were found in **tables-attr.xsl**.

Attribute set	Purpose	Called from
relcell	Formats the `<relcell>` element of `<reltable>`.	tables.xsl , map-elements.xsl
relcolspec	Formats the `<relcolspec>` element of `<reltable>`.	tables.xsl , map-elements.xsl
relheader	Formats the `<relheader>` element of `<reltable>`.	tables.xsl, map-elements.xsl
relrow	Formats the `<relrow>` element of `<reltable>`.	tables.xsl, map-elements.xsl
reltable	Formats the `<reltable>` element.	tables.xsl , map-elements.xsl
reltable__title	Formats the `<title>` element within `<reltable>`.	tables.xsl , map-elements.xsl

Reference attribute sets

The attribute sets that you use to format the appearance of elements specific to reference topics are found in **reference-elements-attr.xsl**. Here is a list of those attributes along with their purpose and what XSLT stylesheets they are called from.

> **Note** In previous versions of the DITA Open Toolkit, these attribute sets were found in **tables-attr.xsl** and **commons-attr.xsl**.

Attribute set	Purpose	Called from
properties	Formats the `<properties>` table.	reference-elements.xsl
properties__body	Formats the `<properties>` table. Also in **tables-attr_axf.xsl**	reference-elements.xsl
property	Formats content in the `<property>` element of `<properties>`.	reference-elements.xsl
property.entry	Formats the table cells of the `<property>` element of `<properties>`.	reference-elements.xsl
property.entry__content	Formats content in the body cells of a `<properties>` table when those cells are not in the column marked as the key column.	reference-elements.xsl
property.entry__keycol-content	Formats content in the body cells of a `<properties>` table when those cells are in the column marked as the key column.	reference-elements.xsl
prophead	Formats content in the `<prophead>` element of `<properties>`. Also in **tables-attr_axf.xsl**	reference-elements.xsl
prophead.entry	Formats the table cells of the `<prophead>` element of `<properties>`.	reference-elements.xsl

Attribute set	Purpose	Called from
prophead.entry__content	Formats content in the `<prophead>` cells of a `<properties>` table when those cells are not in the column marked as the key column.	reference-elements.xsl
prophead.entry__keycol-content	Formats content in the `<prophead>` cells of a `<properties>` table when those cells are in the column marked as the key column.	reference-elements.xsl
prophead__row	Formats the `<prophead>` row of the `<properties>` table.	reference-elements.xsl
refbody	Formats content in the `<refbody>` element.	reference-elements.xsl
reference	Formats content in the `<reference>` element.	reference-elements.xsl
refsyn	Formats content in the `<refsyn>` element.	reference-elements.xsl

Static content attribute sets

The attribute sets that you use to format the appearance of static content (largely in headers and footers) are found in **static-content-attr.xsl**. Here is a list of those attributes along with their purpose and what XSLT stylesheets they are called from.

Attribute set	Purpose	Called from
__body__even__footer	Formats the footer on even body pages	static-content.xsl
__body__even__footer__heading	Formats the running heading in the footer on even body pages	static-content.xsl
__body__even__footer__pagenum	Formats the page number in the footer on even body pages	static-content.xsl

Attribute set	Purpose	Called from
__body__even__header	Formats the header on even body pages	static-content.xsl
__body__even__header__ heading	Formats the running heading in the header on even body pages	static-content.xsl
__body__even__header__ pagenum	Formats the page number in the header on even body pages	static-content.xsl
__body__first__footer	Formats the footer on first body pages	static-content.xsl
__body__first__footer__ heading	Formats the running header in the footer on first body pages	static-content.xsl
__body__first__footer__ pagenum	Formats the page number in the footer on first body pages	static-content.xsl
__body__first__header	Formats the header on first body pages	static-content.xsl
__body__first__header__ heading	Formats the running heading in the header on first body pages	static-content.xsl
__body__first__header__ pagenum	Formats the page number in the header on first body pages	static-content.xsl
__body__footnote__ separator	Formats the separator that separates body text from footnotes	static-content.xsl
__body__last__footer	Formats the footer on last body pages	static-content.xsl
__body__last__header	Formats the header on last body pages	static-content.xsl
__body__odd__footer	Formats the footer on odd body pages	static-content.xsl
__body__odd__footer__ heading	Formats the running heading in the footer on odd body pages	static-content.xsl
__body__odd__footer__ pagenum	Formats the page number in the footer on odd body pages	static-content.xsl

Attribute set	Purpose	Called from
__body__odd__header	Formats the header on odd body pages	static-content.xsl
__body__odd__header__heading	Formats the running heading in the header on odd body pages	static-content.xsl
__body__odd__header__ pagenum	Formats the page number in the header on odd body pages	static-content.xsl
__chapter__frontmatter__ name __container	Format the chapter label (e.g. "Chapter")	commons.xsl
__chapter__frontmatter__ number __container	Formats the chapter number	commons.xsl
__glossary__even__footer	Formats the footer on even glossary pages	static-content.xsl
__glossary__even__footer__ pagenum	Formats the page number in the footer on even glossary pages	static-content.xsl
__glossary__even__header	Formats the header on even glossary pages	static-content.xsl
__glossary__even__header__ pagenum	Formats the page number in the header on even glossary pages	static-content.xsl
__glossary__odd__footer	Formats the footer on odd glossary pages	static-content.xsl
__glossary__odd__footer__ pagenum	Formats the page number in the footer on odd glossary pages	static-content.xsl
__glossary__odd__header	Formats the header on odd glossary pages	static-content.xsl
__glossary__odd__header__ pagenum	Formats the page number in the header on odd glossary pages	static-content.xsl
__index__even__footer	Formats the footer on even Index pages	static-content.xsl
__index__even__footer__ pagenum	Formats the page number in the footer on even Index pages	static-content.xsl

Attribute set	Purpose	Called from
__index__even__header	Formats the header on even Index pages	static-content.xsl
__index__even__header__pagenum	Formats the page number in the header on even Index pages	static-content.xsl
__index__odd__footer	Formats the footer on odd Index pages	static-content.xsl
__index__odd__footer__pagenum	Formats the page number in the footer on odd Index pages	static-content.xsl
__index__odd__header	Formats the header on odd Index pages	static-content.xsl
__index__odd__header__pagenum	Formats the page number in the header on odd Index pages	static-content.xsl
__toc__even__footer	Formats the footer on even TOC pages	static-content.xsl
__toc__even__footer__pagenum	Formats the page number in the footer on even TOC pages	static-content.xsl
__toc__even__header	Formats the header on even TOC pages	static-content.xsl
__toc__even__header__pagenum	Formats the page number in the header on even TOC pages	static-content.xsl
__toc__odd__footer	Formats the footer on odd TOC pages	static-content.xsl
__toc__odd__footer__pagenum	Formats the page number in the footer on odd TOC pages	static-content.xsl
__toc__odd__header	Formats the header on odd TOC pages	static-content.xsl
__toc__odd__header__pagenum	Formats the page number in the header on odd TOC pages	static-content.xsl

Attribute set	Purpose	Called from
even__footer	Establishes common formatting for even page footers that is called by other attribute sets such as *__body__even__footer* or *__index__even__footer.*	static-content.xsl
even__header	Establishes common formatting for even page headers that is called by other attribute sets such as *__body__even__header* or *__index__even__header.*	static-content.xsl
odd__footer	Establishes common formatting for odd page footers that is called by other attribute sets such as *__body__odd__footer* or *__index__odd__footer.*	static-content.xsl
odd__header	Establishes common formatting for even page headers that is called by other attribute sets such as *__body__odd__header* or *__index__odd__header.*	static-content.xsl
pagenum	Established common formatting for page numbers that is called by other attribute sets such as *__body__odd__footer__pagenum.*	static-content.xsl

Task element attribute sets

The attribute sets you use to format the appearance of task elements are found in **task-elements-attr.xsl**. Here is a list of those attributes along with their purpose and what XSLT stylesheets they are called from.

> **Note** This attribute set file consists of task-related attribute sets that were previously found in **lists-attr.xsl**, **tables-attr.xsl**, and **commons-attr.xsl**.

Attribute set	Purpose	Called from
chhead	Formats the `<chhead>` element of `<choicetable>`. Also in **tables-attr_axf.xsl**.	task-elements.xsl
chhead.chdeschd	Formats content in the `<chdeschd>` element within the `<chhead>` element of `<choicetable>`	task-elements.xsl
chhead.chdeschd__content	Formats content in the `<chdeschd>` cells of a `<choicetable>`.	task-elements.xsl
chhead.choptionhd	Formats content in the `<choptionhd>` element within the `<chhead>` element of `<choicetable>`	task-elements.xsl
chhead.choptionhd__content	Formats content in the `<choptionhd>` cells of a `<choicetable>`.	task-elements.xsl
chhead__row	Formats the `<chhead>` row of a `<choicetable>`.	task-elements.xsl
choices	Formats the entire `<choices>` block; generally used to control spacing above and below block	task-elements.xsl
choices.choice	Formats the entire `<choice>` block; generally used to control spacing above and below block	task-elements.xsl

Attribute set	Purpose	Called from
choices.choice__body	Formats the text of a `<choice>`, excluding the numbering; overridden by attributes in *choices.choice__content*	task-elements.xsl
choices.choice__content	Formats all content within a `<choice>` element except labels; can be overridden by properties of lower-level element such as `<p>`	task-elements.xsl
choices.choice__label	Formats the entire label block in a choice; generally used to control keep with next and indent properties, also to create space between label and text	task-elements.xsl
choices.choice__label__ content	Formats the unordered choice number/letter/bullet	task-elements.xsl
choicetable	Formats the `<choicetable>` element.	task-elements.xsl
choicetable__body	Formats the `<choicetable>` element. Also in **tables-attr_axf.xsl**	task-elements.xsl
chrow	Formats the `<chrow>` element of `<choicetable>`	task-elements.xsl
chrow.chdesc	Formats a `<chdesc>` cell of a `<choicetable>`.	task-elements.xsl
chrow.chdesc__content	Formats content in the `<chdesc>` cells of a `<choicetable>` when those cells are not in the column marked as the key column. Also in **tables-attr_axf.xsl**	task-elements.xsl
chrow.chdesc__keycol-content	Formats content in the `<chdesc>` cells of a `<choicetable>` when those cells are in the column marked as the key column.	task-elements.xsl

Attribute set	Purpose	Called from
chrow.choption	Formats a `<choption>` cell of a `<choicetable>`.	task-elements.xsl
chrow.choption__content	Formats content in the `<choption>` cells of a `<choicetable>` when those cells are not in the column marked as the key column.	task-elements.xsl
chrow.choption__keycol-content	Formats content in the `<choption>` cells of a `<choicetable>` when those cells are in the column marked as the key column.	task-elements.xsl
cmd	Formats text in the `<cmd>` element.	task-elements.xsl
context	Formats text in the `<context>` element; can be overridden by properties of lower-level element such as `<p>`. Uses attributes in the *section* attribute set in addition to any entered here.	task-elements.xsl
context__content	Formats content within the `<context>` element.	task-elements.xsl
info	Formats text in the `<info>` element; can be overridden by properties of lower-level element such as `<p>`	task-elements.xsl
postreq	Formats text in the `<postreq>` element; can be overridden by properties of lower-level element such as `<p>`. Uses attributes in the *section* attribute set in addition to any entered here.	task-elements.xsl
postreq__content	Formats content within the `<postreq>` element.	task-elements.xsl

Attribute set	Purpose	Called from
prereq	Formats text in the `<prereq>` element; can be overridden by properties of lower-level element such as `<p>`. Uses attributes in the *section* attribute set in addition to any entered here.	task-elements.xsl
prereq__content	Formats content within the `<prereq>` element.	task-elements.xsl
result	Formats text in the `<result>` element; can be overridden by properties of lower-level element such as `<p>`. Uses attributes in the *section* attribute set in addition to any entered here.	task-elements.xsl
result__content	Formats content within the `<result>` element.	task-elements.xsl
stepresult	Formats text in the `<stepresult>` element; can be overridden by properties of lower-level element such as `<p>`	task-elements.xsl
steps	Formats the entire ordered `<steps>` block; generally used to control spacing above and below block	task-elements.xsl
steps.step	Formats the entire ordered `<step>` block; generally used to control spacing above and below block	task-elements.xsl
steps.step__body	Formats the text of a `<step>`, excluding the numbering; overridden by attributes in *steps.step__content*	task-elements.xsl

Attribute set	Purpose	Called from
steps.step__content	Formats all content within an ordered `<step>` element except labels; can be overridden by properties of lower-level element such as `<cmd>`	task-elements.xsl
steps.step__label	Formats the entire label block in an ordered step; generally used to control keep with next and indent properties, also to create space between label and text	task-elements.xsl
steps.step__label__content	Formats the ordered step number/letter/bullet	task-elements.xsl
steps-unordered	Formats the entire `<steps-unordered>` block; generally used to control spacing above and below block	task-elements.xsl
steps-unordered.step	Formats the entire unordered `<step>` block; generally used to control spacing above and below block	task-elements.xsl
steps-unordered.step__body	Formats the text of a `<step>`, excluding the numbering; overridden by attributes in *steps-unordered.step__content*	task-elements.xsl
steps-unordered.step__ content	Formats all content within an unordered `<step>` element except labels; can be overridden by properties of lower-level element such as `<cmd>`	task-elements.xsl
steps-unordered.step__label	Formats the entire label block in an unordered step; generally used to control keep with next and indent properties, also to create space between label and text	task-elements.xsl
steps-unordered.step__label __content	Formats the unordered step number/letter/bullet	task-elements.xsl

Attribute set	Purpose	Called from
stepsection	Formats the entire `<stepsection>` block; generally used to control spacing above and below block	task-elements.xsl
stepsection__label	Formats the entire label block in a stepsection item; generally used to control keep with next and indent properties, also to create space between label and text	task-elements.xsl
stepsection__label__content	Formats all content within a `<substep>` item except labels	task-elements.xsl
stepsection__body	Formats the text of a `<stepsection>` item, excluding the label; overridden by attributes in *stepsection__content*	task-elements.xsl
stepsection__content	Formats all content within a `<stepsection>` item except labels	task-elements.xsl
stepxmp	Formats text in the `<stepxmp>` element; can be overridden by properties of lower-level element such as `<p>`	task-elements.xsl
substeps	Formats the entire `<substeps>` block; generally used to control spacing above and below block	task-elements.xsl
substeps.substep	Formats the entire `<substep>` block; generally used to control spacing above and below block	task-elements.xsl
substeps.substep__body	Formats the text of a `<substep>`, excluding the numbering; overridden by attributes in *substeps.substep__content*	task-elements.xsl

Attribute set	Purpose	Called from
substeps.substep__content	Formats all content within a `<substep>` element except labels; can be overridden by properties of lower-level element such as `<cmd>`	task-elements.xsl
substeps.substep__label	Formats the entire label block in a substep; generally used to control keep with next and indent properties, also to create space between label and text	task-elements.xsl
substeps.substep__label__content	Formats the substep number/letter/bullet	task-elements.xsl
task	Formats the entire `<task>` block	task-elements.xsl
taskbody	Formats text in the `<taskbody>` element; can be overridden by properties of lower-level element such as `<context>`, `<cmd>` or `<p>`. Uses attributes in the *body* attribute set in addition to any entered here.	task-elements.xsl
task.example	Formats the title of the `<example>` element.	task-elements.xsl
task.example__content	Formats content within the `<example>` element.	task-elements.xsl
tutorialinfo	Formats text in the `<tutorialinfo>` element; can be overridden by properties of lower-level element such as `<p>`	task-elements.xsl

Table attribute sets

The attribute sets that you use to format the appearance of tables are found in **tables-attr.xsl** and **tables-attr_fop.xsl**. Here is a list of those attributes along with their purpose and what XSLT stylesheets they are called from.

> **Note** Although <dl> is rendered as a list in most XML editors, the PDF plugin renders it as a table by default; therefore, its related attribute sets are included in **tables-attr.xsl**.

Attribute set	Purpose	Called from
__tableframe__bottom	Formats the rule along the bottom edge of table cells.	attr-set-reflection.xsl, tables.xsl
__tableframe__left	Formats the rule along the left edge of table cells.	tables.xsl
__tableframe__none	Formats the rules on table cells when no rule is specified.	tables.xsl
__tableframe__right	Formats the rule along the right edge of table cells.	attr-set-reflection.xsl, tables.xsl
__tableframe__top	Formats the rule along the top edge of table cells.	attr-set-reflection.xsl, tables.xsl
dl	Formats the <dl> element. Also in **tables-attr_axf.xsl** and **tables-attr_fop.xsl**.	tables.xsl
dl.dlhead	Formats a <dlhead> element in a <dl> table. Also in **tables-attr_axf.xsl**	tables.xsl
dl.dlhead__row	Formats the <dlhead> row of a <dl> table.	tables.xsl
dl__body	Has an alternative in tables-attr_axf.xsl. Also in **tables-attr_axf.xsl**	tables.xsl

Attribute set	Purpose	Called from
dlentry	Formats a `<dlentry>` element in a `<dl>` table.	tables.xsl
dlentry.dd	Formats a `<dd>` cell in a `<dl>` table.	tables.xsl
dlentry.dd__content	Formats the contents of a `<dd>` cell in a `<dl>` table. Also in **tables-attr_axf.xsl**	tables.xsl
dlentry.dt	Formats a `<dt>` cell in a `<dl>` table.	tables.xsl
dlentry.dt__content	Formats the contents of a `<dt>` cell in a `<dl>` table. Also in **tables-attr_axf.xsl**	tables.xsl, tables_fop.xsl
dlhead.ddhd__cell	Formats a `<ddhd>` cell in a `<dl>` table.	tables.xsl
dlhead.ddhd__content	Formats the contents of a `<ddhd>` cell in a `<dl>` table.	tables.xsl
dlhead.dthd__cell	Formats a `<dthd>` cell in a `<dl>` table.	tables.xsl
dlhead.dthd__content	Formats the contents of a `<dthd>` cell in a `<dl>` table.	tables.xsl
simpletable	Formats the `<simpletable>` element. Also in **tables-attr_axf.xsl** and **tables-attr_fop.xsl**.	tables.xsl
simpletable__body	Formats each `<strow>` in `<simpletable>`. Also in **tables-attr_axf.xsl**	tables.xsl
sthead	Formats the `<sthead>` element of `<simpletable>`. Also in **tables-attr_axf.xsl**	tables.xsl
sthead.stentry	Formats each `<stentry>` element in the `<sthead>` element of `<simpletable>`.	tables.xsl

Attribute set	Purpose	Called from
sthead.stentry__content	Formats content in the `<stentry>` cells of the `<sthead>` element of `<simpletable>` when those cells are not in the column marked as the key column.	tables.xsl
sthead.stentry__keycol- content	Formats content in the `<stentry>` cells of the `<sthead>` element of`<simpletable>` when those cells are in the column marked as the key column.	tables.xsl
sthead__row	Formats the table row in the `<sthead>` element of `<simpletable>`.	tables.xsl
strow	Formats the `<strow>` element of `<simpletable>`.	tables.xsl
strow.stentry	Formats the table cells of the `<strow>` element of `<simpletable>`.	tables.xsl
strow.stentry__content	Formats content in the `<stentry>` cells of the `<strow>` element of `<simpletable>` when those cells are not in the column marked as the key column.	tables.xsl
strow.stentry__keycol- content	Formats content in the `<stentry>` cells of the `<strow>` element of `<simpletable>` when those cells are in the column marked as the key column.	tables.xsl
table	Formats the `<table>` element.	tables.xsl
table.tgroup	Formats the `<tgroup>` element of `<table>`.	tables.xsl
table.title	Formats the `<title>` element within `<table>`.	tables.xsl

Attribute set	Purpose	Called from
table__tableframe__all	Formats the frame on all side of tables.	attr-set-reflection.xsl
table__tableframe__ bottom	Formats the frame on the bottom side of tables.	attr-set-reflection.xsl
table__tableframe__left	Formats the frame on the left side of tables.	tables-attr.xsl
table__tableframe__right	Formats the frame on the right side of tables.	tables-attr.xsl
table__tableframe__sides	Formats the frame on the left and right sides of tables.	attr-set-reflection.xsl
table__tableframe__top	Formats the frame on the top side of tables.	attr-set-reflection.xsl
table__tableframe__ topbot	Formats the frame on the top and bottom sides of tables.	attr-set-reflection.xsl
tbody.row	Formats the table rows within the `<tbody>` element of `<table>`.	attr-set-reflection.xsl, tables.xsl
tbody.row.entry	Formats the table cells within the `<tbody>` element of `<table>`.	attr-set-reflection.xsl, tables.xsl
tbody.row.entry__ content	Formats the content of each table cell within the `<tbody>` element of `<table>`.	attr-set-reflection.xsl, tables.xsl
tbody.row.entry__firstcol	Formats the content of each table cell within the first column of tables.	tables.xsl
tfoot.row	Formats table footers.	attr-set-reflection.xsl
tfoot.row.entry	Formats cells in table footers.	attr-set-reflection.xsl
tfoot.row.entry__content	Formats the text in table footer cells.	attr-set-reflection.xsl

Attribute set	Purpose	Called from
tgroup.tbody	Formats the `<tbody>` element within `<table>`. Also in **tables-attr_axf.xsl**	tables.xsl
tgroup.tfoot	Formats the `<tfoot>` element within `<table>`. Also in **tables-attr_axf.xsl**	none
tgroup.thead	Formats the `<thead>` element within `<table>`. Also in **tables-attr_axf.xsl**	tables.xsl
thead.row	Formats the table rows within the `<thead>` element of `<table>`.	attr-set-reflection.xsl, tables.xsl
thead.row.entry	Formats the table cells within the `<thead>` element of `<table>`.	attr-set-reflection.xsl, tables.xsl
thead.row.entry__content	Formats the content of each table cell within the `<thead>` element of `<table>`.	attr-set-reflection.xsl, tables.xsl
thead__tableframe__ bottom	Formats the rule along the bottom edits of table header cells.	attr-set-reflection.xsl

Table of Contents attribute sets

The attribute sets that you use to format the appearance of the TOC are found in **toc-attr.xsl**, **toc-attr_axf.xsl**, and **toc-attr_fop.xsl**. (There is no XEP-specific TOC attributes file.) Here is a list of those attributes along with their purpose and what XSLT stylesheets they are called from.

Attribute set	Purpose	Called from
__toc__appendix__ content	Formats TOC entries generated from `<appendix>` elements in a bookmap. Uses *__toc__topic__content* attribute set for font size, etc. based on hierarchy.	toc.xsl
__toc__chapter__ content	Formats TOC entries generated from `<chapter>` elements in a bookmap. Uses *__toc__topic__content* attribute set for font size, etc. based on hierarchy.	toc.xsl
__toc__content	Renamed to *__toc__topic__content*	none
__toc__header	Formats the header of the TOC.	toc.xsl
__toc__indent	Controls the amount of indent between each level of the TOC	toc.xsl
__toc__leader	Formats the leader between the topic names and the page numbers in the TOC	toc.xsl
__toc__link	Formats linked page numbers in the TOC	toc.xsl, lot-lof.xsl
__toc__mini	Formats the contents of each cell in the mini-TOC table	commons.xsl
__toc__mini__body	Not used	none
__toc__mini__header	Formats the heading of each topic in the mini-TOC	commons.xsl
__toc__mini__label	Not used	none

Attribute set	Purpose	Called from
__toc__mini__list	Formats the list of topics in the mini-TOC	commons.xsl
__toc__mini__summary	Formats the shortdesc of subtitle of each topic in the mini-TOC	commons.xsl
__toc__mini__table	Formats the mini-TOC table. Also in **commons-attr_fop.xsl**.	commons.xsl
__toc__mini__table__body	Formats the body of the mini-TOC table. Also in **toc-attr_axf.xsl**.	commons.xsl
__toc__mini__table__column_1	Formats the first column of the mini-TOC table	commons.xsl
__toc__mini__table__column_2	Formats the second column of the mini-TOC table	commons.xsl
__toc__notices__content	Formats TOC entries generated from items in the `<notices>` section of a bookmap. Uses *__toc__topic__content* attribute set for font size, etc. based on hierarchy.	toc.xsl
__toc__part__content	Formats TOC entries generated from `<part>` elements in a bookmap. Uses *__toc__topic__content* attribute set for font size, etc. based on hierarchy.	toc.xsl
__toc__preface__content	Formats TOC entries generated from items in the `<preface>` section of a bookmap. Uses *__toc__topic__content* attribute set for font size, etc. based on hierarchy.	toc.xsl
__toc__title	Formats the TOC title	toc.xsl
__toc__topic__content	Formats entries in the TOC that are not formatted by a more specific set, like *__toc__chapter__content*	toc.xsl

Attribute set	Purpose	Called from
__toc__topic__content__glossary	Formats TOC entries generated from `<glossarylist>` elements in a bookmap.	toc.xsl
__toc__indent__glossary	Controls the amount of indent for TOC entries generated from `<glossarylist>` elements in a bookmap.	toc.xsl
__toc__indent__booklist	Establishes common indent levels for booklist files such as the glossary, list of tables, and list of figures. Called by other attribute sets such as _toc__indent__lot.	toc.xsl
__toc__indent__lof	Formats the indent levels for the list of figures.	toc.xsl
__toc__indent__lot	Formats the indent levels for the list of tables.	toc.xsl
__toc__item__right	Not used.	none
__toc__page-number	Formats page numbers in the TOC.	toc.xsl
__toc__topic__content__booklist	Establishes common formatting for the titles of booklist files such as the glossary, list of tables, and list of figures. Called by other attribute sets such as _toc__topic__content__lot.	toc.xsl
__toc__topic__content__lof	Formats the TOC entry text for the list of figures.	toc.xsl
__toc__topic__content__lot	Formats the TOC entry text for the list of tables.	toc.xsl

Basic settings variables

The variables that you use to format page masters are found in **basic-settings.xsl**. Although these are variables and not attribute sets, the **basic-settings.xsl** file is included with the attribute set files. Here is a list of those variables along with their purpose and what XSLT stylesheets they are called from.

Attribute set	Purpose	Called from
default-font-size	Defines the default size of type for text blocks.	commons-attr.xsl, toc-attr.xsl
default-line-height	Defines the default line height (or leading) for text blocks. By default is 12pt but usually better expressed as a percentage, such as 120%.	commons-attr.xsl
generate-front-cover	Indicates whether or not to generate a front cover.	front-matter.xsl
generate-back-cover	Indicates whether or not to generate a back cover.	layout-masters.xsl, front-matter.xsl
generate-toc	Indicates whether or not to generate a table of contents.	bookmarks.xsl, commons.xsl, glossary.xsl, toc.xsl
mirror-page-margins	Defines whether (true) or not (false) to automatically create mirrored page margins for left and right verso pages. The default is false, which requires manually creating mirrored regions for left and right pages.	layout-masters.xsl, static-content.xsl
page-width	Defines the total width of page masters in the PDF.	layout-masters-attr.xsl
page-height	Defines the total height of page masters in the PDF.	layout-masters-attr.xsl
page-margins	Defines the default margins of page masters in the PDF.	layout-masters.xsl, static-content.xsl
page-margin-inside	Defines the inside margin of page masters in the PDF. Lends itself to mirrored pages.	layout-masters-attr.xsl

Attribute set	Purpose	Called from
page-margin-outside	Defines the outside margin of page masters in the PDF. Lends itself to mirrored pages.	layout-masters-attr.xsl
page-margin-top	Defines the top margin of page masters in the PDF.	layout-masters-attr.xsl
page-margin-bottom	Defines the bottom margin of page masters in the PDF.	layout-masters-attr.xsl
side-col-width	Defines the width of the side column area.	commons-attr.xsl, glossary-attr.xsl, links-attr.xsl, toc-attr.xsl, task-elements-attr.xsl

Appendix C

Localization variables list

The following variables are included in the localization variables and strings files by default. (Variable categories are mine.) Variables whose descriptions end with (•) are found in the localization variables files. Variables whose descriptions end with (▶) are found in the strings files.

> **Note** There are also a number of variables in these files that are specific to online help formats. Those variables are not included in this list.

The names in the Variable column are the actual name of the variable as it appears in the localization variables files. Notice that variable names are not consistent. Some contain spaces, some are sentence case, some are mixed case, some are lower case, some contain special characters. This is another relic of the many hands that have touched the DITA Open Toolkit over the years.

Admonition variables

Variable	Description	Used in
Note	The prefix that precedes text inside `<note type="note">`. (▶)	commons.xsl
Notice	The prefix that precedes text inside `<note type="notice">`. (•)	commons.xsl
Tip	The prefix that precedes text inside `<note type="tip">`. (▶)	commons.xsl
Important	The prefix that precedes text inside `<note type="important">`. (▶)	commons.xsl
Fastpath	The prefix that precedes text inside `<note type="fastpath">`. (▶)	commons.xsl

Variable	Description	Used in
Remember	The prefix that precedes text inside `<note type="remember">`. (►)	commons.xsl
Attention	The prefix that precedes text inside `<note type="attention">`. (►)	commons.xsl
Caution	The prefix that precedes text inside `<note type="caution">`. (►)	commons.xsl
Danger	The prefix that precedes text inside `<note type="danger">`. (►)	commons.xsl
Restriction	The prefix that precedes text inside `<note type="restriction">`. (►)	commons.xsl
Warning	The prefix that precedes text inside `<note type="warning">`. (►)	commons.xsl
Trouble	The prefix that precedes text inside `<note type="warning">`. (►)	commons.xsl
note Note Image Path	Previously, the path and file name of the image that precedes text inside `<note type="note">`. No longer references an image. (•)	commons.xsl
attention Note Image Path	The path and file name of the image that precedes text inside `<note type="attention">`. (•)	commons.xsl
caution Note Image Path	The path and file name of the image that precedes text inside `<note type="caution">`. (•)	commons.xsl
danger Note Image Path	The path and file name of the image that precedes text inside `<note type="danger">`. (•)	commons.xsl
fastpath Note Image Path	Previously, the path and file name of the image that precedes text inside `<note type="fastpath">`. No longer references an image. (•)	commons.xsl

Variable	Description	Used in
important Note Image Path	Previously, the path and file name of the image that precedes text inside `<note type="important">`. No longer references an image. (•)	commons.xsl
remember Note Image Path	Previously, the path and file name of the image that precedes text inside `<note type="remember">`. No longer references an image. (•)	commons.xsl
restriction Note Image Path	Previously, the path and file name of the image that precedes text inside `<note type="restriction">`. No longer references an image. (•)	commons.xsl
warning Note Image path	The path and file name of the image that precedes text inside `<note type="warning">`. (•)	commons.xsl
notice Note Image path	Previously, the path and file name of the image that precedes text inside `<note type="notice">`. No longer references an image. (•)	commons.xsl
tip Note Image Path	Previously, the path and file name of the image that precedes text inside `<note type="tip">`. No longer references an image. (•)	commons.xsl
other Note Image Path	Previously, the path and file name of the image that precedes text inside `<note type="other">`. No longer references an image. (•)	commons.xsl
#note-separator	The character that follows the label. The default is a colon. (•)	commons.xsl

Cross-reference variables

Variable	Description	Used in
On the page	Text that precedes page number of a reference. (•)	links.xsl
Page	Text that precedes page number of a reference when reference has no title. (•)	links.xsl
This link	Text that precedes page number of a reference in chm/html, when reference has no title. (•)	Not used
Cross-Reference	Text that precedes a cross-reference when no cross-reference text is specified. (•)	Not used
Content-Reference	Text that precedes a content-reference (conref) when the target cannot be resolved. (•)	commons.xsl

Glossary variables

Variable	Description	Used in
Glossary odd footer	The footer that appears on odd-numbered glossary pages. (•)	static-content.xsl
Glossary even footer	The footer that appears on even-numbered glossary pages. (•)	static-content.xsl
Glossary odd header	The header that appears on odd-numbered glossary pages. (•)	static-content.xsl
Glossary even header	The header that appears on even-numbered glossary pages. (•)	static-content.xsl
Glossary	Text used as the heading for the glossary. (•)	bookmarks.xsl, glossary.xsl, toc.xsl

Header/footer variables

Variable	Description	Used in
Body odd footer	The footer that appears on odd-numbered body pages. (•)	static-content.xsl
Body even footer	The footer that appears on even-numbered body pages. (•)	static-content.xsl
Body first footer	The footer that appears on first body pages. (•)	static-content.xsl
Body first header	The header that appears on first body pages. (•)	static-content.xsl
Body odd header	The header that appears on odd body pages. (•)	static-content.xsl
Body even header	The header that appears on even body pages. (•)	static-content.xsl
Preface odd footer	The footer that appears on odd-numbered preface pages. (•)	static-content.xsl
Preface even footer	The footer that appears on even-numbered preface pages. (•)	static-content.xsl
Preface first footer	The footer that appears on first preface pages. (•)	static-content.xsl
Preface odd header	The header that appears on odd-numbered preface pages. (•)	static-content.xsl
Preface even header	The header that appears on even-numbered preface pages. (•)	static-content.xsl
Preface first header	The header that appears on first preface pages. (•)	static-content.xsl
Toc odd footer	The footer that appears on odd-numbered TOC pages. (•)	static-content.xsl
Toc even footer	The footer that appears on even-numbered TOC pages. (•)	static-content.xsl
Toc odd header	The header that appears on odd TOC pages. (•)	static-content.xsl

Variable	Description	Used in
Toc even header	The header that appears on even TOC pages. (•)	static-content.xsl
Index odd footer	The footer that appears on odd-numbered index pages. (•)	static-content.xsl
Index even footer	The footer that appears on even-numbered index pages. (•)	static-content.xsl
Index odd header	The header that appears on odd index pages. (•)	static-content.xsl
Index even header	The header that appears on even index pages. (•)	static-content.xsl

List variables

Variable	Description	Used in
Ordered List Number	The numbering format used in `<ol>`. (•)	links.xsl, lists.xsl, task-elements.xsl, ut-domain.xsl
Unordered List bullet	The character used as the bullet in `<ul>`. (•)	commons.xsl, lists.xsl, task-elements.xsl, links.xsl

Related links variables

Variable	Description	Used in
Related Links	Text used as the label in related links. (•)	links.xsl
Related concepts	Text used to label links to related concepts. (▶)	links.xsl
Related tasks	Text used to label links to related tasks. (▶)	links.xsl
Related references	Text used to label links to related references. Not used. (•)	none

Variable	Description	Used in
Related reference	Text used to label links to related references. (▶)	links.xsl
Related information	Text used to label links to related items that cannot be identified as concepts, tasks or references. (▶)	links.xsl
Untitled section	Text used for reference text in sections with no explicit content or title. Not used. (•)	none
List item	Text used for reference text in list items with no explicit content. (•)	links.xsl
Foot note	Text used for reference text in footnotes with no explicit content. (•)	links.xsl

Task variables

Variable	Description	Used in
Optional Step	The text that precedes steps with @importance = "optional". (•)	task-elements.xsl
Task Prereq	The label that precedes text inside `<prereq>`. (•)	task-elements.xsl
task_prereq	The label that precedes text inside `<prereq>`. (▶)	task-elements.xsl
Task Context	The label that precedes text inside `<context>`. (•)	task-elements.xsl
task_context	The label that precedes text inside `<context>`. (▶)	task-elements.xsl
Task Steps	The label that precedes text inside `<steps>`. (•)	task-elements.xsl
task_procedure	The label that precedes text inside `<steps>`. (▶)	task-elements.xsl
Task Result	The label that precedes text inside `<result>`. (•)	task-elements.xsl
task_results	The label that precedes text inside `<result>`. (▶)	task-elements.xsl

Variable	Description	Used in
Task Example	The label that precedes text inside `<example>`. (•)	task-elements.xsl
task_example	The label that precedes text inside `<example>`. (▶)	task-elements.xsl
Task Postreq	The label that precedes text inside `<postreq>`. (•)	task-elements.xsl
task_postreq	The label that precedes text inside `<postreq>`. (▶)	task-elements.xsl
#steps-unordered-label	The label that precedes text inside `<steps-unordered>`. (•)	task-elements.xsl
task_procedure_unordered	The label that precedes text inside `<steps-unordered>`. (▶)	task-elements.xsl
Option	The column heading for the first column in a `<choicetable>`. (▶)	tables.xsl
Description	The column heading for the second column in a `<choicetable>`. (▶)	tables.xsl

Title variables

Variable	Description	Used in
Preface title	Text that precedes titles in prefaces. (•)	commons.xsl
Notices title	Text that precedes titles in notices. (•)	commons.xsl
Figure.title	Text that precedes `<title>` in `<fig>`. (▶)	commons.xsl, links.xsl, lot-lof.xsl
Figure Number	Text that precedes the autonumbering of the `<title>` in `<fig>`. (▶)	links.xsl
Table.title	Text that precedes `<title>` in `<table>`. (▶)	links.xsl, lot-lof.xsl, tables.xsl

Variable	Description	Used in
Table Number	Text that precedes the autonumbering of the `<title>` in `<table>`. (►)	links.xsl
Navigation title	The prefix that appears before text in `<navtitle>`. (•)	commons.xsl
Search title	The prefix that appears before text in `<searchtitle>`. (•)	commons.xsl
Chapter with number	Text that precedes titles in chapters. (•)	commons.xsl
Appendix with number	Text that precedes titles in appendices. (•)	commons.xsl
Part with number	Text that precedes titles in parts. (•)	commons.xsl
List of Tables	Text used as the heading for the List of Tables. (•)	bookmarks.xsl, lot-lof.xsl, toc.xsl
List of Figures	Text used as the heading for the List of Figures. (•)	bookmarks.xsl, lot-lof.xsl, toc.xsl

TOC and Index variables

Variable	Description	Used in
Table of Contents	Text used as the heading for the Table of Contents. (•)	bookmarks.xsl, toc.xsl
Table of Contents Chapter	Text that precedes entries for `<chapter>` in the TOC. (•)	toc.xsl
Table of Contents Appendix	Text that precedes entries for `<appendix>` in the TOC. (•)	toc.xsl
Table of Contents Part	Text that precedes entries for `<part>` in the TOC. (•)	toc.xsl

Variable	Description	Used in
Table of Contents Preface	Text that precedes entries for `<preface>` in the TOC. (•)	toc.xsl
Table of Contents Notices	Text that precedes entries for `<notice>` in the TOC. (•)	toc.xsl
Mini Toc	Text used as the heading of the mini-TOC at the beginning of a chapter. (•)	commons.xsl
Index	Text used as the heading for the index. (►)	bookmarks.xsl, index.xsl, toc.xsl
Index Continued String		index.xsl
Index See String	Text that precedes the `<see>` element in an index. (•)	index.xsl
Index See Also String	Text that precedes the `<see-also>` element in an index. (•)	index.xsl

Miscellaneous variables

Variable	Description	Used in
Product Name	Product name that appears in headers or footers. (•)	root-processing.xsl
Required-Cleanup	Text used for a label generated from `<required-cleanup>`.	commons.xsl
#menucascade-separator	The separator (such as >) used to separate `<uicontrol>` elements within a `<menucascade>`. (•)	ui-domain.xsl
#quote-start	The character used at the beginning of quotes indicated by `<q>`. (•)	commons.xsl
#quote-end	The character used at the end of quotes indicated by `</q>`. (•)	commons.xsl

Appendix D

DITA OT changes for DITA 1.3

The 2.x series of the DITA Open Toolkit provides a lot of nice new functionality regardless of the version of DITA you're using. If you move to DITA 1.3—to take advantage of scoped keys or ditaval scope, for example—then you have to move to DITA OT 2.x as well to process those features. Because DITA 1.3 is such a major driver for DITA OT 2.x adoption, it's useful to know exactly what changes have been implemented for the new DITA 1.3 features and what this book covers.

Here's a list of the major changes and new features.

Scoped keys

DITA 1.3 includes new functionality to specify which key values apply to specific nodes in a map. Users make this specification via the new *keyscope* attribute.

The Open Toolkit processes key scopes in the **maprefImpl.xsl** file, which is part of DITA OT preprocessing. Because key scopes are processed before the Open Toolkit invokes the `org.dita.pdf2` plugin, the plugin does not include any processing for scoped keys, and this book doesn't cover this feature.

Branch filtering

DITA 1.3 includes a new element, `<ditavalref>` that allows a ditaval to be referenced from any node within a map.

The Open Toolkit processes branch filtering in the **maprefImpl.xsl** file, which is part of DITA OT preprocessing. As with key scopes, branch filtering is handled before the Open Toolkit invokes the `org.dita.pdf2` plugin is invoked. Therefore, the plugin does not include any processing for branch filtering, and this book doesn't cover this feature.

Troubleshooting

DITA 1.3 includes a new troubleshooting topic type, new `<tasktroubleshooting>` and `<steptroubleshooting>` elements for use in tasks, and a new value, "trouble", for the *type* attribute on the `<note>` element.

As of publication, the DITA OT doesn't include any processing specific to the troubleshooting topic type, `<tasktroubleshooting>` or `<steptroubleshooting>`. It does include processing for the new `<note>` type. Customizing the new `<note>` type is the same as customizing any other `<note>` type and multiple exercises cover doing so (**Notes formatting** *(p. 214)*). This book also covers how to add templates and attribute sets specific to the troubleshooting topic type and elements:

- **Add task troubleshooting attribute sets and variables** *(p. 273)*
- **Add task troubleshooting templates** *(p. 272)*

Future releases of the DITA OT will likely include these items out of the box.

Release management

DITA 1.3 includes new elements that allow writers to record information such as release notes directly in a topic prolog. While you can use these elements in output if you want, they are not intended to be part of output, so the DITA OT does not include any processing for them. This book does not cover customizing these elements.

Context-sensitive help

DITA 1.3 includes new elements to enable writers to specify the context-sensitive id's and display windows that a topic should use within online help. This information can be specified at the topic or map level.

These elements are likely to require highly specialized customization based on individual need. For that reason, the DITA OT does not include any processing for them. This book does not cover customizing these elements.

XML Mention

DITA 1.3 includes a new XML Mention domain, which consists of elements to mark up named XML constructs, including elements, attributes, entities, processing instructions, and document-type declaration components.

The DITA OT includes a new attribute set file (**xml-domain-attr.xsl**) and stylesheet (**xml-domain.xsl**) to process these elements. There is nothing unusual about customizing these elements for PDF, so this book does not address them. However, the exercises here should provide enough guidance for you to easily handle this new domain.

Highlight domain

DITA 1.3 includes new elements, `<line-through>` and `<overline>`, in the highlighting domain.

The DITA OT includes new attribute sets for these elements in (**hi-domain-attr.xsl**) and templates (**hi-domain.xsl**) to process these elements. There is nothing unusual about customizing these elements for PDF, so this book does not address them. However, the exercises here should provide enough guidance for you to easily handle these new elements.

Table accessibility and output

DITA 1.3 includes new attributes (*scope, headers*) to identify the column name when a cell is read out loud by Braille or spoken word browsers. It also includes the new attribute *orient* to print tables in landscape mode, and *rotate* to rotate the content in table cells 90 degrees counterclockwise.

The DITA OT includes some processing for *scope* and *headers*, but the implementation of these attributes in output and by screen readers or other applications is likely to be highly specialized and application-specific. This book does not address these attributes.

The *orient* attribute is covered in **Output a table in landscape orientation** *(p. 298)*. The exercise **Rotate text** *(p. 228)* explains how to work with the *rotate* attribute.

MathML and SVG

The MathML and SVG XML standards are now included in the standard DITA 1.3 DTDs and do not have to be separately integrated. The implementation of these standards in output is likely to be highly specialized and user-specific. This book does not address them.

<div>

DITA 1.3 includes a new <div> element to enable arbitrary grouping of content within a topic. It is primarily intended as a way to group content for content referencing or as a basis for specialization.

The **commons.xsl** stylesheet includes a template for processing <div>. Because it's a container designed to support user-specific processing, this template has no defined output. Implementations will be highly specialized for each user. This book does not address <div>.

deliveryTarget attribute

DITA 1.3 includes a new conditional processing attribute, *deliveryTarget*, to replace the now-deprecated *print* attribute. Users can set controlled values for the *deliveryTarget* attribute by using a subjectScheme map.

This attribute is likely to require highly specialized customization based on individual need. For that reason, the DITA OT does not include default processing for it. This book does not address this element.

Appendix E

Notable differences between the org.dita.pdf2 plugin in Open Toolkit 1.8.5 and 2.4

This is not a comprehensive list of changes by far. There are many changes (major and minor) that do not directly affect PDF plugin creation. Most of these changes have to do with various pre-processing steps that you are not likely to need to change.

To see the full release notes for each 2.x version of the DITA Open Toolkit, go to **www.dita-ot.org**, click the Docs menu, select the version, and then select Release Notes.

Separate plugins for each PDF renderer

Processing for all three PDF renderers (Antenna House, FOP, XEP) was previously combined into org.dita.pdf2. Processing unique to each renderer has been broken out into separate plugins: `org.dita.pdf2.axf`, `org.dita.pdf2.fop`, and `org.dita.pdf2.xep`.

The build files **build_axf.xml**, **build_fop.xml**, and **build_xep.xml** have been moved from `org.dita.pdf2` to the new plugins. and the `org.dita.pdf2/fop` folder has been moved to `org.dita.pdf2.fop`. The following files have been moved from `org.dita.pdf2/cfg/fo/attrs` to the new plugins:

- commons-attr_fop.xsl
- commons-attr_xep.xsl
- index-attr_axf.xsl
- index-attr_xep.xsl
- layout-masters-attr_xep.xsl
- tables-attr_axf.xsl
- tables-attr_fop.xsl
- toc-attr_axf.xsl
- toc-attr_fop.xsl

The following files have been moved from `org.dita.pdf2/xsl/fo` to the respective new plugins:

- flagging_fop.xsl
- index_axf.xsl
- index_fop.xsl
- index_xep.xsl
- root-processing_axf.xsl
- root-processing_fop.xsl
- root-processing_xep.xsl
- topic2fo_shell_axf.xsl
- topic2fo_shell_fop.xsl
- topic2fo_shell_xep.xsl
- topic2fo_shell_axf_template.xsl
- topic2fo_shell_fop_template.xsl
- topic2fo_shell_xep_template.xsl

New DITA 1.3 domains

There are two new files in the `org.dita.pdf2/cfg/fo/attrs` folder, **markup-domain-attr.xsl** and **xml-domain-attr.xsl**, which contain attribute sets to format the elements in the new DITA 1.3 Markup and XML Mention domains. Likewise, there are two new files in the `org.dita.pdf2/xsl/fo` folder, **markup-domain.xsl** and **xml-domain.xsl**, which contain templates to process these elements.

hi-domain-attr.xsl includes attribute sets to format the new DITA 1.3 `<line-through>` and `<overline>` elements. **hi-domain.xsl** includes new templates to process these elements.

Back cover generation

front-matter-attr.xsl includes two new attribute sets, *back-cover* and *__back-cover*, which enable you to define the pagination for the back cover.

layout-masters-attr.xsl includes new attribute sets for the odd and even back cover body regions.

static-content.xsl includes new templates for the odd and even back cover headers and footers.

Additional language handling

Variables, index processing, and i18n files have been added for many new languages. These new files are found in `org.dita.pdf2/cfg/common/index`, `org.dita.pdf2/cfg/common/vars` and `org.dita.pdf2/cfg/fo/i18n`.

In `org.dita.pdf2/cfg/fo/`**font-mappings.xsl**, new default fonts have been added for better non-European language handling.

Table formatting

tables-attr.xsl includes a new attribute set, *table__container*, to format a table rotated using the new DITA 1.3 *orient* attribute.

The same file includes another new attribute set, *tbody.row.entry__firstcol*, which enables you to format the first column of a table separately from the others.

Troubleshooting

commons-attr.xsl includes a new attribute set, *note__label__trouble*, to format the new "trouble" value for the *type* attribute on `<note>` elements. **commons.xsl** includes templates for processing the new value.

Linking

links.xsl includes new code to select the appropriate variable (*Figure Number* or *Figure.title*) based on the value of the *args.figurelink.style* parameter in the build file. It also includes new code to select the appropriate variable (*Table Number* or *Table.title*) based on the value of the *args.tablelink.style* parameter in the build file.

Other changes

In `org.dita.pdf2/resource/`**messages.xml**, many new, clearer error messages have been added and some obsolete messages removed.

There is a new file in the `org.dita.pdf2/xsl/fo` folder: **ut-domain.xsl**, to process elements in the Utilities domain.

basic-settings.xsl now includes three new variables to enable you to easily specify whether to include a front cover, back cover, or TOC in your PDF.

The ANT build parameters that were previously specified in **topic2fo.xsl** are now specified in **basic-settings.xsl**, to make them more easily accessible.

pr-domain.xsl includes new templates to autonumber lines in `<codeblock>` elements. **pr-domain-attr.xsl** includes a new attribute set, *codeblock.line-number*, to format those line numbers.

commons.attr.xsl includes a new attribute set, *div*, to format the new DITA 1.3 `<div>` element. **commons.xsl** includes a new template to process the same element.

commons.attr.xsl includes several new attribute sets, *__expanse__page*, *__expanse__column*, *__expanse__textline*, and *__expanse__spread*, which fully implement the existing *expanse* attribute. **commons.xsl** includes improved processing for *expanse*.

commons.attr.xsl includes a new attribute set, *image.artlabel*, which formats the image file name labels that are created when the *args.artlbl* parameter is set to "yes" for a build. **commons.xsl** tests for the value of this parameter when processing images.

commons.xsl includes processing for the *scalefit* attribute, to correctly automatically scale large images to the page size.

The templates for page generation in **root-processing.xsl** have changed significantly.

In **commons.xsl**, topic processing (including chapter, appendix, part, frontmatter) has changed significantly.

commons-attr.xsl includes new page master attribute sets for each of the page-sequences that the PDF plugin creates by default. They are called by templates in **commons.xsl**, **toc.xsl**, **index.xsl**, **root-processing.xsl**, **preface.xsl**, **lot-lof.xsl**, and **glossary.xsl**. These attribute sets are empty but provide a convenient way to define specific characteristics for each page-sequence.

The *insertVariable* template has been replaced by *getVariable*. The two parameters frequently used with *insertVariable*, *theVariableID* and *theParameters* have been replaced by *id* and *params*, respectively.

Tests for the DITA version found in **bookmarks.xsl**, **front-matter.xsl**, **index.xsl**, **toc.xsl**, and **commons.xsl** no longer exist in OT 2.4.

Appendix F

Sample ANT build file

Previous versions of the DITA OT included a `samples/ant_sample` folder that contained sample ANT build files for all of the out-of-the-box outputs. This folder has gone the way of the dodo, so for your convenience, here is a copy of the **sample_pdf.xml** file from that folder. You can use it as a starter for creating your own ANT build files.

```xml
<?xml version="1.0" encoding="UTF-8" ?>
<!-- This file is part of the DITA Open Toolkit project hosted on
     Sourceforge.net. See the accompanying license.txt file for
     applicable licenses.-->
<!-- (c) Copyright IBM Corp. 2004, 2006 All Rights Reserved. -->

<project name="sample_pdf" default="samples.pdf" basedir=".">

  <property name="dita.dir" location="${basedir}/../.."/>

  <target name="samples.pdf" description="build the samples as PDF" depends="clean.samples.pdf">

    <ant antfile="${dita.dir}/build.xml">
      <property name="args.input" location="${dita.dir}/samples/sequence.ditamap"/>
      <property name="output.dir" location="${dita.dir}/out/samples/pdf"/>
      <property name="transtype" value="pdf"/>
    </ant>
  </target>

  <target name="clean.samples.pdf" description="remove the sample PDF output">
    <delete dir="${dita.dir}/out/samples/pdf"/>
  </target>

</project>
```

Appendix G

GetChapterPrefix template

This template is one of many that members of the Yahoo! DITA-Users list have generously donated to the group. This template was originally developed by Kyle Schwamkrug for National Instruments Corporation.

> **Note** Many thanks to Kyle for donating this code to the DITA Users group and for his generous permission to include it in this book. Thanks also to Steve Fogel for giving me some additional information about it.

```
<!-- This template donated by Kyle Schwamkrug, from the dita-users Yahoo group.-->
<xsl:template name="getChapterPrefix">
    <xsl:variable name="topicType">
        <xsl:call-template name="determineTopicType"/>
    </xsl:variable>
    <!-- Looks back up the document tree to find which top-level topic I'm nested in. -->
    <xsl:variable name="containingChapter" select="ancestor-or-self::
    ►*[contains(@class, ' topic/topic')][position()=last()]"/>
    <!-- And get the id of that chapter. I'll need it later. -->
    <xsl:variable name="id" select="$containingChapter/@id"/>
    <!-- Get the chapters and appendixes from the merged map because, at this point, I don't
        know whether the topic I'm in is inside a chapter or an appendix or a part. -->
    <xsl:variable name="topicChapters">
        <xsl:copy-of select="$map//*[contains(@class, ' bookmap/chapter')]"/>
    </xsl:variable>
    <xsl:variable name="topicAppendices">
        <xsl:copy-of select="$map//*[contains(@class, ' bookmap/appendix')]"/>
    </xsl:variable>
    <xsl:variable name="topicParts">
        <xsl:copy-of select="$map//*[contains(@class, ' bookmap/part')]"/>
    </xsl:variable>
    <!-- Figure out the chapter number. -->
    <xsl:variable name="chapterNumber">
        <xsl:choose>
            <!-- If there's something in $topicChapters with an id that matches the id of the
                context node, then I'm inside a chapter. -->
            <xsl:when test="$topicChapters/*[@id = $id]">
                <xsl:number format="1" value="count($topicChapters/*[@id =$id]/
                ►preceding-sibling::*) + 1"/>
            </xsl:when>
            <!-- If there's something in $topicAppendices with an id that matches the id of the
                context node, then I'm inside an appendix. -->
            <xsl:when test="$topicAppendices/*[@id = $id]">
                <xsl:number format="A" value="count($topicAppendices/*[@id =$id]/
                ►preceding-sibling::*) + 1"/>
            </xsl:when>
```

```
<!-- If there's something in $topicParts with an id that matches the id
     of the context node, then I'm inside a part. -->
          <xsl:when test="$topicParts/*[@id = $id]">
            <xsl:number format="I" value="count($topicParts/*[@id =$id]/
            ▶preceding-sibling::*) + 1"/>
          </xsl:when>
          <xsl:otherwise></xsl:otherwise>
        </xsl:choose>
    </xsl:variable>
    <!-- If $chapterNumber is defined, return it.-->
    <xsl:choose>
        <xsl:when test="$chapterNumber != ''">
            <xsl:value-of select="$chapterNumber"/>
  </xsl:when>
    </xsl:choose>
</xsl:template>
```

Appendix H

GetChapterPrefixForIndex template

This template is a variation of one originally developed by Kyle Schwamkrug for National Instruments Corporation. The original template is one of many that members of the Yahoo! DITA-Users list have generously donated to the group

> **Note** Many thanks to Kyle for donating this code to the DITA Users group and for his generous permission to include it in this book. Thanks also to Steve Fogel for giving me some additional information about it.

```xsl
<xsl:template name="getChapterPrefixForIndex">
  <xsl:param name="currentNode"/>
      <xsl:variable name="containerNode" select="$map//*[@id=normalize-space($currentNode)]"/>

      <xsl:variable name="chapterNumber">
          <xsl:choose>
              <xsl:when test="$containerNode/ancestor-or-self::*[contains
              ▶(@class, ' bookmap/chapter')]">
                  <xsl:number format="1" value="count($containerNode/ancestor-or-self::
                  ▶*[contains(@class, ' bookmap/chapter')]/preceding-sibling::
                  ▶*[contains(@class, ' bookmap/chapter')]) + 1"/>
              </xsl:when>
              <xsl:when test="$containerNode/ancestor-or-self::*[contains
              ▶(@class, ' bookmap/appendix')]">
                  <xsl:number format="A" value="count($containerNode/ancestor-or-self::
                  ▶*[contains(@class, ' bookmap/appendix')]/preceding-sibling::
                  ▶*[contains(@class, ' bookmap/appendix')]) + 1"/>
              </xsl:when>
              <xsl:when test="$containerNode/ancestor-or-self::*[contains
              ▶(@class, ' bookmap/part')]">
                  <xsl:number format="A" value="count($containerNode/ancestor-or-self::
                  ▶*[contains(@class, ' bookmap/part')]/preceding-sibling::
                  ▶*[contains(@class, ' bookmap/part')]) + 1"/>
              </xsl:when>
              <xsl:otherwise></xsl:otherwise>
          </xsl:choose>
      </xsl:variable>
      <!-- If $chapterNumber is defined, return it.-->
      <xsl:choose>
          <xsl:when test="$chapterNumber != ''">
              <xsl:value-of select="$chapterNumber"/>
      </xsl:when>
      </xsl:choose>
  </xsl:template>
```

Appendix I

Specialized element template creation

While a detailed discussion of specialization is outside the scope of this book, here's a brief discussion of how to create new templates to process specialized elements.

Let's say you've specialized an inline element called `<em>`, for emphasis. Its full class is the following: `topic/ph my-hi-d/em`. Say also that you modeled `<em>` on the existing `<b>` element, so you can use the existing template for the `<b>` element as a starting point. The `<b>` element is part of the highlighting domain. You know that all the highlighting domain elements are processed by templates in the **hi-domain.xsl** stylesheet, so you can add **hi-domain.xsl** to your PDF plugin if it isn't already there.

Next, you copy the template that processes the `<b>` element and edit it to reflect the class for your new `<em>` element, as well as to call a new attribute set:

```
<xsl:template match="*[contains(@class,' my-hi-d/em ')]">
    <fo:inline xsl:use-attribute-sets="em">
        <xsl:call-template name="commonattributes"/>
        <xsl:apply-templates/>
    </fo:inline>
</xsl:template>
```

> **Note** If you prefer, you can also create this new template in the file `DITA-OT/plugins/com.company.pdf/cfg/fo/xsl/`**custom.xsl**.

Of course, now you also have to create that new *em* attribute set, which you should already know how to do. If not, take a look at **Create a new attribute set** *(p. 65)*.

For more information on specialization, refer to Eliot Kimber's specialization tutorial, his book *DITA for Practitioners, Volume I: Architecture and Technology*, or his online tutorial, which can be found at **http://dita4practitioners.github.io/dita-specialization-tutorials/**.

Appendix J

Paragraph and character formatting: word processing applications to XSL-FO match-up

It's likely that you're migrating from a common word processing application to DITA and that you want to migrate your template as well. This match-up will get you started customizing attribute sets to reproduce the paragraph and character formatting in your current templates.

These are the closest equivalent FO attributes for the properties found in most common word processing applications. In some cases, the match is exact. In some cases, the match is approximate. In other cases, there is no equivalent.

For details on using each attribute, check your XSL-FO reference.

Alignment and spacing

Property name	FO attribute
alignment	text-align
first indent	text-indent
fixed line space	line-height-shift-adjustment
left indent	margin-left
line space	line-height
right indent	margin-right

Property name	FO attribute
spacing above/before	margin-top/space-above
spacing below/after	margin-bottom/space-below
special indentation	text-indent

Font formatting

Property name	FO attribute
all caps	font-variant
angle	font-style
color	color
double strikethrough	text-decoration
family	font-family
font	font-family
font color	color
font-style	font-style
language	xml:lang
overline	text-decoration
position	baseline-shift
scale	font-stretch
size	font-size
small caps	font-variant

Property name	FO attribute
spacing	letter spacing
spread	letter-spacing
stretch	font-stretch
strikethrough	text-decoration
subscript	baseline-shift
superscript	baseline-shift
underline	text-decoration
variation	text-transform
weight	font-weight

Pagination

Property name	FO attribute
across all columns	span
across all columns and side heads	span
first baseline	relative-align
keep lines together	keep-together
keep with next	keep-with-next.within-page/keep-together
keep with previous	keep-with-previous.within-page/ keep-together
page break before	page-break-before
start	break-before; page-break-before

Property name	FO attribute
side head-alignment	fo:float*
widow/orphan lines/control	orphans, widows

* `<fo:float>` is not an attribute. It's an FO element that you have to include in your XSLT stylesheets to create floating text.

Other properties

Property name	FO attribute
frame above	border-top
frame below	border-bottom
hyphenate/don't hyphenate	hyphenate; hyphenation-keep
maximum # adjacent	hyphenation-ladder-count
maximum word spacing	word-spacing
minimum word spacing	word-spacing
optimum word spacing	word-spacing
shortest prefix	hyphenation-remain-character-count
shortest suffix	hyphenation-push-character-count

Index

CPSIA information can be obtained
at www.ICGtesting.com
Printed in the USA
BVOW09s2223270317

479156BV00004B/82/P

9 781937 434540